FOOD SCIENCE AND TECHNOLOGY

MILK PRODUCTION

Food Science and Technology

Additional books in this series can be found on Nova's website under the Series tab.

Additional E-books in this series can be found on Nova's website under the E-book tab.

FOOD SCIENCE AND TECHNOLOGY

MILK PRODUCTION

BOULBABA REKIK
EDITOR

Nova Science Publishers, Inc.
New York

For permission to use material from this book please contact us:
Telephone 631-231-7269; Fax 631-231-8175
Web Site: http://www.novapublishers.com

Library of Congress Cataloging-in-Publication Data

Milk production / editor, Boulbaba Rekik.
p. cm.
Includes index.
ISBN 978-1-62100-061-7 (hardcover)
1. Milk yield. 2. Milk in human nutrition. I. Rekik, Boulbaba.
SF251.M477 2011
637'.1--dc23

2011030124

Published by Nova Science Publishers, Inc. † New York

Contents

Preface

Milk in different species is probably the world most perfect food because of its contents in valuable nutrients from lipids to vitamins. Ruminant milk is consumed by humans of all ages. During the last decades, we have witnessed substantial changes in milk production levels, milk quality and its perception, and transformation techniques. In one hand, selection programs within species, breeds or populations have resulted in sizeable genetic gains in milk yields to meat over changing market demands. Causes of variation of produced milk quantities and constituents are varied, the genetic makeup of animals, the production environment (resources and climate conditions), and social and economic status in the region or country of production. On the other hand, revolutionary transformation techniques, of which the outcome may be affected by the quality of the raw milk essentially the bacteriological content, of milk to products added horizons to ways of milk valorization. Even though, emphasis is on versatile high quality milk products within countries with leading dairy industries, e.g., North America and some European countries, milk production will need to nearly double in the world over the next decade to follow population and income growth. The strongest demand for milk and milk products are anticipated for developing countries where an important population growth is expected.In recent years, biological and product authenticity in addition to animal welfare are emerging topics affecting the dynamic of the milk industry worldwide. The main focus of these concepts is providing healthy products searched for by consumers.

This book is organized in five parts. These are outlined below:

Physiology of Lactation

Potential functional role of aquaporin water channels in the transport of water and small solutes across endothelial and epithelial barriers in the mammary gland are discussed. And the roles of these membrane proteins in milk production are explored.

Milk for Humans of All Ages

Colstrum and milk nutritional properties are highlighted. In particular up-to-date information of the importance of fatty acids of milk in human nutrition are reviewed with emphasis on education commitment for more efficient future human healthbenefits from milk and dairy products in general.

Milk Production by Different Species in Varied Geographical Area

Milk production varies with species, breeds and the production environment. In this part, are summarized results of different studies focusing on the role of genetic factors (breed and genetic polymorphism) on quantitative and qualitative characteristics of milk. Furthermore, health issues, essentially last advances in mastitis control are presented. The search for feeding resources is also investigated through two experiments on feeding peripartal diets in sheep and the incorporation of fish oil in the dietary lipid supplement of a diet based on alfalfa hay are discussed.

Milk Products and Valorization Techniques

The role of milk content in proteins on physico-chemical and dairy-technological characteristic of milk were investigated in relation to genetic types from breeds present in the production area of the Parmigiano-Reggiano cheese, in comparison with those of the Italian Friesian cow milk. For this aim, the following four comparisons were carried out: Italian Brown *vs* Italian Friesian, Jersey *vs* Italian Friesian; Ayrshire *vs* Italian Friesian.

Animal Well Being

Under the prevailing moral discourse, human beings predominate over other species and the nature. It is high time for us to revisit this taken-for-granted attitude to emphasize the awareness of ethical animal use for a durable secured development of both animal welfare and dairy industries.

Rekik, Boulbaba
Development Editor

List of Contributors

Ali Mobasheri
A. Nikkhah
H. Salmon
Rey Gutiérrez Tolentino
Salvador Vega y León
Claudia Radilla Vázquez
María Radilla Vázquez
Samuel Coronel Nuñez
Marta Coronado Herrera
A. Sacco
D. Sacco
G. Casiello
V. Mazzilli
A. Ventrella
F. Longobardi
María Elena Fátima Nader-Macías
Cristina Bogni
Fernando Juan Manuel Sesma
María Carolina Espeche
Matías Pellegrino
Lucila Saavedra
Ignacio Frola
A. Ben Gara
B. Jemmali
H. Hammami
H. Rouissi
M. Bouallegue
B. Rekik
M. Karam-Babaei
K. L. Jacobsen
S. Harris
E. J. DePeters

S. J. Taylor
Andrea Summer
Massimo Malacarne
Paolo Formaggioni
Piero Franceschi
Primo Mariani
Zenobia C.Y. Chan
Wing-Fu Lai

Part I: Physiology of Lactation

In: Milk Production
Editor: Boulbaba Rekik, pp. 1-11
ISBN 978-1-62100-061-7

Chapter I

Mammary Glands, Aquaporins and Milk Production

Ali Mobasheri*
Division of Veterinary Medicine, School of Veterinary Medicine and Science, Faculty of Medicine and Health Sciences, University of Nottingham, Sutton Bonington Campus, Leicestershire LE12 5RD, United Kingdom

Abstract

The mammary gland is a specialized, enlarged sudoriferous or sweat (apocrine) gland that produces and secretes milk during lactation. Milk consists of simple sugars, lipids, proteins, vitamins and minerals dissolved in water, which accounts for up to 88% per unit volume of milk. The water content of milk varies depending on the animal species under investigation and the physiological state of the lactating animal. Current knowledge suggests that water is secreted across the mammary epithelium in a transcellular manner, in response to an osmotic gradient produced largely by the lactose content of milk. Milk yield and quality are important criteria for the dairy industry. Despite the economic importance of milk yield, little is known about the physiological mechanisms responsible for water transport in the bovine mammary gland. Recent studies suggest that several aquaporin proteins are present in the rodent, bovine and human mammary gland. Aquaporins (AQP) play fundamental roles in water and small solute transport across epithelial and endothelial barriers. Immunohistochemical techniques have confirmed the presence of AQP1 and AQP3 water channels in rat, mouse, bovine and human mammary glands. Studies from our laboratory suggest that in addition to AQP1 and AQP3 the AQP4, AQP5 and AQP7 proteins are expressed in different epithelial and endothelial locations in the mammary gland. This chapter discusses the potential functional role of aquaporin water channels in the transport of water and small solutes across endothelial and epithelial barriers in the mammary gland and explores how these membrane proteins may be involved in milk production.

*Corresponding author: E-mail: ali.mobasheri@nottingham.ac.uk

Keywords: Mammary Gland; Milk Production; Aquaporin; Water Channel; Immunohistochemistry.

Introduction

The mammary gland is a unique and dynamic tissue that undergoes epithelial expansion and invasion during puberty and cycles of branching and lobular morphogenesis, secretory differentiation, and regression during pregnancy, lactation, and involution [1]. Unlike most mammalian organs, the mammary gland undergoes the majority of its development in the adult organism.Embryonically, the mammary gland's development begins with invasion of the underlying fat pad by a rudimentary ductal structure [2]. Postnatal growth occurs in two phases: ductal growth and early alveolar development during estrous cycles, and cycles of proliferation, differentiation, and death that occur with each pregnancy, lactation, and involution.

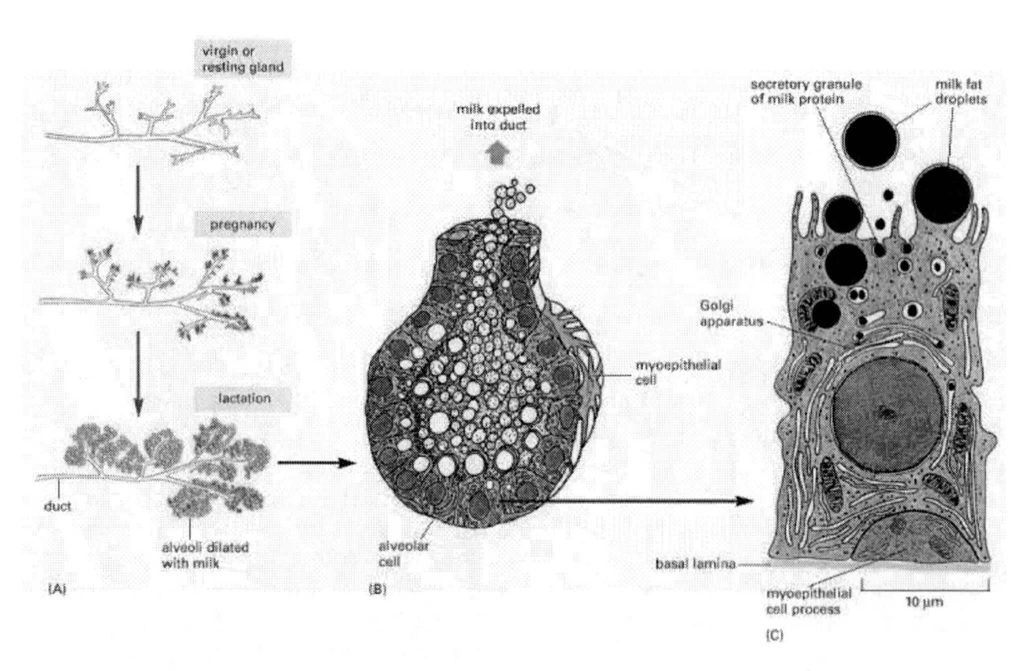

Figure 1. The mammary gland. (A) The growth of alveoli from the ducts of the mammary gland during pregnancy and lactation. Only a small part of the gland is shown. The "resting" gland contains a small amount of inactive glandular tissue embedded in a large amount of fatty connective tissue. During pregnancy an enormous proliferation of the glandular tissue takes place at the expense of the fatty connective tissue, with the secretory portions of the gland developing preferentially to create alveoli. (B) One of the milk-secreting alveoli with a basket of contractile myoepithelial cells (green) embracing it. (C) A single type of secretory alveolar cell produces both the milk proteins and the milk fat. The proteins are secreted in the normal way by exocytosis, while the fat is released as droplets surrounded by plasma membrane detached from the cell. Figure reproduced from Molecular Biology of the Cell. (4th edition) with the permission of the publishers. Bruce Alberts, Dennis Bray, Julian Lewis, Martin Raff, Keith Roberts, and James D. Watson .New York: Garland Science; 2002. Information obtained from the NCBI Bookshelf. A service of the National Library of Medicine, National Institutes of Health (NIH).

Therefore, its distinctive developmental aspects and complex regulation by hormones and growth factors has made the mammary gland a focus of research for biologists from many different disciplines.The variety of epithelial structures and stromal changes that occur throughout the life of a mammary gland have made it a challenging organ to study [2]. The development, proliferation and differentiation of the mammary gland involve the concerted actions of a variety of hormones and growth factors such as estrogen, progesterone, and prolactin.

In addition to these regulatory endocrine factors, normal mammary development and lactation require the cellular communication and cell-cell interactions between the stromal and parenchymal elements of the mammary gland [3]. Figure 1 outlines the growth of alveoli from the ducts of the mammary gland during pregnancy and lactation. It highlights the differences between "resting", "pregnancy" and "lactating" states and illustrates the functional unit consisting of milk-secreting alveoli with a basket of contractile myoepithelial cells embracing it.

Anatomy, Histology and Physiology of the Mammary Gland

The mammalian mammary gland is a specialized, enlarged sudoriferous or sweat (apocrine) gland that produces and secretes milk. Milk provides nourishment to the newborn mammalian offspring, which supports and enhances life.

It is a vital source of nourishment for the neonate and growing mammal from birth to weaning. It contains vital nutrients and colostrum which provides passive mucosal immunity [4, 5]. Milk consists of simple sugars, lipids, proteins, vitamins and minerals dissolved in water [5]. Water accounts for up to 88% per unit volume of milk (although this percentage will vary depending on the species under investigation and the physiological state of the lactating animal [6]. Current understanding is that water is secreted across the mammary epithelium in a transcellular manner, in response to an osmotic gradient produced largely by the lactose content of the milk (Shennan & Peaker, 2000; McManaman et al., 2006).

In terms of anatomical architecture, mammary glands are essentially modified and enlarged sweat glands. Mammary lobes are comprised of secretory acini, which are formed from cuboidal epithelial cells, responsible for the synthesis and secretion of milk, surrounded by myoepithelial cells. The latter contract during the milk ejection reflex, increase the pressure in acini and stimulate the expulsion of milk. Interstitial spaces between acini, ducts and lobes are filled with connective tissue and for the most part, adipose tissue. In lactating animals, active acini secrete milk, which drains into small intralobular excretory ducts (Figures 2 and 3).

Ducts are also lined with epithelial cells and may have an outer layer of myoepithelial cells derived from the same lineage [7, 8]. Intralobular ducts join to form a network of interlobular ducts leading into a complex system of lactiferous ducts. In the cow, lactiferous ducts converge into a single lactiferous cistern within each quarter where milk collects before being "let down" through the teat canal during the milking process or suckling.

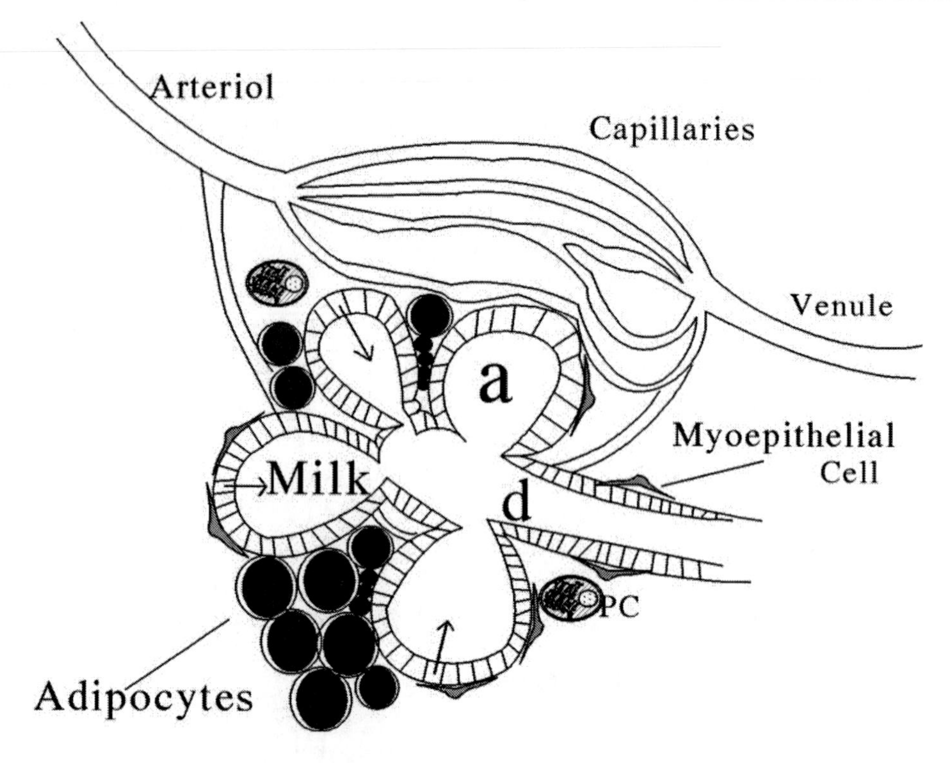

Figure 2. Model alveolus (a) with subtending duct (d) showing blood supply, adipocyte stroma, myoepithelial cells, and plasma cells (PC). Image from the Biology of the Mammary Gland website, reproduced with the permission of Margaret C. Neville.

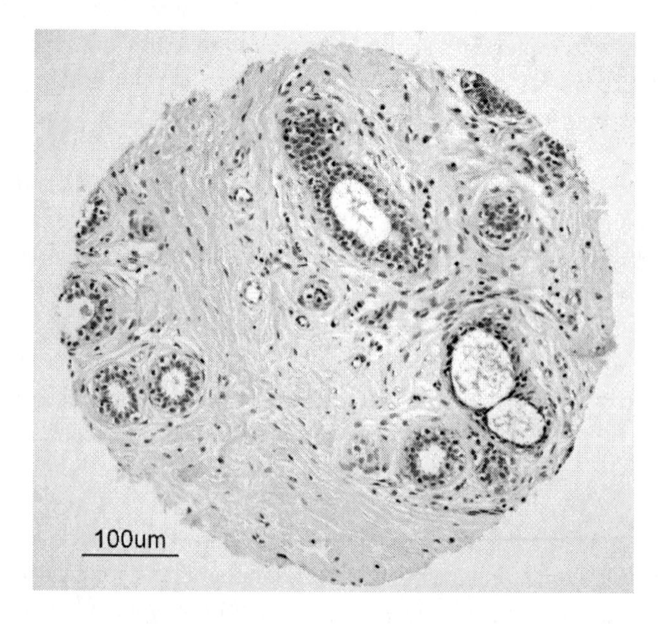

Figure 3. Histology of human mammary gland. The sample shown in this figure is from a tissue microarray developed by the Cooperative Human Tissue Network (CHTN) of the National Cancer Institute (http://www.chtn.nci.nih.gov/). Mammary gland alveoli and ducts form an extensive network of interconnected structures.

Milk Production

Five distinct processes are involved in the mammary epithelium in the secretion of milk (Figure 4).

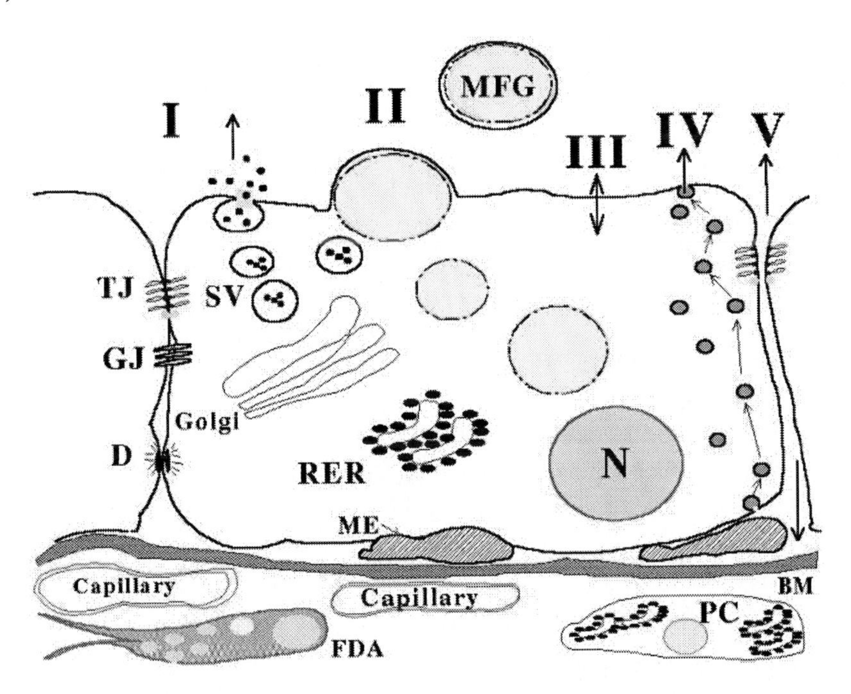

Figure 4. Alveolar Cell from lactating mammary gland. N, nucleus; TJ, tight junction; GJ, gap junction; D, desmosome; SV, secretory vesicle; FDA, fat-depleted adipocyte; PC, Plasma Cell; BM, basement membrane; ME, cross section through process of myoepithelial cell; RER, rough endoplasmic reticulum. See text for explanation of secretory pathways I (exocytosis), II (lipid), III (apical transport), IV (transcytosis) and V (paracellular pathway). Reproduced with permission from Milk Secretion: An Overview by Margaret C. Neville.

These pathways operate in parallel to transform precursors derived from the blood or interstitial fluid into milk constituents. Although the biochemical processes involved are fundamentally the same in all mammals, differences in their relative rates and, in some cases, in the nature of the products synthesized result in milks whose composition differs widely from one species to another. Some of the milk secretion pathways, e.g., exocytosis of protein-containing vesicles and transcytosis of immunoglobulins, are similar to processes in many exocrine organs. In contrast, the mechanism for fat secretion is unique to the mammary gland. Four secretory processes are synchronized in the mammary epithelial cell of the lactating mammary gland: exocytosis, lipid synthesis and secretion, transmembrane secretion of ions and water and transcytosis of extra-alveolar proteins such as immunoglobulins, hormones and albumin from the interstitial space. A fifth pathway, the paracellular pathway, allows the direct transfer of materials between the milk space and the interstitial space. This pathway is open in the pregnant gland and allows the transfer of molecules at least as large as intact immunoglobulins. It is closed in the fully lactating gland providing a tight barrier between the milk and interstitial spaces. This barrier opens again in the presence of mastitis and during involution [9, 10].

Aquaporin Water Channels

Aquaporins are a family of membrane bound proteins that are extensively distributed in microorganisms[11], animals [12-14] and plants[15-17]. They are small integral membrane proteins that are expressed in a variety of epithelial tissues where they are responsible for regulating rapid water movement across epithelial barriers driven by osmotic gradients. Aquaporins play fundamental roles in water and small solute transport across epithelial and endothelial barriers [18, 19].

In mammals, aquaporins are located at strategic membrane sites in endothelia and a variety of epithelia, most of which have well-defined physiological functions in fluid absorption or secretion[22].To date, 13 members of the aquaporin gene family have been identified in humans: AQP0-AQP12 (20). Animal genome projects have also confirmed the presence of multiple aquaporin genes encoding distinct protein isoforms. The proteins encoded by aquaporin genes have been classified into two major groups based on their substrate permeabilities: 1) the classical water permeable aquaporins are permeated by water and include AQP1, AQP2, AQP4, AQP5 and AQP8 [12]; 2) the water and small solute permeable aquaglyceroporins exhibit permeability to water and a range of small neutral solutes. These may include substances such as glycerol and urea. Aquaglyceroporins include AQP3, AQP7, AQP9 and AQP10 [12, 21].

Immunolocalization of Aquaporins in Rat, Mouse, Human and Bovine Mammary Glands

Until recently nothing was known about the expression of aquaporins in the mammary glands of rodents, humans and economically important milk producing animals. Recent immunohistochemical studies of aquaporins in the rat and mouse mammary gland have confirmed the presence of AQP1 and AQP3 proteins in both species(23). Matsuzaki and co-workers also used RT-PCR to study the expression of AQP1, AQP2, AQP3, AQP4, AQP5, AQP6 and AQP7 in the rat mammary gland [23]. In addition to AQP1 and AQP3 they found evidence for the presence of AQP4, AQP5, AQP7 and AQP9 transcripts in the rat mammary gland. However, they did not use antibodies to localize these proteins in the rodent mammary gland.

Our own work using human tissue microarrays has provided immunohistochemical evidence for AQP1 and AQP3 in human mammary glands [24, 25](Figures 5 and 6). AQP1 has been located in both the apical and basolateral membranes of capillary endothelia in the rodent mammary gland[23].

AQP3 has been located immunohistochemically in basolateral membranes of secretory epithelial cells and intralobular and interlobular duct epithelial cells in rat and mouse mammary tissue [23]. Although the expression of mRNA transcripts encoding AQP4, AQP5, AQP7 and AQP9 was demonstrated in rodent mammary by RT-PCR, the presence of the corresponding proteins was not investigated and the presence of AQPs 4, 5 and 7 in the rodent mammary gland was not studied using immunohistochemical methods [23].

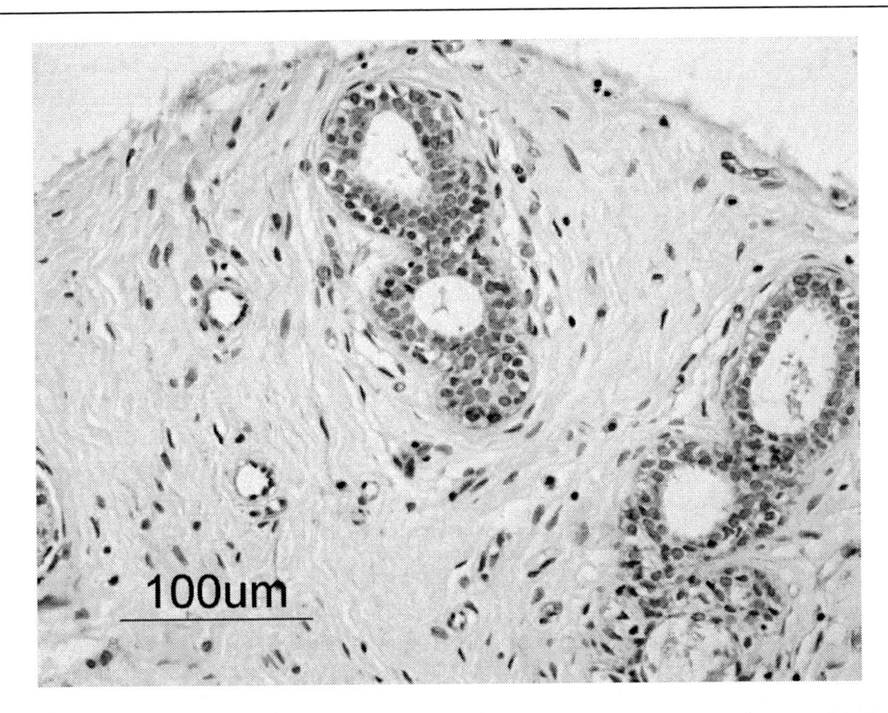

Figure 5. Expression of AQP1 in the humanmammary gland. The tissue section shown in this figure is from a tissue microarray developed by the Cooperative Human Tissue Network (CHTN) of the National Cancer Institute (http://www.chtn.nci.nih.gov/). The section was immunostained using affinity purified polyclonal antibodies to rat AQP1. The antibody used has been shown to exhibit broad mammalian cross-reactivity.

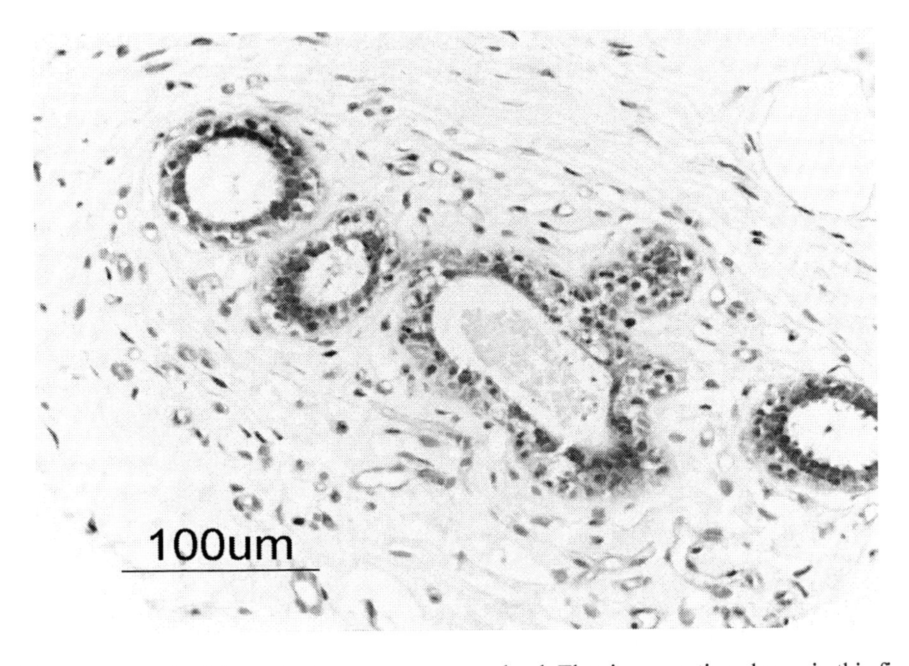

Figure 6. Expression of AQP3 in the human mammary gland. The tissue section shown in this figure is from a tissue microarray developed by the Cooperative Human Tissue Network (CHTN) of the National Cancer Institute (http://www.chtn.nci.nih.gov/). It was immunostained using a commercially produced affinity purified polyclonal antibodies to rat AQP3.

Physiological Relevance of Aquaporins to Milk Production

Milk is a secreted body fluid that consists of water plus lipids, electrolytes, vitamins, sugars and specific milk proteins. The delivery of water to the mammary gland by the circulatory system and the movement of water across endothelial and epithelial barriers are critical for milk synthesis and secretion in lactating animals. There is published information on sodium and chloride transport in mammary epithelia[26]. However, the functional anatomy of the mammary gland has not been extensively studied in the context of water transport across endothelial and epithelial barriers and bulk fluid movement in lactating animals. Consequently, very little published information has been available about aquaporin expression, distribution and function in mammary tissues. AQP1 and AQP3 proteins have been shown to be present in the capillary endothelia and epithelial cells of mouse and rat mammary glands r [23]. Similar observations from our laboratory suggest that AQP1 and AQP3 proteins are also present in non-lactating human mammary glands [24, 25]. The data we have recently published on aquaporins in the bovine mammary gland [27] confirms some of the recent observations made by [23] in rodent mammary glands and our own observations in human mammary glands [24, 25]. In addition, we have made a number of novel observations in support of the RT-PCR findings made by Matsuzaki and co-workers[23] in rodent mammary glands.

Our bovine data [27] has confirmed the presence of these proteins in the bovine mammary gland, although in distinct cellular locations. We also observed a striking and previously unreported abundance of AQP1 in myoepithelial cells underlying teat duct epithelia [27]. The physiological reason for this is at present unknown, but high AQP1 levels in this location may contribute to increased permeability of teat duct epithelia in the lactating bovine mammary gland. The relative abundance of the protein also appears to represent a distinct aspect of differentiation of myoepithelial cells in this location, compared with those surrounding the acini.

Goats' milk is known to contain numerous cell fragments known as "christiesomes" which originate from secretory epithelial cells of the mammary gland [28, 29]. These cell fragments are known to contain intact and well-preserved endoplasmic reticulum, mitochondria and lipid droplets. Furthermore, they are capable of triglyceride synthesis. Although cows' milk has been shown to contain very few cellular fragments, it does contain different and denser particles containing fewer vesicles and numerous microvillus-like protrusions on one side – known as "sunbursts" [28]. Although it has been suggested that these particles are residues of dead cells, it is possible that these fragmented cytoplasm-containing entities contain membrane proteins, which are targeted to the apical membranes of mammary secretory epithelial cells. Recent studies have provided evidence for urinary excretion of AQP2 water channel proteins during pregnancy [30, 31]. Biochemical analysis of "christiesomes" and "sunbursts" should reveal if these cellular fragments contain membrane proteins of the apical membranes of alveolar epithelial cells.

Comparing our data with published data from rodent studies clearly indicates that some of the AQPs expressed in the lactating mammary tissues were found in expected anatomical locations; immunohistochemical labelling in the rat mammary gland suggests that AQP1 is localized to the capillaries and AQP3 is localized to the basolateral membranes of the alveolar secretory cells. These results suggest that aquaporins are present in lactating mammary glands and may be participants in the control of milk water content by diluting the sugar, protein and lipid contents of milk to an isotonic solution as it descends through the teat duct system.

Published data has provided interesting new information about the possible regulation of water homeostasis in the mammary gland. Milk yield is of major economic importance to the dairy industry. Therefore, it is necessary to understand the underlying molecular physiology involved in fluid movement in lactation. Economically important production traits within the dairy industry such as milk yield traits have large phenotypic variation and high heritability (Brotherstone et al., 1997; Jamrozik and Schaeffer, 1997; Royal et al., 2002; Wall et al., 2003) and thus are largely influenced by the genotype of the cow. However, the underlying physiological mechanisms responsible for the genetic variation between animals and indeed within a lactation period have not been clearly identified. Gaining further information about the underlying genetic regulation of aquaporins and their production and activity may have the potential to help improve the accuracy of breeding value prediction for yield traits allowing producers to make more informed selection decisions. Clearly this is important work that needs to be done.

Conclusion

A number of studies have attempted to identify specific genes required for the functional development of mammary epithelium [32]. However, very few studies have examined the potential contribution of aquaporins to milk production. The altered expression of a number of molecular markers, including the progesterone, estrogen, and prolactin receptors, the Na/K/2Cl cotransporter proteins (NKCC1) and aquaporin 5 (AQP5), and several markers of skin differentiation (Sprr2A and keratin 6) has been reported in breast cancer[33]. The identification of AQP5 expression in neoplastic breast epithelium preceded subsequent reports of other aquaporins in rat, mouse [23] and bovine mammary glands [27]. The studies reviewed in this chapter suggest that several aquaporin proteins collaborate in the production of milk in the mammary gland. Future studies will need to determine if additional aquaporins are involved at different stages of lactation and correlate the expression of these with the expression of milk protein genes and proteins. A better understanding of the molecular mechanism involved in milk production will have significant benefits for animal breeding programmes. Further work is required to develop a larger bank of mammary tissue samples for more comprehensive immunohistochemical analysis using custom designed tissue micro-arrays. It will also be useful to obtain further information about aquaporin expression in mammary glands of non-lactating animals, those undergoing mammary development during first pregnancy, and those recovering from mastitis.

Acknowledgements

This work received financial support from BSAS/GenesisFaraday (http://www.genesis-faraday.org/) through a Vacation Scholarship to Miss Bryony Kendall. I am grateful to Judith

Maxwell, Ami Sawran, Alexander German, Melissa Royal and David Marples for their ongoing support and collaboration.

References

[1] Masso-Welch, P. A., Darcy, K. M., Stangle-Castor, N. C.& Ip, M. M. (2000). A developmental atlas of rat mammary gland histology. *J Mammary Gland Biol Neoplasia*. Apr, 5(2),165-85.

[2] Richert, M. M., Schwertfeger, K. L., Ryder, J. W., &Anderson, S. M. (2000). An atlas of mouse mammary gland development. *J Mammary Gland Biol Neoplasia,*Apr5(2), 227-41.

[3] Rillema, J. A. (1994). Development of the mammary gland and lactation. Trends. *Endocrinol Metab*, 5(4), 149-54.

[4] Salmon, H. (1999). The mammary gland and neonate mucosal immunity. *Vet Immunol Immunopathol*, 72(1-2), 143-55.

[5] Shennan, D. B.,& Peaker, M. (2000). Transport of milk constituents by the mammary gland. *Physiol Rev,* 80(3), 925-51.

[6] McManaman, J. L., &Neville, M. C. (2003). Mammary physiology and milk secretion. *Adv Drug Deliv Rev*, 55(5), 629-41.

[7] Pechoux, C., Gudjonsson, T., Ronnov-Jessen, L., Bissell, M. J., &Petersen, O. W. (1999). Human mammary luminal epithelial cells contain progenitors to myoepithelial cells. *Dev Biol*, 206(1), 88-99.

[8] Gudjonsson, T., Adriance, M. C., Sternlicht, M. D., Petersen, O. W., &Bissell, M. J. (2005). Myoepithelial cells: their origin and function in breast morphogenesis and neoplasia. *J Mammary Gland Biol Neoplasia,* 10(3), 261-72.

[9] Neville, M. C. (1995). Sampling and storage of human milk. In: Jensen RG, editor. Handbook of Milk Composition. San Diego: Academic Press, pp. 63-79.

[10] Neville, M. C. (1995). Lactogenesis in women: A cascade of events revealed by milk composition. San Diego: Academic Press.

[11] Calamita, G. (2000). The Escherichia coli aquaporin-Z water channel. *Mol Microbiol*, 37(2), 254-62.

[12] Agre, P., King, L. S., Yasui, M., Guggino, W. B., Ottersen, O. P., Fujiyoshi, Y., Engel, A., &Nielsen, S.Aquaporin water channels--from atomic structure to clinical medicine. *J Physiol*, 542(Pt 1), 3-16.

[13] King, L. S., Kozono, D., &Agre, P. (2004). From structure to disease: the evolving tale of aquaporin biology. *Nat Rev Mol Cell Biol*, 5(9), 687-98.

[14] Agre, P.,& Kozono, D. (2003). Aquaporin water channels: molecular mechanisms for human diseases. *FEBS Lett,*555(1), 72-8.

[15] Schaffner, A. R. (1998). Aquaporin function, structure, and expression: are there more surprises to surface in water relations? *Planta*, 204(2), 131-9.

[16] Chrispeels, M. J., & Maurel, C. (1994). Aquaporins: the molecular basis of facilitated water movement through living plant cells? *Plant Physiol*, 105(1), 9-13.

[17] Johansson, I., Karlsson, M., Johanson, U., Larsson, C., &Kjellbom, P. (2000). The role of aquaporins in cellular and whole plant water balance. *Biochim Biophys Acta*, 1465(1-2), 324-42.

[18] Verkman, A. S., &Mitra, A. K. (2000). Structure and function of aquaporin water channels. *Am J Physiol Renal Physiol*, 278(1), F13-28.

[19] Verkman, A. S. (2002). Aquaporin water channels and endothelial cell function. *J Anat*, 200(6), 617-27.

[20] Castle, N. A. (2005). Aquaporins as targets for drug discovery. *Drug Discov Today*, 10(7), 485-93.

[21] Hibuse, T., Maeda, N., Nagasawa, A., &Funahashi, T. (2006). Aquaporins and glycerol metabolism.*Biochim Biophys Acta*, 1758(8), 1004-11.

[22] Brown, D., Katsura, T., Kawashima, M., Verkman, A. S., &Sabolic, I. (1995). Cellular distribution of the aquaporins: a family of water channel proteins. *Histochem Cell Biol*, 104(1), 1-9.

[23] Matsuzaki, T., Machida, N., Tajika, Y., Ablimit, A., Suzuki, T., Aoki, T., Hagiwara, H., &Takata, K. (2005). Expression and immunolocalization of water-channel aquaporins in the rat and mouse mammary gland. *Histochem Cell Biol*,123(4-5), 501-12.

[24] Mobasheri, A., Wray, S., &Marples, D. (2005). Distribution of AQP2 and AQP3 water channels in human tissue microarrays. *J Mol Histol*,36(1-2), 1-14.

[25] Mobasheri, A., &Marples, D. (2004). Expression of the AQP-1 water channel in normal human tissues: a semiquantitative study using tissue microarray technology. *Am J Physiol Cell Physiol*, 286(3), C529-37.

[26] Blaug, S., Hybiske, K., Cohn, J., Firestone, G. L., Machen, T. E., &Miller, S. S. (2009). ENaC- and CFTR-dependent ion and fluid transport in mammary epithelia. *Am J Physiol Cell Physiol*, 281(2), C633-48.

[27] Mobasheri, A., Kendall, B. H., Maxwell, J. E., Sawran, A. V., German, A. J., Marples, D., Luck, M. R., &Royal, M. D. (2009). Cellular localization of aquaporins along the secretory pathway of the lactating bovine mammary gland: An immunohistochemical study. *Acta Histochem,* Oct 21.

[28] Wooding FB, Morgan G, Craig H. "Sunbursts" and "christiesomes": cellular fragments in normal cow and goat milk. Cell Tissue Res. 1977 Dec 28;185(4):535-45.

[29] Clegg, R. A. (1978). Lipoprotein lipase in "Christiesomes" from goats' milk: a membrane-bound enzyme [proceedings]. *Biochem Soc Trans*, (6),1205-7.

[30] Buemi, M., D'Anna, R, Di Pasquale G, Floccari F, Ruello A, Aloisi C, Leonardi I, Frisina N, Corica F. (2001). Urinary excretion of aquaporin-2 water channel during pregnancy. *Cell Physiol Biochem*, 11(4), 203-8.

[31] Schrier, R. W., Fassett, R. G., Ohara, M., &Martin, P. Y. (1998).Pathophysiology of renal fluid retention. *Kidney Int Suppl*, 67, S127-32.

[32] Shillingford, J. M., Miyoshi, K., Robinson, G. W., Bierie, B., Cao, Y., Karin, M.,& Hennighausen, L. (2003). Proteotyping of mammary tissue from transgenic and gene knockout mice with immunohistochemical markers: a tool to define developmental lesions.*J Histochem Cytochem*, 51(5), 555-65.

[33] Grimm, S. L.,& Rosen, J. M. (2003). The role of C/EBPbeta in mammary gland development and breast cancer. *J Mammary Gland Biol Neoplasia*, 8(2), 191-204.

Part II: Milk for Humans of All Ages

In: Milk Production
Editor: Boulbaba Rekik, pp. 15-20
ISBN 978-1-62100-061-7

Chapter II

Science and Pseudo-Science of Milk Implications for Human Health

A. Nikkhah [*]
Department of Animal Sciences, Faculty of Agricultural Sciences,
University of Zanjan, Zanjan 313-45195 Iran

Abstract

Milk production in ruminants is a bio-complex process that necessitates conversion and renovation of least available plant materials (e.g., cellulose) into most enriched available nutrients. This development involves pregastric rumen fermentation of plant cells by microbes and production of volatile fatty acids (VFA) amongst others as substrates for mammary milk lactose, fat and protein synthesis. As a result, milk contains numerous bioactive substances that function beyond solely their nutritive value. Essential amino acids, specialized protein such as casein, lactalbumins and globulins, peptides, nucleosides, nucleotides, unsaturated fatty acids such as conjugated linoleic acids, sphingomyelins, fat soluble vitamins and minerals of mainly calcium are principal examples. Milk sufficiency for neonatal brain, nervous and immune systems, and bone development and supportive tissue growth for even up to two years without a major need for alternative foods is an evolutionary proof for its irreplaceable role in human nutrition. However, the increasing concerns of cardiovascular disorders, hypertension, and related complexities in modern populations due to improper nutrition seem to have contributed to forming a fallacious public perception about milk in general and milk fat in particular as a possible risk factor. Despite several recent emphases on the functional nature of milk as a whole in improving health status in different populations, the confusion exists where education is suboptimal. Insightful education on the science of milk ought to accompany research evidence to enable the public to discern a pseudo-science that unconsciously disregards milk as a saturated fat animal food with serious risks to human health. It is reasoned that animal fat in the form of milk should not be considered biologically similar to animal fats supplied by non-milk sources. With limited saturated fat intake from non-milk sources, increased milk consumption could bear a multitude of positive impacts on

[*]E-mail: nikkhah@znu.ac.ir

health even with high fat content. Milk is a collection of bioactive substances with unique nutritional properties that synergistically optimize the health of mind and physics in different age groups.

Keywords: Science, Pseudo-science, Milk, Health, Education.

Ruminant Milk, Man, and the Nature

Milk production in ruminants is a bio-complex process that necessitates conversion and renovation of least available plant materials, such as cellulose, into most invaluable nutrients namely lactose, amino and fatty acids, and energizing water soluble vitamins (Nikkhah, 2010). These developments involve pregastric rumen fermentation of plant cells to produce volatile fatty acids (VFA) and microbial proteins, supplying most of the substrates for milk secretion (Gordon et al., 1947; Nikkhah, 2010).

As a result, milk contains many bioactive substances that function beyond their sole nutritive value (Donnet-Hughes et al., 2000). Hence, milk is considered the most natural functional food. Essential amino acids, specialized proteins such as casein and peptides, lactalbumins and immunoglobulins, nucleosides, nucleotides, unsaturated fatty acids of mainly conjugated linoleic acids, sphingomyelins, fat soluble vitamins and calcium are vital examples (Donnet-Hughes et al., 20003; Stelwagen et al., 2009).

Crucially, milk sufficiency for neonatal developments of brain network, nervous and immune systems, and skeletal frame all occur during the first few months of life without demands for alternatives foods. This provides an evolutionary attestation for the irreplaceable milk role in creating next generations (Nikkhah, 2010; Figure 1). The objective is to enlighten milk nutritional properties and emphasize public educationcommitment for more efficient future human healthbenefits from milk and dairy products in general.

Figure 1.Continued on next page.

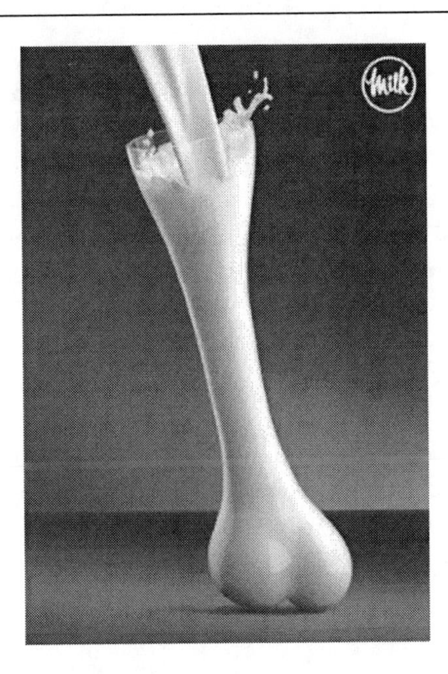

Figure 1. Ruminant milk and human wellbeing represents a historical relationbetween man and the nature.

Critical Evaluation of Literature on Milk and Human Health

The increasing modern era's concerns from cardiovascular, cognitive and aging complexities due to improper nutrition of non-milk ingredients appear to have alarmingly contributed to forming a fallacious public acuity about milk as a risk factor to optimum health (Lanou, 2009). For instance, very early (< 3 mo after birth) neonatal cow milk intake has been related to insulin-dependent diabetes (Gerstein, 1994; Scott, 2005). Apart from bioreasons for such a relationship, non-breast milk consumption during very early stages of life is unarguably uncommon. Modern nutrition does in no standard circumstances authorize feeding such quite young neonates' non-breast milks.

Population studies with compelling scientific application to real life scenarios demonstrated that unlike unsubstantiated beliefs, cow milk is not responsible for the major gastrointestinal related immune malfunctions (Knip et al., 2005). Moreover, consuming cow milk by lactating mothers has proved to enrich breast milk beta-lactoglobulin and ovalbumin components (Fukushima et al., 1997). Furthermore, youngsters consuming cow milk develop more standard body frame structure, whereas children deprived of milk intake exhibit poor bone health (Okada, 2004; Black, 2002). In complementary addition, what make ruminant milk exclusive are its processing and manipulation opportunities to better meeting human nutrient requirements with no major health compromises. Enrichment with vitamin E and skim milk production from immunized cows for lowering blood cholesterol in hypercholesterolemic patients are amongst key examples (Golay, 1990; Hayes et al., 2001).

Most recently, conjugated linoleic acid(CLA) isomers of about 20, notably CLA $_{cis9\ trans11}$ (75-90%) and CLA $_{trans10\ cis12}$ from cow and goats milk have been emerging as novel agents with anti-carcinogenic, anti-atherogenic, anti-inflammatory, and anti-lipogenic effects

(McGuire & McGuire, 2000; Lock&Bauman, 2004; Haenlein, 2004). These effects protect against hypertension, cardiovascular disease (CVD) and progressive obesity. Accordingly, considerable research has focused on altering modern dairy cows and somewhat goats nutrition programs to improve milk nutrient profile, and enhance components with human healthimplications (Lock&Bauman, 2004). These are besides milk immunoglobulins beneficial impacts against Enterotoxigenic Escherichia coli (Tacket et al., 1988). Another group of milk bioactives include casomorphins, immunostimulating peptides, and ACE-inhibitory peptides. These substances form prime structures of milk proteins and act as plausible physiological modulators during milk digestion along the gut. Thus, such modulators can provide essential sources for designing commercial bioactive features or 'functional foods' (Xu, 1998).

Recent observational studies report no increased CVD with increased milk or other dairy products intake (Pfeuffera & Schrezenmeira, 2000, 2006). Dairy consumption has in several cases reduced the incidence of one or more metabolic syndromes (Pfeuffera & Schrezenmeira, 2000). Whey proteins are insulinotropic, medium chain fatty acids improve insulin sensitivity, and alongside calcium may favorably influence body weight and fat distribution. Peptides and calcium can reduce blood pressure and blood cholesterol. Additionally, dairy consumption benefits folate availability and thereby lowers circulating homocystein levels and heart attack risks (Pfeuffera & Schrezenmeira, 2000).

Notwithstanding such recent emphases on the functional nature of milk in improving human health status in different populations, major confusions exist where education is suboptimal. Insightful education must commit to distinguishing science from pseudo-science of ruminant milk implications for the 21'st century's stressful times. Such science embodied with research evidence will enable the public to discern the pseudo-science that has likely unconsciously or deliberately disregarded milk and introduced it only as an overly saturated animal fat food with risks to human health. It will be reasoned that animal fat in the form of milk should not be considered biologically similar to non-milk animal fats, simply because milk fats are accompanied by numerous various substances, each with distinctive benefits (Donnet-Hughes et al., 2000; Pfeuffera & Schrezenmeira, 2000, 2006; Xu, 1988). Even, evidence exists that medium- and short-chain fatty acids of milk specifically butyric acid, publically considered possibly unhealthy, improve intestinal fat and nitrogen assimilation and insulin sensitivity (Gao et al., 2009; Tantibhedhyangkul & Hashim, 1975).

Conclusion

Milk is considered the most natural functional food. The increasing concerns from cardiovascular, cognitive and aging complexities in modern populations, mostly due to improper nutrition and lifestyle have wrongly contributed to forming a fallacious public perception about milk as a risk factor to optimal health. With limited saturated fat intake from non-milk sources, increased lower fat milk intake should lead to a multitude of health benefits. Substantive emphasis on public education of science and pseudo-science of milk intake is inevitable. Proper education is the gold missing piece of an accurate civic perception of milk intake relationship with the fitness of mind and physics in different human age groups.

Acknowledgments

The University of Zanjan (Zanjan, Iran) is acknowledged for supporting the author's teaching and research programs of improving science education worldwide.

References

Black, R. E., Williams, S. M., Jones, I. E. & Goulding, A. (2002). Children who avoid drinking cow milk have low dietary calcium intakes and poor bone health. *Am. J. Clin. Nutr.* 76, 675-680.

Donnet-Hughes, A., Duc, N., Serrant, P., Vidal, K. & Schiffrin, E. J. (2000). Bioactive molecules in milk and their role in health and disease: The role of transforming growth factor- β . Immun. *Cell Biol.* 78, 74-79.

Fukushima, Y., Kawata, Y., Onda, T. & Kitagawa, M. (1997). Consumption of cow milk and egg by lactating women and the presence of beta-lactoglobulin and ovalbumin in breast milk. *Am. J. Clin. Nutr.* 65, 30-35.

Gao, Z., Yin, J., Zhang, J., Ward, R. E., Martin, R. J. & Lefevre, M. (2009). Butyrate improves insulin sensitivity and increases energy expenditure in mice diabetes. *Diabet.* 58, 1509-15.

Gerstein, H. C. (1994). Cow's milk exposure and type I diabetes mellitus: a critical overview of the clinical literature. *Diabet. Care*17, 13-19.

Golay, A. (1990). Cholesterol-lowering effect of skim milk from immunized cows in hypercholesterolemic patients. *Am. J. Clin. Nutr.* 52, 1014-1019.

Gordon, H. H., Levine, S. Z. & McNamara, H. (1947). A Comparison of Human and Cow's Milk. *Am. J. Dis. Child.* 73(4), 442-452.

Haenlein, G. N. W. (2004). Goat milk in human nutrition. *Small Rum. Res.* 51, 155-163.

Hayes, K. C., Pronczuk, A. & Perlman, D. (2001). Vitamin E in fortified cow milk uniquely enriches human plasma lipoproteins. *Am. J. Clin. Nutr.* 74, 211-218.

Knip, M., Vaarala, Q. & Kokkonen, J. (2005). Cow milk is not responsible for most gastrointestinal immune-like syndromes—evidence from a population-based study. *Am. J. Clin. Nutr.* 82, 1327-1335.

Lanou, A. J. (2009). Should dairy be recommended as part of a healthy vegetarian diet? Counterpoint. *Am. J. Clin. Nutr.* 89, 1638S-1642S.

Lock, A. L. &Bauman, D. E. (2004). Modifying milk fat composition of dairy cows to enhance fatty acids beneficial to human health. *Lipids* 39, 1197-1206.

McGuire, M. A. & McGuire, M. K. (2000). Conjugated linoleic acid (CLA): A ruminant fatty acid with beneficial effects on human health. *J. Anim. Sci.* 77, 1-8.

Nikkhah, A. (2010). Ruminant milk and human wellbeing: a multi-species review. In Milk Production. NOVA Science Publishers, Inc., NY, USA.

Okada, T. (2004). Effect of cow milk consumption on longitudinal height gain in children. *Am. J. Clin. Nutr.* 80, 1088-1089.

Pfeuffera, M. & Schrezenmeira, J. (2000). Bioactive substances in milk with properties decreasing risk of cardiovascular diseases. *Br. J. Nutr.* 84, 155-159.

Pfeuffer, M. & Schrezenmeir, J. (2006). Milk and the metabolic syndrome. *Obes. Rev.* 8, 109-118.

Scott, F. W. (2005). Cow milk and insulin-dependent diabetes mellitus: is there a relationship? *Am. J. Clin. Nutr.* 51, 489-491.

Stelwagen, K., Carpenter, E., Haigh, N., Hodgkinson, A. & Wheeler, T. T. (2009). Immune components of bovine colostrum and milk. *J. Anim. Sci.* 87(13 suppl), 3-9.

Tacket, C. O., Losonsky, G., Link, H., Hoang, Y., Guesry, P., Hilpert, H. & Levine, M. M. (1988). Protection by milk immunoglobulin concentrate against oral challenge with Enterotoxigenic Escherichia coli. *N. Engl. J. Med.* 318, 1240-1243.

Tantibhedhyangkul, P. & Hashim, S. A. (1975). Medium-chain triglyceride feeding in premature infants: effects on fat and nitrogen absorption. *Pediatrics* 55, 359-370.

Xu, R-J. (1998). Bioactive peptides in milk and their biological and health implications. *Food Rev. Int.*14, 1-16.

In: Milk Production
Editor: Boulbaba Rekik, pp. 21-35

ISBN 978-1-62100-061-7

Chapter III

Ruminant Milk and Human Wellbeing: A Multi-Species Review of the Natural Relations

A. Nikkhah[*]

Department of Animal Sciences, Faculty of Agricultural Sciences,
University of Zanjan, Zanjan 313-45195 Iran

Abstract

Milk is unarguably the most nutritious bio-fluid in the nature. Ruminant milk provides an ideal food for humans of all ages. Thus, proper appreciation of milk nutritional value by the public requires thorough education of its biological descriptions in major ruminant species. The chapter assembles nutritional characteristics for humans of milk components and production by cattle, buffaloes, sheep, goats, and camels. Qualitative exclusivities and quantitative contributions of milk products from different ruminants to modern and post-modern human nutrition are discussed. Unique beneficial effects of different milk fats, proteins, peptides, minerals, vitamins, and other bioactive substances on optimal meeting of essential nutrients requirements and reducing the incidence of metabolic and cardiovascular complexities, cancers, traumas, and cognitive issues are also described. The advanced ongoing education and knowledge dissemination will be an obligation for unbiased milk natural health properties to be most deservingly realized by the new millennium's humans.

Keywords: Milk, Ruminant, Education, Nutrition, Human, Health.

[*]E-mail: nikkhah@znu.ac.ir

Introduction

Yield and composition of ruminant milk vary due to inter-breed and inter-family genetic differences. These genetic versatilities have major influence on milk products quality and thus on human digestion capacity and health. In addition, lactation stage, daily and hour-based fluctuations, season, parity, diet properties, udder dynamics and health, and processing techniques affect milk contents and human health implications. Approximates for world milk production are given in Table 1 and Figure 1. Asia, Europe, and North America are the major milk and accordingly meat producers (Figures 1 and 2). World milk production appears to have been more consistent for cows than for buffalos, with sheep and goats milk fluctuating considerably from year to year (Figure 3).

Table 2 demonstrates the different milk composition among species. The highest per capita dairy products intakes in the Western Europe, Scandinavians as well as Australia and Canada (Table 3) highlight a necessity of continued and further education in these and other world regions on human health implication of ruminant milk.

Table 1. World ruminants' milk production approximates (million tons)

Continent	Cow	Sheep	Goats	Buffalo	Total
Africa	13	1.5	2.0	1.5	18
North America	85	-	0.5	-	85.5
South America	30	0.04	0.2	-	31
Asia	49	4	4	37	94
Europe	173	3.8	1.7	0.1	179
World	494	8	13	70	585
% (world)	84.4	1.4	2.2	12	100

Table 2. Average milk composition per 100 g milk in different species

	Cow	Buffalo	Sheep	Goats	Camel	Human
Fat	4.0	7.0	6.0	4.5	4.2	3.5
Protein, g	3.4	4.0	4.8	3.8	3.5	1.9
Lactose, g	4.8	5.1	5.0	4.7	4.4	6.5
Minerals, g	0.7	0.8	0.9	0.8	0.7	0.2
Solids-non-fat, g	9.0	9.8	10.3	9.0	8.6	7.3
Total solids, g	13.3	16.7	16.3	13.5	12.8	12.1
Cholesterol, mg	14	9	11	10	37	16
Calcium, IU	120	195	170	100	143	32
Phosphorus, mg	93	174	158	111	116	14
Saturated FA, g	2.4	4.2	3.8	2.3	2.6	1.8
Monounsaturated FA, g	1.1	1.7	1.5	0.8	1.6	1.6
Polyunsaturated FA, g	0.1	0.2	0.3	0.1	0.6	0.5

FA = fatty acids.

Table 3. Top 10 per capita cow milk products consumers (2006)

Country	Milk (liters)	Cheese (kg)	Butter (kg)
Finland	184	19.1	5.3
Sweden	146	18.5	1.0
Ireland	130	10.5	2.9
Netherlands	123	20.4	3.3
Norway	117	16.0	4.3
Spain	119	9.6	1.0
Switzerland	113	22.2	5.6
Britain	111	12.2	3.7
Australia	106	11.7	3.7
Canada	95	12.2	3.3

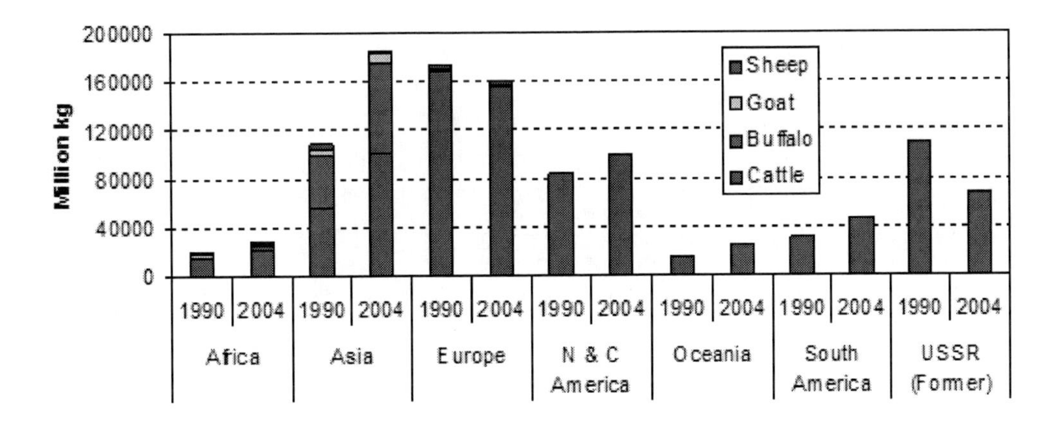

Figure 1. Total world ruminant milk production.

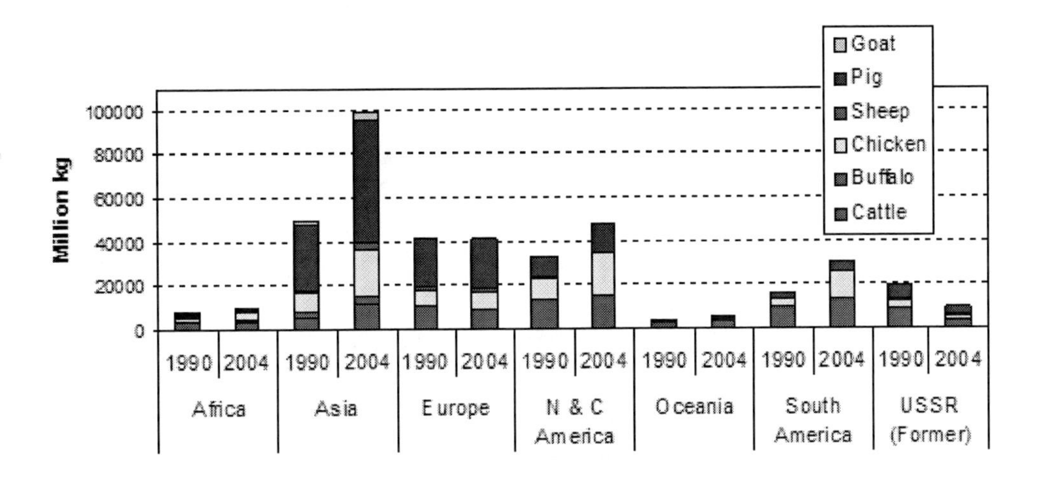

Figure 2. Total world meat production.

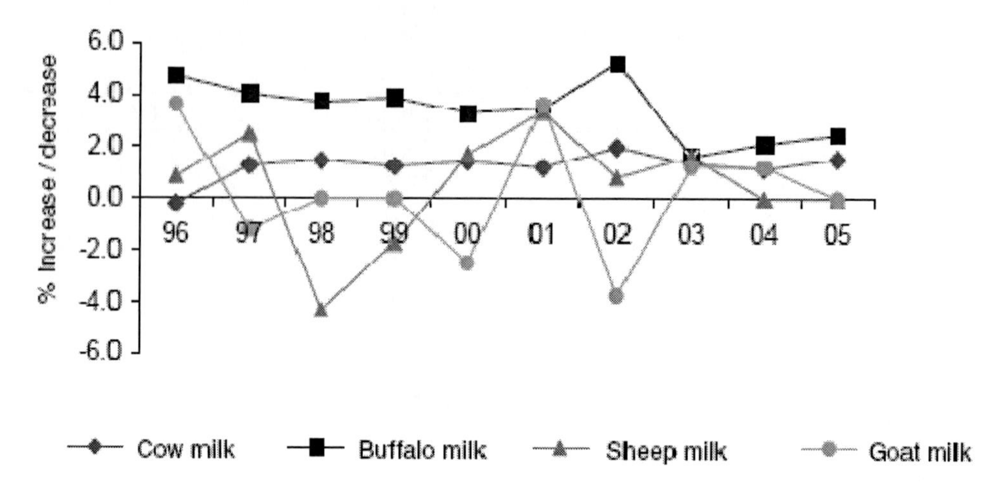

Figure 3. Milk production trends between 1995 and 2005 in different species.

Relative to all milk from cows, buffaloes, ewes and goats, ewe milk contributes 1.5% and goat milk 2.0% to the world total. In the Mediterranean, they are respectively 11.7% and 3.3%. In many regions such as Greece, small ruminants milk is > 20% of all milk produced. In some regions, the combined production of ewe and goat milk is > 20% of all regional milk (FAO, 1994). Some countries, such as Iran have about 1 ewe and goat per person. Some countries with high ewe and goat populations have even more than 1 ewe and goat per hectare permanent pasture land. The contributions of ewe and goat milk to human nutrition vary largely from < 5 to > 450 kg per individual annually. The protein and calcium supply from animal foods including dairy products also varies between countries. Averages in many regions are much below minimum daily requirements of about 70 g protein and 800 mg calcium (DRI, 2007). Buffalo milk is commercially more viable than cow milk in terms of fat-based and SNF-based milk products e.g., butter, ghee, and milk powders. Buffalo milk has lower water and higher fat content than average cow milk. The lower cholesterol content of buffalo milk makes it more popular for humans of the new era. Camel milk is somewhat saltier than cow milk. While produced by animals living in the toughest environments, it is three times richer in ascorbic acid or vitamin C than cow milk. In some countries including Iran, Russia, Kazakhstan, and India, camel milk is prescribed for hastening recovery from a multitude of diseases, as it is rich in vitamin C, iron, unsaturated fatty acids, and B family vitamins. This chapter describes underestimated exclusive nutritional characteristics of ruminants' milk in providing healthier foods for humans of the modern and post-modern times. Cow milk will be regarded as a basis for comparisons.

Sheep Dairy Products

Depending on economics, ewe milk is mostly processed into cheeses and yogurt. Sheep's milk is richer in essential vitamins and minerals than cow milk. However, it is not yet as popular in serving home and restaurants tables. That is to some extent due to high fat content of ewe milk that is a hit with people searching for low-fat, natural dairy products. Because of the nutritional advantages over cow milk, sheep milk should be able to contribute further to the health market as a major source of essential fatty and amino acids as well as vitamins and minerals.

Besides the greater fat content (6.7 versus 2.5-5%), sheep milk is much richer in riboflavin (B2) (4.3 vs. 2.2 mg/l), thiamine (1.2 vs. 0.5 mg/l), niacin (5.4 vs. 1.0mg/l), pantothenic acid (5.3 vs. 3.4mg/l), vitamin B6 (0.7 vs. 0.5mg/l), vitamin B12 (0.09 vs. 0.03mg/l) and biotin (5.0 vs. 1.7 mg/l) than cow milk. Folate level is almost similar between sheep and cow milks (0.5 mg/l). Calcium in sheep milk is between 162 and 259 mg/100g versus 110 mg/100g for cow milk. Phosphorous, sodium, magnesium, zinc and iron levels are also higher in sheep than in cow milk. As a result, total milk solids percentages are about 18.3 and 12.1% in sheep and cow milk, respectively. Hence, for producing ice cream and related products, sheep milk becomes more viable and economical because of its concentrated substance, thus eliminating a need for adding nutritional supplements.

The Mediterranean countries produce 66% of the world's ewe milk. Sheep husbandry in Mediterranean countries has a lamb-suckling period of about 1 mo followed by intensive milking post-weaning.

In Spain, for instance, ovine milk makes up to 5% of the total milk production from cows, sheep, and goats. This amount reaches up to 20% in some regions of Spain. The average annual cheese intake in several Mediterranean regions is > 20 kg/person. Milk profiles of fatty acids (FA) are more critical in evaluating milk nutritional benefits to humans, as FA have been related to the incidence of a variety of metabolic disorders. Anticarcinogenic and antiatherogenic properties are attributed to the intake of conjugated linoleic cid (CLA) isomers, notably C18:2 cis-9, trans-11 isomer or rumenic acid (Lee et al., 2005; Bauman et al., 2006). Linolenic acid (C18:3 cis-9, cis-12, cis-15) is another FA of interest in cardiovascular complexities (Metka et al., 2006; Mesa et al., 2007). The proportions of monounsaturated and polyunsaturated FA in foods are of determining importance for human health, as well.

The ongoing research evaluates FA profiles and contents of milk from different ruminants in varying regions, to quantify effects of genetics and recent genetic improvements (Bobe et al., 2008; Carta et al., 2008), food source and other environmental factors (Metka et al., 2006; Gómez-Cortés et al., 2008).

Information on factors affecting variations in FA profiles of milk in commercial dairy sheep is limited. Factors such as flock, lactation stage, age, parity, and testing day within flocks considerably contribute to variations in milk composition of fat, protein, and somatic cell count (Gonzalo et al., 1994; El-Saied et al., 1998; Othmane et al., 2002).

Season especially under extensive and semi-extensive managements has dramatic effects on milk FA profiles and content of polyunsaturated FA of small ruminants (Chilliard et al., 2003). This is because fresh grass quality and feeding strategy vary in spring and summer. Similar to cow milk, antimicrobial residues in ewe milk can seriously compromise the health of consumers through development of antibiotic resistance or different types of immune system allergies (EMEA, 1999).

Such risks have recently led to the establishment of the National Residue Monitoring Plan for substances and residues in non-bovine milk by the European Union. Screening tests for antibiotic residues are conducted to detect milk antibiotic contaminations. An example is the Delvotest SP test of ewe milk for the growth of Bacillus stearothermophillus var. calidolactis (Althaus et al., 2002, 2003). More research on development of further identification assays for antibiotic residues in ewe milk is warranted (Yamaki et al., 2004).

Goats Dairy Products

Quantitatively, goats' milk makes a minor contribution to total ruminant milk production, when compared to cow milk. Goat milk is consumed in some regions as whole fluid milk as well as in some other regions as dairy products. The Mediterranean regions produce 18% of the world's goats milk (Table 2) (FAO, 1994).Goats and cows do not only greatly differ anatomically; they are also highly distinct in physiology and biochemistry (Haenlein, 1980a; 1992a), requiring specific strategies for their differential husbandry. For instance, goats and cows are very dissimilar in water requirements, basic metabolism, reproduction cycles and efficiency, feed preferences and feeding behavior, endocrinology, forestomach digestion, milk bio-synthesis and secretion and milk composition. The goat milk has almost no alpha-s-1-casein, beta-carotene, agglutinin; while being lower in citric acid, Na, Fe, S, Zn, Mo, ribonuclease, alkaline phosphatase, lipase, xanthine oxidase, N-acetylneuraminic acid, orotic acid, pyridoxine, folate, vitamin B12, and vitamin C, compared with cow milk (Parkash and Jenness, 1968; Jenness, 1980). Goats' milk also has lower freezing point, pH and folic acid but greater Ca, K, Mg, P, Cl, Mn, vitamin A, vitamin D, nicotinic acid, choline, inositol, medium-chain length fatty acids, small diameter fat globules, and somatic cell counts than cow milk (Droke et al., 1993). Considerable differences exist between goats and cows in mineral metabolism (Lengemann, 1970; Haenlein, 1980b; 1991a; 1992b; Ademosun et al., 1992). The lower orotic acid content of goat milk has been notified in preventing fatty liver and associated syndromes (Robinson, 1980). Moreover, the more fragile fat globule membranes in goat milk have the control over off-flavors (Patton et al., 1980; Bakke et al., 1977). Furthermore, the greater glycerol ethers in goat milk than in cow milk make the former a long-lasting palatable food for newborns (Ahrne et al., 1980).

The metabolism of acetate derived from rumen fermentation differs between cow and goats. Acetate use by the mammary gland in goats leads to fats richer in medium-chain fatty acids (Parkash and Jenness, 1968). In addition, fatty acids selection for glycerol attachment and triacylglycerol biosynthesis differs between cow and goats (Jenness, 1980). These reflect fundamental differences in cell biochemistry, genomics and enzymatic reactions between goats and cows.

As for milk proteins, beta-caseins are more prevailing than alpha-caseins in goat's milk compared to cow milk (Jenness, 1980). More recent evidences suggest that previously assumed lack of alpha-s-1 casein in goat milk is untrue (Boulanger et al., 1984; Ambrosoli et al., 1988; Mora-Gutierrez et al., 1991; Haenlein, 1991b). Certain breeds and strains have either no or low or high alpha-s-1 casein levels. Low levels hasten cheese coagulation and lessen curd firmness, cheese yield and its heat resistance that all will affect products digestibility for humans. Transgenic goat milk has been considered in treating intestinal diseases in children. Milk from transgenic goats carries a gene for an antibacterial enzyme and lysozymes, present in human breast milk, which altered intestinal bacteria in young goats and pigs fed the transgenic milk. These data may lead to milk that can protect infants and youngsters against gut complexities, which claim > 2 million young lives annually worldwide (World Health Organization). Lysozyme is a protein found in mammals' body fluids e.g., tears, saliva and milk, and in high levels in human breast milk. Goat milk contains 0.06% as much lysozyme as do breast milk. Lysozymes prevent bacteria growth via destroying cell wall and leaking out cell contents. Because of such effects, lysozymes stimulate the growth of beneficial intestinal bacteria. It is for these reasons that human milk feeding is compulsory and contributes immensely to the optimal well-being of infants.

Gene-transfer technology has been utilized to develop transgenic dairy goats with the genes responsible for human lysozymes to produce lysozyme-rich milk. Young goats and pigs were fed pasteurized, lysozyme-rich milk of the transgenic goats. Control groups received pasteurized milk from non-transgenic goats. The pigs were used due to their similar digestive systems to humans. The kid goats were used as a model for neonate ruminants including goats, sheep and cows with multi-chambered stomachs. In both animal models, the results showed that the transgenic goat milk affected gut bacteria growth, notably in opposite fashions. The young pigs fed the lysozyme-rich milk had reduced coliform bacteria in the small intestine, such as fewer Escherichia coli than the control groups. Some strains of E. coli cause severe intestinal illness. However, the kid goats on lysozyme-rich milks had increased coliform bacteria levels and approximately similar E. coli numbers, compared with the control group. All control and treatment goats and pigs were healthy and had normal growth performance. These findings clearly indicated that consuming human lysozyme-enriched milk impacts upon the bacterial population and composition of the gut in two distinct animal models. Thus, potential exists for biotechnology to improve health criteria for ruminants' milk through unique human milk constituents introduction into ruminants' milk. Producing lysozyme-rich solid milk and even transgenic dairy goat herds is a gold goal where intestinal disorders claim lives of human infants. The technology could further be applied to dairy cattle whose quantitative contribution to human nutrition is much larger than goat milk.

Last but not certainly least, whilst being a viable alternative for cow milk for optimal nutrition of growing children (Razafindrakato et al., 1994), goat milk could be a more gainful target for nutritional strategies that aim to enrich milk with conjugated linoleic acidisomers. Such specialized fatty acids have been implicated as potential protective substances against cancer and immune malfunction in humans (Bauman et al., 2006). The increasing trend in goats milk consumption in many regions including Britain, Australia, and New Zealand as well as Mediterranean and Middle East countries promises a more representative image of its nutritional significant role in human health.

Camel Dairy Products

Research results show that camel milk has on average 11.7% total solids, 3.0% protein, 3.6% fat, 0.8% ash, 4.4% lactose, 0.13% acidity and a pH of 6.5. The levels of Na, K, Zn, Fe, Cu, Mn, niacin and vitamin C are greater and thiamin, riboflavin, folacin, vitamin Bt12, pantothenic acid, vitamin A, lysine and tryptophan are relatively lower than those of cow milk. Gas liquid chromatography analysis of milk fat suggests a molar percent of 26.7 for palmitic, 25.5 oleic, 11.4 myristic, and 11.0 palmitoleic. Camel milk is more similar to goats milk and contains less short-chain fatty acids than cow, sheep and buffalo milks. In vitro protein digestibility and protein efficiency ratio values were 81.4% and 2.69, respectively, based on 90.0% and 2.50 for ANRC-Casein (Khan and Iqbal, 2001). Another study revealed that camel milk contained substantially less vitamins A (010 vs. 0.27 mg/L) and B2 (0.57 vs. 1.56 mg/L), similar vitamin E content (0.56 vs. 0.60 mg/L), and about 3 times greater vitamin C (37.4 vs. 11.0 mg/L) than cow milk, respectively (Farah et al., 1992).

In arid regions and drylands, camel is termed to be the gold goal pseudo-ruminant of the 21st century. For example, in Kazakstan, 37% of the total milk comes from camel, with sheep, yak and cows supplying 30%, 23% and 10%, respectively. Camel daily milk production varies from 3.5 up to 40 L under intensive management. Lactation lasts between 9-18 months, with peak yield occurring during the first 2-3 month of lactation. In a recent study, the range of major contents of camel milk were reported as following: fat 2.9-5.5%, protein 2.5-4.5%, lactose 2.9-5.8%, ash 0.35-0.95%, water 86.3-88.5% and SNF 8.9-14.3%, with a mean specific gravity of 1.03 (Khan and Iqbal, 2001).

The camel milk is getting more recognition as a global product supporting human health. The FAO predicts that camel dairy products to appear on European supermarket shelves. However, logistic challenges such as those of manufacturing and processing must be overcome. Despite the increasing demands from Sahara to Mongolia, the annual 5.4 million tones camel dairy products are still greatly inadequate. Sector and local investments must escalate to meet demands and create profitable markets both in the Middle East and the Western world. There are about 300 million potential customers in the Middle East and millions more in Africa, Europe and the Americas for camel dairy products. Although somewhat saltier than cow milk, camel milk represents and results from cost-effective animal husbandry under hard environmental conditions. Camel milk is 3 times richer in Vitamin C than cow milk. In many regions of Iran, Russia, Kazakhstan and India, camel milk is traditionally regarded and prescribed as a cure for disease recovery (Khan and alBukhari, 1974; Shabo et al., 2005; Yagil, 1982). Oral camel milk administration has recently proved protective against cadmium induced toxicity in rats as a model for humans (Al-Hashem et al., 2009). Camel milk is also known for its rich iron, unsaturated fatty acids and B vitamins contents. A main challenge in camel milk processing is its incompatibility with the Ultra High Temperature (UHT) exposure needed to make milk long-lasting. Another main challenge is the most camel producers' nomad nature. An additional challenge is animal related. Camels are reputedly stubborn animals. Unlike cows and other docile ruminants that store all their milk in udders, camels store their milk further up the body. Besides all, camel milk production is generally considered an inferior-tech business, such that a scanty five liters daily secretion is regarded a decent yield. Even with remaining extensive systems, improved feeding, husbandry and veterinary services should enable daily yields of up to 20-40 kg. There seems to be at least 17 million camels (Camelus dromedarius) in the world including about 12 million in Africa and 5 million in Asia (FAO). Milk production varies between 1500 and 15000 kg during a lactation cycle of between 9 and 18 months (Yagil, 1982).

Camel milk has 21 different amino acids (Osorhaan, 1979). One kg of camel milk supplies 100% of daily human requirements for calcium and phosphorus, 57.6% for potassium, 40% for iron, copper, zinc and magnesium, and 24% for sodium (Batsuh, 1994). Saturated acids comprise about 61% of milk fatty acids, including palmitin (27%), stearin (14.7%), myristic acid (12.4%), with kapron and capric acid being only 0.8% (Indra, 1989). Thus, camel milk fat is made of 39% unsaturated fatty acid. Camel milk can help treat some liver problems, lower bilirubin output, lighten vitamin inadequacy and nutrient deficiency, and augment immunity (Jadambaa, Batsuh, Baigalmaa, 2000). Fermented camel milk contains some minor amount of alcohol, and 12.4% of dry matter including 4.8% fat, 3.6% protein, 0.12 mg/kg carotene and 56 mg/kg vitamin C. One kg of fermented milk has 766 kcal energy. From each 3 kg of fermented milk, about 1 kg of curd is produced, almost half of the curd yield in sheep, goat and cow milk. Processing 100 kg of camel milk in temperatures between 50-550 C can yield 9-10 kg of sour cream with 66% 6 of fat. Processing the same amount of camel milk in temperatures between 45-550 C can generate 12-14 kg of sour cream with 35-40% of fat (Indraa and Osorhaan, 1987). One kg of curds is derived from six liters of camel milk, one kg of dried curd is produced from 9.06 l of milk, one kg of melted butter with 98% of fat is made from 15 liters of milk with 5.66% of fat, one kg of sour cream with 76.8% of fat is made from 13.5 l of milk with 5.66% of fat. Fermented camel milk is made of pure starter culture. It must be stirred 5000-6000 times by electrical beaters every 3 hours. Fermentation starts at temperatures of 20-25 degrees for the first 24 hours, after that the temperature should be cooled down to 18-200, and fermentation will be complete after 6 hours. The fermented milk can be kept for 60 days in dark bottle at a temperature of 4-50C and becomes ready for consumption. Carbonated fermented milk can be preserved for up to 90 days.

A major use of camel milk is in allergy caused by various foods, especially ruminant-born foods of mainly cow milk and milk products. Such allergies may cause anaphylactic reactions. These include 1) immediate reaction within 45 min of drinking cow milk that leads to urticaria, angioedema and likely true anaphylactic reactions; 2) medium reactions occurring 45 min to 20 hours that manifests with pallor, vomiting and diarrhea; and 3) long-term reactions happening > 20 hours with mixed reactions in the skin and respiratory and gastrointestinal tracts. Importantly, camel is a non-ruminant, despite ruminating, so called Tylopode. Camel milk is exclusive by its rather low milk fat of about 2% that is made mainly from polyunsaturated fatty acids. These fatty acids tend to be fully homogenized and form smooth white milk. In addition, camel milk proteins are the most determining constituents in preventing/curing food-born allergies. This is because camel milk lacks beta-lactoglobulin (Merin et al., 2001) and possesses a different beta-casein (Beg et al., 1986) than that of cow milk. These proteins make cow milk allergenic. Moreover, camel milk contains many immunoglobulins, being compatible with human counterparts. Camel milk is also rich in vitamin C, calcium and iron, thus empowering immune function. Camel milk was recently shown to cure allergies in children of 8 months to 10 years old (Shabo et al., 2005).

In several regions of Iran, Pakistan and India, camel milk has long been used therapeutically against dropsy, Jaundice, spleen and liver complexities, tuberculosis, asthma, anaemia, and piles (Akundov et al., 1972; Urazakov and Bainazarov, 1974; Sharmanov et al., 1978). Camel milk is healthier when it is drunk as a cool drink (Gast et al., 1969). The milk also apparently has slimming properties (Yasin and Wahid, 1957). The belief by the Bedouin of the Sinai Peninsula exists that any internal disease can be cured by drinking camel milk through driving from the body all the bacteria. Such milk properties must additionally be attributed to what camel eats. For instance, milk from camel consuming straw might not exhibit the favorable health effects. By the rule of thumb considerations, generally camel milk nutritional value would be considered the lowest following milk from ewe, goat and cow. However, 4–5 kg of camel milk and milk products are sufficient to meet nutrient requirements of a man in terms of energy, lipids, proteins and calcium.

Following immediate use, the left-over camel milk is curdled and soured. The casein prepared from this milk product is called 'industrial casein' (Pant and Chandra, 1980), as it is not very fir for human intake, and is rather used for glue and gums making. The industrial casein from cow milk is however rich in proteins. The whey proteins of camel milk are richer in nitrogen than that of cow milk (Pant and Chandra, 1980). The high proteins and amino acidproportions of camel milk casein make camel milk an appropriate food supplement for humans. The unfavorable odor and taste however, could affect its popularity. Thus, it is recommended to purify the camel industrial casein to maintain its competitive status in the human healthmarkets.

Buffalo Dairy Products

Buffalo milk is higher in total solids, fat, proteins and vitamins and lower in cholesterol compared to cow milk (Table 2). Buffalo milk also contains less cholesterol and more tocopherol or a natural antioxidant (Talpur et al., 2007). The peroxidase activity is 2 to 4 times greater in buffalo milk than in cow milk, suggesting that buffalo milk has superior natural qualities (Chantalakhana and Falvey, 1999).

Buffalo milk appears whiter than cow milk, since it lacks the yellow pigment carotene, a vitamin A precursor, although buffalo milk contains more vitamin A than cow milk. In addition, due to its greater opacity of casein miscelles and greater colloidal proteins, calcium and phosphorus, buffalo milk looks whiter with superior whitening properties than cow milk. Therefore, unlike the pale-creamish yellow cow milk and the golden yellow cow milk fat, buffalo milk is characteristically whiter.

The UHT-processed buffalo milk and cream are essentially whiter and more viscous than cow milk, partly owing to conversion of more calcium and phosphorus into the colloidal form. As a result, buffalo milk is more suitable for tea and coffee whiteners' production than is cow milk. It is also an economical alternative to cow milk for casein, caseinates, whey protein concentrates, and fat-rich dairy products production.

Buffalo milk generally possesses 58% more calcium, 40% more protein, and 43% less cholesterol than cow milk. Buffalo milk is utilized for production of butter, butter oil (clarified butter or ghee), soft and hard cheeses, condensed milks, ice cream, yoghurt and buttermilk. The chemophysical properties of buffalo milk ease processing. For instance, 1 kg of cheese requires about 8 kg of cow milk, but only about 5 kg of buffalo milk. In regions like India, about 30% of the total milk production is converted into ghee and about 20% is converted into products such as dahi (curd), khoa (dehydrated milk) and milk sweets (Chantalakhana and Falvey, 1999).

Figure 4. Modern dairy buffalo farming (right) and Mozzarella made of buffalo milk (left).

Traditions and regimes for buffalo milk intake are quite varied in different parts of the world. Nonetheless, demand is increasing for buffalo milk and dairy products. Dairy buffalo milk production has increased to about 13-15% of the total world ruminant milk production. As such, buffalo populations have been steadily and consistently rising in major buffalo raising countries. The production systems move towards intensification for improved nutrient efficiency and economics (Figure 4). Table 4 shows the top 10 buffalo milk producers in the world, led by India and Pakistan. Except Italy, all other major buffalo milk producers are in Middle East and Asia (Table 4).

Table 4. Top 10 buffalo milk producers in 2007

Country	Milk production in tons
India	59,211,000
Pakistan	20,372,000
China	2,900,000
Egypt	2,300,000
Nepal	958,603
Iran	241,500
Myanmar	220,462
Italy	200,000
Vietnam	32,000
Turkey	30,375
Total	86,574,539

Mozzarella

Mozzarella cheese is synonymous with buffalo and Italy. Buffalo production came Italy in the 17th century, while in stretched curd cheeses were made from cow milk. In the 1950's the soft stretched curd from cow milk called "Fior di Latte" was made distinct from that of buffalo milk or "Mozzarella". Mozzarella cheese intake has increased worldwide, with Italy exporting 15% of its mozzarella (Borghese, 2005). The mozzarella from Campana - close to Naples - in southern Italy has a Denomination Origin Protected (DOP) registration in the European Union as "Mozzarella di Bufala Campana". Buffalo mozzarella is a chewy textured cheese with the porcelain-white color, a very thin coat, and shaped small and round (Figure 4). Besides mozzarella are other products namely tereccia, ricotta, crescenza, robiola, caciocavallo, butter and yoghurts. Fresh mozzarella from buffalo milk is clearly distinctive from its industrial type obtained from cow milk or mixed buffalo and cow milk. Buffalo mozzarella has a shorter shelf life mainly because of greater fat and presence of natural yeasts and microbes, being maximum 3-5 days without refrigeration. The cheese is quite soft and tasty, juicy and creamy, and rich in milk and flavours and live yeasts and microbes. The industrial mozzarella from cow milk or a blend of cow and buffalo mills has a longer shelf life of about 25-30 days with refrigeration. However, it is free of live yeast and microbes, and is thus less soft and not juicy, used for making pizzas and caprese. In regions such as India, Pakistan and Iran with extensive buffalo production, making mozzarella and ricotta chesses is a promising technology for optimal human nutritional habits.

Paneer

Paneer is a special type of cheese produced in Iran, India and Pakistan. There are products similar to paneer in different parts of the world, including paneer-khiki in Iran, kareish in Egypt, armavir in the Western Caucasus, zsirpi in the Himalayas, feta in the Balkans and queso criollo, queso pais, and queso llanero in Latin America. Paneer is obtained via heat and acid coagulation of buffalo milk. A typical paneer is marble white with somewhat spongy form, firm texture and a sweetish-acidic-nutty flavor. Based on Indian standards, paneer should contain a maximum of 70% moisture and at least 50% fat in dry matter (Aneja etal., 2002).

Conclusion

As the most nutritious natural fluid, milk contains many bioactive substances that function beyond their sole nutritional properties. Thus, milk provides an ideal food for humans of all ages. The most proper appreciation by the public of such nutritional uniqueness requires thorough education and dissemination of species-wise milk functional roles in human nutrition. The chapter underlined comparative nutritional characteristics for humans of milk components and production by cows, buffaloes, sheep, goats and camels. Qualitative exclusivities and quantitative significance of ruminants' milk dairy products to modern man nutrition were highlighted. Profound knowledge into specific constituents of milk from different species (e.g., cholesterol, saturated, monounsaturated and polyunsaturated fatty acids; vitamins; minerals, favorable and unfavorable modern farming residues) will aid in adopting the most appropriate milk for different individuals. Thorough education and knowledge dissemination will be an ongoing obligation for unbiased milk natural health properties to be most deservingly realized by the new millennium's humans.

Acknowledgments

The Ministry of Science, Research and Technology of Iran and The University of Zanjan (Zanjan, Iran) are gratefully acknowledged for their supports of the author's programs towards improving science education in the new millennium.

References

Ademosun, A. A., Bosman, H. G., Haenlein, G. F. W. & Adebowale, E. A. (1992). Recent advances in nutrient requirements of goats. Proc. Fifth Intern. Conf. Goats, New Delhi, India, ICAR.

Ahrne, L., Bjoerck, L., Raznikiewicz, T. & Claesson, O. (1980). Glycerol ether in colostrum and milk from cow, goat pig, and sheep. *J. Dairy Sci.* 63, 741-745.

Akhundov, A. A., Dyrdyev, B. & Serebryakov, E.R. (1972). Effect of combined treatment on water electrolyte exchange in pulmonary TBC patients. *Zdravookhr. Turkm.* 16, 40–44.

Al-Hashem, F., Dallak, M., Bashir, N., Abbas, M., Elessa, R., Khalil, M. & Al-Khateeb, M. (2009). Camel's milk protects against cadmium chloride induced toxicity in white albino rats. *Am. J. Pharmacol. Toxicol.* 4 (3), 107-117.

Althaus, R. L., Torres, A., Montero, A., Balasch, S. & Molina, M. P. (2003). Detection limits of antimicrobials in ewe milk by Delvotest photometric measurements. *J. Dairy Sci.* 86, 457–463.

Ambrosoli, R., Di Stasio, L. & Mazzocco, P. (1988). Content of alpha-s-1 casein and coagulation properties in goat milk. *J. Dairy Sci.* 71, 24-28.

Aneja, R. P., Mathur, B. N., Chandan, R. C. & Banerjee, A. K. (2002). Principles of processing, Section 2. Technology of Indian Milk and Milk Products, Dairy India Yearbook. (Ed. and pub. P.R. Gupta), Delhi, India: p. 50.

Althaus, R. L., Peris, C., Montero, A. & Molina, M. P. (2002). Detection limits of antimicrobials in ewe milk by Delvotest. *Milchwissenschaft* 57, 660–664.

Bauman, D. E., Mather, I. H., Wall, R. J. & Lock, A. L. (2006). Major advances associated with the biosynthesis of milk.*J. Dairy Sci.* 89, 1235–1243.

Beg, O. U., von-Bahr-Lindststrom, H., Zaidi, Z. H. & Jornvall, H. (1986). Characterisation of camel milk protein rich proline identifies a new beta casein fragment. *Regul. Pept.* 15, 55–62.

Bobe, G., Minick Bormann, J. A., Lindberg, G. L., Freeman, A. E. & Beitz, D. C. (2008). Estimates of genetic variation of milk fatty acids in US Holstein cows. *J. Dairy Sci.* 91, 1209–1213.

Borghese, A. (2005). Buffalo cheeses and milk industry. Buffalo production and research (Ed. A. Borghese) REU Technical series 67, FAO regional office for Europe. 185 pp.

Borghese A. & Mazzi, M. (2005). Buffalo populations and strategies in the world. Buffalo production and research, (Ed. A. Borghese) REU Technical series 67, FAO regional office for Europe. 19 pp.

Carta, A., Casu, S., Usai, M. G., Addis, M., Fiori, M., Fraghí, A., Miari, S., Mura, L., Piredda, G., Schibler, L., Sechi, T., Elsen, J. M. & Barillet, F. (2008). Investigating the genetic component of fatty acids content in sheep milk. *Small Rumin. Res.* 79, 22–28.

Chantalakhana, C. & Falvey, L. (1999). Smallholder dairying in the tropics. ILRI (International Livestock Research Institute), Nairobi, Kenya: 462 pp.

Chilliard, Y., Ferlay, A., Rouel, J. & Lambere, G. (2003). A review of nutritional and physiological factors affecting goat milk synthesis and lipolysis. *J. Dairy Sci. 86*, 1751–1770.

CVMP/342/99 Final Report. Available: http://www.emea.eu.int/pdfs/vet/regaffair/034299.ENC.pdf.

De La Fuente, L. F., Barbosa, E., Carriedo, J. A., Gonzalo, C., Arenas, R., Fresno, J. M. & San Primitivo, F. (2009). Factors influencing variation of fatty acid content in ovine milk *J. Dairy Sci.* 92, 3791–3799.

DRI: Dietary Reference Intakes. (2007). Recommended Intakes for Individuals, Food and Nutrition Board, Institute of Medicine, National Academies. National Academy of Sciences, Washington, D.C. USA.

Droke, E. A., Paape, M. J. & Di Carlo, A. L. (19930. Prevalence of high somatic cell counts in bulk tank goat milk. *J. Dairy Sci.* 76, 1035-1039.

El-Saied, U. M., Carriedo, J. A. & San Primitivo, F. (1998). Heritability of test day somatic cell counts and its relationship with milk yield and protein percentage in dairy ewes. *J. Dairy Sci.* 81, 2956–2961.

European Medicines Evaluation Agency (EMEA). (1999). Antibiotic resistance in the European Union associated with therapeutic use of veterinary medicines: Report and qualitative risk assessment by the committee for veterinary medicinal products. EMEA/

Farah, Z., Rettenmaier, R., Atkins, D. (1992). Vitamin content of camel milk. *Internat. J.Vit. Nutr. Res.* 62, 30-33.

FAO. (1994). Production Yearbook 1993. Food & Agric. Organiz. of the United Nations, Statistical Series No. 117, vol. 47, 254.

Gast, M., Mauboisj, L. and Adda, J. (1969). Le lait et les produits laitiers en Ahaggar. Centr. Rech. Anthr. Prehist. Ethn.

Gómez-Cortés, P., Frutos, P., Mantecón, A. R., Juarez, M., de la Fuente, M. A. & Hervás, G. (2008). Addiction of olive oil to dairy ewe diets: Effect on milk fatty acid profile and animal performance. *J. Dairy Sci.* 91, 3119–3127.

Gonzalo, C., Carriedo, J. A., Baro, J. A. & San Primitivo, F. (1994). Factors influencing variation of test day milk yield, somatic cell count, fat, and protein in dairy sheep. *J. Dairy Sci.* 77, 1537–1542.

Haenlein, G. F. W. (1980a). Goats: Are they physiologically different from other domestic food animals? Intern. *Goat and Sheep Res.* 1, 173-175.

Haenlein, G. F. W. (1980b). Mineral nutrition of goats. J. Dairy Sci. 63, 1729-1748.

Jenness, R. (1980). Composition and characteristics of goat milk: Review 1968-1979. *J. Dairy Sci.* 63, 1605-1630.

Kakar, M. A., Sadullah, M-U-H., Nasrullah, J., Akhtar, K. & Ehsanullah. (2009). Buffalo Milk Processing and its Future Prospects in Balochistan, Pakistan. Pakistan *J. Zool. Suppl. Ser.*, 9, 529-532.

Khan, M. M. & al-Bukhari, S. (1974). Translation of the Meanings of the Koran. Saudi Arabia: Al-Medina Islamic University.

Khan, B. B. & Iqbal, A. (2001). Production and composition of camel milk: Review. *Pak. 1. Agri. Sci.* 38, 3-4.

Lee, K. W., Lee, H. J., Cho, H. Y. & Kim, Y. J. (2005). Role of the conjugated linoleic acidin the prevention of cancer. *Crit. Rev. Food Sci. Nutr.* 45, 135–144.

Lengemann, F. W. (1970). Metabolism of radioiodine by lactating goats given iodine-131 for extended periods. *J. Dairy Sci.* 53, 165-170.

Merin, U., Bernstein, S. D., Bloch-Damti, N., Yagil, R & van Creveld, C., Lindner, P. A. (2001). Comparative study of milk proteins in camel (Camelus dromedarius) and bovine colostrum. *Livest. Product. Sci.* 67, 297–301.

Mesa, M. D., Aguilera, G. M. & Gil, A. (2007). Efectos saludables de los lípidos en la dieta. *Aliment. Nutr. Salud* 14, 12–26.

Metka, Z., Vekoslava, S. & Rogelj, I. (2006). Milk fatty acid composition of goats grazing on alpine pasture. *Small Rumin. Res.* 64, 45–52.

Othmane, M. H., De La Fuente, L. F., Carriedo, J. A. & San Primitivo, F. (2002b). Factors affecting test-day milk composition in dairy ewes, and relationships amongst various milk components. *J. Dairy Res.* 69, 53–62.

Pant, R. & Chandra, P. (1980). Composition of cow and camel milk proteins and industrial casein. Milchwissenschaft 35, 91–93.

Parkash, S. & Jenness, R. (1968). The composition and characteristics of goats' milk: A review. *J. Dairy Sci. Abstr.* 30, 67-87.

Patton, S., Long, C. & Sokka, T. (1980). Effect of storing milk on cholesterol and phospholipid of skim milk. *J. Dairy Sci.* 63, 697-700.

Razafindrakato, O., Ravelomanana, N., Rasolofo, A., Rakotarimanana, R. D., Gourgue, P., Coquin, P., Briend, A. & Desjeux, J-F. (1994). Goat's milk as a substitute for cow's milk for undernourished children: A randomized double blind clinical trial. *Pediatrics.* 94, 64-69.

Shabo, Y., Barzel, R., Margoulis, M. & Yagil, R. (2005). Camel milk for food allergies in children. *Immunol. Allerg.* 7, 796-798.

Talpur, F. N., Memon, N. N. & Bhanger, M. I. (2007). Comparison of fatty acid and cholesterol content of Pakistani water buffalo breeds. *Pak. J. Anal. Environ. Chem.* 8, 15-20.

Thomas, C. S. (2008). Efficient dairy buffalo production. Buffalo Milking Product Area – Milking © DeLaval International AB, Tumba, Sweden.

Urazakov, N. U. & Bainazarov, S. H. (1974). The 1st clinic in history for the treatment of pulmonary tuberculosis with camel's sour milk. *Probl. Tuberk.* 2, 89–90.

Yamaki, M., Berruga, M. I., Althaus, R. L., Molina, M. P. & Molina, M. (2004). Occurrence of Antibiotic Residues in Milk from Manchega Ewe Dairy Farms. *J. Dairy Sci.* 87, 3132–3137.

Yagil R. (1982). Camels and Camel Milk. Italy: FAO (Food and Agricultural Organization of the UN).

Yagil, R. (1982). Camel Milk, FAO Production and Health Paper. 26, 9.

Yasin, S. A. & Wahid, A. (1957). Pakistan camels. A preliminary survey. *Agric. Pakist.* 8, 289–297.

In: Milk Production
Editor: Boulbaba Rekik, pp. 37-73
ISBN 978-1-62100-061-7

Chapter IV

Colostral and Lactogenic Maternal Immunity: Humoral and Cellular Factors of Induction and Transmission to the Neonate

H. Salmon*
INRA, IASP, Lymphocyte et Immunité des Muqueuses,
37380 Nouzilly, France

Abstract

Colostrum and milk, secretions of mammary gland (MG) are the two components of the post-natal delivery of maternal immunity to the neonate. In monogastrics, sIgA is the predominant colostrum or milk immunoglobulins depending upon to the degree of prenatal Ig transfer, whereas in ruminant IgG1 predominate. In ungulate as swine and ruminant absence of prenatal Ig transfer is compensated for by IgG enriched colostrum. These immunoglobulins, typically absorbed within the first 36 hours of life, by entering in blood, provide the newborn with the antibodies that arose from antigenic stimulation of the mother's systemic immune system and sustain the systemic protection of the neonate against invasive pathogens. In contrast, the passive mucosal protection of neonate mammals is dependent on the continuous supply until weaning of maternally secretory dimeric IgA (Monogastric) and IgG1 (ruminants), the so-called lactogenic immunity.

The plasma cells (cIgA+ cells) of the mammary gland produces sIgA, which is, then secreted into the milk via the poly-Ig receptor of mammary epithelial cells. These antibodies are produced in response to intestinal and respiratory antigens, including pathogens and commensal organisms likely to be encountered by the neonate shortly after birth. Protection is also mediated by cellular immunity, which is transferred via maternal cells present in mammary secretions. The mechanisms underlying the various immunological links, that is cell (cIgA+ cells) trafficking from inductor sites (mucosal surfaces) to MG, involve hormonally regulated addressins and chemokines specific to

* Email : Henri.Salmon@tours.inra.fr

these compartments. As the same pattern of adhesion molecules -vascular adressin/homing receptor- MadCAM-1/α4β7 in one hand, VCAM-1/α4β1 in another and CCL20 chemokine was observed in small gut and nasal mucosa indicate the existence of cellular link between upper respiratory tract and MG in addition to the cellular entero-mammary link. By comparison, absence of MadCAM-1 in ruminant MG, in agreement with the absence of cellular link with the intestinal immune system together with the low blood sIgA translocation to MG explain the low levels of IgA in bovine mammary secretions. The enhancement of colostrogenic immunity therefore depends on the stimulation of systemic immunity, whereas the enhancement of lactogenic immunity depends among other means of appropriate stimulation at induction sites, gut and upper respiratory tract. In addition, mammary secretions provide factors other than immunoglobulins that protect the neonate and regulate the development of mucosal immunity.

Introduction

The mammary gland (MG) is a complex highly specialized organ that has evolved to provide nutrition for the young. The production of milk by the female parent following parturition is the only feature by which mammals are distinguished from other vertebrates. The mammary secretory cells of mammals are similar with milk containing lactose and casein. There is evidence that the MG appeared before the placenta as in monotreme mammals that display reptilian features have well developed MG but no placenta (Goldman et al., 1998).

In placental mammals (eutherians), gestation has replaced lactation to varying degrees and lactation has far greater capacity to transfer nutrients than gestation. However, depending on the mammal and on the level of pre-natal transfer of immunoglobulins (Ig), the MG is more or less involved in two different mechanisms of protection: one providing the neonate with passive protection and the other protecting the MG itself. Among eutherian mammals, unlike humans, lagomorphs and rodents, which transfer some serum antibodies via the placenta, ungulates (Perissodactyla, Artiodactyla, and Cetacea) have a epitheliochorial placenta impermeable to immunoglobulins, and are therefore born hypo- or agammaglobulinemic, as shown earlier in swine (Kim, 1975). Although the foetus is immunocompetent systemically (it elaborates an immune response after intra-muscularly antigen injection) (Binns, 1967) and probably, mucosaly as evidenced by an immune response in the gut after colonization by virus (Redman et al., 1978), the newborn piglet is immunologically underdeveloped at birth so that these responses are primary immune with a long lag period (King et al., 2003), (Hammerberg et al., 1989), (Bianchi et al., 1999).

Many infectious agents in the fields, including *E. coli* and *Salmonella* spp., transmissible gastro-enteritis virus (TGE) and rotavirus, may proliferate rapidly, before the animal has had time to develop antibodies, and this proliferation may result in neonatal colibacillosis, septicemia or fatal enteritis. However, this does not preclude the presence of natural antibodies, produced in fetal mice by B1 lymphocytes, often polyreactive and binding with low affinity to multiple antigens, including microbial surfaces, thus providing protection against common environmental pathogens; but as these “innate” antibodies do not bind with high affinities, their protectice capabilities are limited (Manz et al., 2005; Bianchi et al.,

1999). Thus the survival of neonate piglets depends upon their ingestion of colostrum during the first few hours of life.

The colostrum provides the newborn with maternal serum antibodies generated through antigenic stimulation of the mother's systemic immune system. This immunity to infection, mediated by IgG and IgM, results in pathogen neutralization. Furthermore, due to the absence of previous antigenic sensitization, the neonate cannot produce their own local immune responses rapidly enough to protect at the intestinal and respiratory mucosa - the first sites to be invaded and challenged by environmental antigens.

Neonates are continuously exposed to new microbes that enter the gastrointestinal tract with food. In humans, this exposure begins with breast milk, which contains up to 10^9 microbes/l in healthy mothers (Moughan et al., 1992) and which may include many potential pathogens.

Protection against local pathogens is mostly conferred by milk-derived immunity (lactogenic immunity, a term coined by Haelterman (Haelterman, 1975)), until weaning. Bohl and Saif demonstrated that lactogenic immunity in pigs is associated with secretory IgA (Bohl et al., 1972), (Bohl et Saif, 1975), (Saif et al., 1994) resulting from antigenic stimulation of mucosal associated lymphoid tissue in the gut -the so-called entero-mammary immunological link- (Hanson et al., 1979), (Hanson et Korotkova, 2002), (Evans et al., 1980a), (Kleinman et Walker, 1979), (Fishaut et al., 1981) or repiratory tract-mammary gland link (Lanza et al., 1995b).

From a theoretical point of view, presence of these IgA antibodies in milk may be due either to the selective transport of dimeric IgA from the inductive site to the blood and from the blood by translocation through the mammary epithelium to the milk or by the migration of sensitized IgA B cells from the inductive site(s) into the MG where they transform locally into plasma cells which deliver IgA which are translocated in the milk (Goldblum et al., 1975); (Saif et Bohl, 1979).

The relationship between lactating mammary gland function in neonatal gastrointestinal function is an example of the parallel evolution of two organs; the MG and gut, which after birth, together undertake functions performed by the placenta during intrauterine life. During mammalian evolution the process of suckling has been adapted to serve a variety of different strategies for feeding and protecting the young (Weaver, 1992). Mammals differ in the selectivity and duration of their intestinal permeability to proteins (Norcross, 1982) and the period for which the intestine is permeable to immunoglobulins (i.e to absorb them intact) also varies among species and among immunoglobulin isotypes.

This chapter focuses on the mechanisms at work in sow and bovine MG and refers to selected mechanisms in other monogastric species, including rodents and human, when also relevant.

As colostrum and milk differ greatly in their immunoglobulin content, we will divide this review into sections dealing with these two secretions separately and covering the humoral and cellular elements involved in the protection of the neonate and of the MG itself. The mechanisms underlying the recruitment of cells to the MG, including the homing of lymphocytes/lymphoblasts to the MG during pregnancy and lactation will be detailed, together with approaches for increasing lactogenic immunity by enhancing immunoglobulin production in the MG and the subsequent secretion of immunoglobulin via the neonatal Fc receptor (FcRn) and pIgR.

1. Transfer of Systemic and/or Mucosal Protection to the Neonate via Colostrum

Colostrum, defined as the "first" secretions from the mammary gland present after giving birth represent the accumulated secretions of the mammary gland over the last few weeks of pregnancy together with protein actively transported from the blood stream.Colostrum has many known and unknown properties, Ig and non-Ig components and depending on the species it includes different concentrated levels of antibodies and isotypes and many of immune cells (Table 1).

1.1. Transfer of Ig across the Wall of Neonatal Gut to the Blood Circulation

1.1.1. Short-Time 24h IgAin Human

In human, the placentation is hemochorial so that transplacental transfer of IgG occurs in utero by an FcRn-mediated process: the IgG level in newborn blood (11mg/ml) is similar to that of its mother in serum (12.4mg/ml), so that there is no need to transfer IgG through the neonatal gut (Baintner, 2007). Therefore, the main immunological function of mammary secretions is therefore to protect the newborn mucosae by providing IgA.: in colostrum, the 1day post-partum, approximately 90% of IgA occurs in the sIgA form composed of equal proportion of dimeric and tetrameric molecules containing 1 molecule of secretory component (Mestecky, 2001) the high level of secretory IgA (140mg/ml) decrease very rapidly to 10 mg/ml at d3 and plateaued at d4 to the milk level , 0.6mg/ml. Small amount of IgA is transferred across the gut of newborn for 18 to 24h after birth (Ogra et al., 1978).

1.1.2. 24-36h Ig Transfer through the Gut Wall in Ungulates.

There exist some difference in cattle and pig in that the major isotype of pig colostrum is IgG, in comparison to IgG1 in bovine, but IgG1 persisted at high level in bovine milk (Salmon, 1999) , in contrast to pig, where IgA predominate in milk.

1.1.2.1. Colostral Ig filtration/production by MG

In sow, estimates of the ratio of IgA to IgG range from 0.16 to 0.22 in colostrum and 2.1 to 6.96 in milk (Berthon et al., 2000). Studies based on the radioactive labeling of immunoglobulin have shown that almost 100% of IgG, 40% of IgA and 85% of IgM of colostrum are derived from sow serum(Bourne et Curtis, 1973); similarly in bovine colostrum 100% IgG, 50% IgA and 60% IgM are serum derived. The selection of isotype is performed in the mammary gland where epithelial cells, through the mediation of FcRn and pIgR receptors on their surface, concentrate serum IgG and IgA and accumulate them in the colostral secretions. In sows, FcRn is expressed in the MG during colostrogenesis, (Schnulle et Hurley, 2003) and similarly in ruminants FcRn in late pregnancy(Mayer et al., 2002), (Sayed-Ahmed et al., 2010).

In ruminants, the Ig transport is highly selective for IgG1 but not for IgG2 (Watson, 1980) which probably diffuse through the tissues (Gitlin et al., 1976).

Table 1. Comparison of immunoglobulin (Ig) concentration in serum, colostrum and milk of neonates in various mammals, orders/species according to the placentae (number of barriers/permeability to Ig), duration of neonatal gut permeability after birth and proportion of dimeric IgA in serum

			Immunoglobulin Concentration in mg/ml					
ORDER *Species*	Placental / gut transmission		*Isotype*	MOTHER *Serum*	*Mammary secretion* Colostrum	Milk	OFFSPRING Neonatal serum At birth	After sucklings (D or Week)
					Day post-partum(c)	Day post-partum(L)		
Primates *Homo sapiens*	3/+/	-	Total IgG	12.4[br]	0.52(c1)/0.41(c2) /0.14(c3)[LJ]	0.06(L4-L18)[LJ]	11[Am]	11 (D4) [Am]
			IgA (20% dimeric)	2.9 (1.4-4)	140(c1)/30(c2)/ 10(c3)[He]	1.8 (c4=L4-L 18)[He]	0	<0.5 (2wk)
					62(c1)/69(c2)/20(c3)[LJ]/	7 (c4)/5(c5)[LJ]		
			IgM	1.2 (0.5-2)	0.6(c1)/0.7(c2)/0.3(c3)/0 .06[LJ]	0.06(c4=L4=L18 mo)[LJ]	0.2	0.3 (2wk)
Ungulata								After colostrum
Swine[1]	6/-	24-36h	IgG	22	64	1.37	-	19
			IgA (50% dimeric)	2	16	3.04	-	4
			IgM	1	4	0.98	-	1
Bovine[1]	5/-	24-36h	IgG1	11	88 (48, Ovine)	0.35	-	44(? h après birth)
			IgG2	8	2	0.06	-	1
			IgA (100% dimeric)	0.3	3	0.05	-	1.7
			IgM	2	9	0.04	-	4

Table 1.(Continued)

			Immunoglobulin Concentration in mg/ml					
Rodents								
Mouse	1/+-	21d	Total Ig	0.5-2.5 G1/0.4-1.5G2a/0.2-0.6G2b/0.1-0.2G3 [HB]	-	0.12-0.24 (L5-L9)[MB] 0.12 (IgG) [deGroot]	G1,G2a,G2b 20-30% Adult[Apple]	G1,G2a,G2b Adult level J7pp G3: 1
			IgA	0.6(L4)/0.6(L8) /0.8(L13)[Ha] -	- 0.05/(c1-L6)[VF]	0.6 (L4)/1.1 (L8) /1.10 (L13)[Ha] 0.05(L5)-0.25(L14)[VF]linear increase	Presence[Apple]	No IgA transfert (synthese propre at >=28d)
			IgM	0.2 [HB]	-	0.59	Presence[Apple] -	No IgM transfert across the gut wall(own synthesis at 5-13d)
Rat	1	26d	IgG2a	7	0.7	1(L2)/1.2(L4) /1(L19)[McG]	1	1 (birth) to 6 (at weaning 21d) [McG]
			IgG2b	0.9	0.05	0.3		0.9
			IgA (50% dimeric)	0.2	1.2	1.3(L2)/1.4(L4) /0.6(L19)	-	0 (D0-D15 then increase to attain adult value -0.15- at D30
			IgM	1	0	0.002	-	0

Superscript correspond to : Appleby (Appleby et Catty, 1983) , 1 : Silim (Silim et al., 1990), br : Brandtzaeg (Brandtzaeg, 1988), Am: Ammann (Ammann et Stiehm, 1966), He=Hennart (Hennart et al., 1991) , LJ=Lewis-Jones (Lewis-Jones et al., 1985), Ha=Halsey (Halsey et al., 1982b) , De Groot (de Groot et al., 2000), MB: Mink (Mink et Benner, 1979) , VF: Van derr Feltz (van der Feltz et al., 2001) , McGhee (McGhee et al., 1975) , WB: Wilson (Wilson et Butcher, 2004), HB: Defresne (Defresne, 1998).

The selective transport of IgG1 from maternal blood into MG secretions during colostral synthesis two to three weeks before parturition results in a marked decrease in serum IgG1 levels at this time (Butler et al., 2006), (Butler, 1998)and then the IgG1 is maintained at a low concentration during lactation. The IgA fraction of sow colostrum contains several molecular forms, with the 6.4S (IgA monomeric) and 9.3S (IgA dimeric) forms predominating over the 11S sIgA form (secretory IgA, sIgA dimeric). The high levels of the non-secretory form of IgA originate from the diffusion of blood IgA (Porter, 1979) through tissues, composed of roughly 50% monomeric IgA and 50% dimeric IgA (Vaerman et al., 1997), whereas in bovine blood, nearly all IgA are polymeric and of gut origin (Sheldrake et al., 1985b). The secretion of polymeric Ig, IgA and IgM into colostrum is mediated by the pIgR in the MG (Kumura et al., 2000), (Le-Jan, 1993).

The production of pIgR by MG epithelial cells seems to limite for IgA translocation, as the overexpression of pIgR in MG results in a doubling of IgA concentration in milk (de Groot et al., 1999 ; de Groot et al., 2000). As pIgR transport occurs continuously and independently of pIgA binding, free secretory component (SC) may be released when unoccupied pIgR are cleaved, and may therefore be found in colostrum and milk. This is the case for human breast (Mestecky, 2001), but apparently not for sow milk (Butler et al., 2006).

Recently, the developmental expression of pIgR gene in sheep mammary gland and its regulation has been studied. It was shown that the pIgR expression was upregulated during the third trimester and intensified 3 days after parturition to reach its highest level of expression during lactation. Further experiments have suggested that the enhancement of the pIgR expression resulted from combined effects of both circulating hormones (prolactin and glucocorticoids) and locally produced cytokines (interferon-gamma) (Rincheval-Arnold et al., 2002).

1.1.2.2. Transfer of Systemic and Mucosal Humoral Immunityacross the Gut

In ungulates the passive acquisition of maternal systemic/mucosal immunity by the young is mediated entirely by the colostrum ingested within the first 24-36 hours after birth. During this period the colostral proteins present in the lumen of the newborn's intestine are absorbed by highly vacuolized immature enterocytes permeable to macromolecules(Baintner, 1994). Effective over the entire length of the small intestine, this mode of transport is not IgG specific, despite the presence of the major histocompatibility complex class I-related molecule (Simister et Mostov, 1989) FcRn in gut epithelial cells in piglet (Stirling et al., 2005) and lamb (Mayer et al., 2002). This process of transcytose into enterocytes, results in high amounts of colostral protein transfer during the first 24-48h and absorption of the majority of colostral protein including IgG, IgM and IgA (with the exception of secretory IgA(Baintner, 2007); however, the detailed mechanisms involved is not fully understood (Danielsen et al., 2006). This process is obstructed in the inflamed preterm gut (Danielsen et al., 2006) and stimulated by colostral factors (Jensen et al., 2001). Once absorbed, IgA (monomeric and dimeric) may undergo reverse transudation from the neonatal blood through the epithelial cells of the respiratory tract, accounting for the observation that colostral IgA protects the upper respiratory tract in calves (Porter, 1979) and piglets (Bradley et al., 1990). Similarily, it has been shown that blood IgG is transudated across the epithelial lining of the gut, restricting the replication of enteric viruses and other pathogens. Provided that blood IgG levels are high enough (Ogra et al., 1971), IgG may also undergo retrotransudation into the gut.

During the colostrum-mediated transfer, several factors are responsible for maintaining the integrity of maternal antibodies. Colostrum has a high buffering capacity and this reduces the denaturation of immunoglobulins that would occur due to gastric proteases and extremes of pH. In the cow, colostral IgG1 seems to be more resistant to proteolysis by chymotrypsin than are other classes of antibodies. Lastly, colostrum contains a trypsin inhibitor which prevents the lytic activity of this enzyme(Watson, 1980). The intestinal absorption of colostral antibodies ceases when the newborn's enterocytes are replaced by mature intestinal epithelial cells that do not exhibit endocytosis of macromolecules. This process of "gut closure" takes place within the 36 hours after birth (Watson, 1980).

1.1.3. 21d Post-Partum Transfer Of IgGthrough Wall Gut to the Circulation in Neonate Rodents by Colostrum and then Milk

1.1.3.1. MG Ig Production

In mice, a small but significant transfer of IgG to the foetus occur *in utero* in the last 5 days of gestation but the bulk of passively acquired immunoglobulin is derived from colostrum/milkafter birth (Appleby et Catty, 1983); the transfer of IgG from milk is quantitatively much more important than via the placenta. At birth, whereas antibody obtained by transmission*in utero* is only 10% of maternal serum level, by 10 days the suckling mouse gain twice level of maternal Ig by gut absorption(Appleby et Catty, 1983); but FcRn is not involved in the secretion of IgG in milk (Israel et al., 1995), and hence colostrum/milk IgG may result from diffusion through the tissues.

1.1.3.2. Passage across Neonatal Gut Wall

Whilst in the ungulates, passage of immunoglobulins across the intestinal wall has largely terminated 36h after birth, in suckling rats and mice transfer continues for up to 16-21 days. This is due to the presence of FcRn receptor onto enterocytes of the proximal intestine (duodenum and jejunum) which bind and transport IgG until weaning. Indeed, neonatal mice homozygous for a targeted disruption of β2 M gene have reduced FcRn expression on their intestinal cells and consequently have much lower serum IgG levels during the first month after birth than littermates with functional FcR; Mainly studied in rodents, the specificity of the mechanism of transport of the IgGs (that persists during lactation) is due to the existence of FcRn receptors on the apical surface of enterocytes located in the duodenum and in the proximal jejunum of the young (He et al., 2008)These cells, which differ from those present in the ileum or in the adult, exhibit an intricate system of vesicles for endocytosis (coated vesicles) originating from the apical surface of the cell at the base of microvilli, and ensure the selective transport of IgG to the basolateral membrane. The FcRn receptors on the apical surface of enterocytes specifically bind IgGs at normal pH of intestinal contents (6.0-6.5), whereas, they do not at pH 7.4. (Rodewald et Kraehenbuhl, 1984) The binding of IgGs to receptors located on the basolateral membrane also is pH-dependent. Thus, it seems that these antibodies are transported into the enterocyte in the form of ligand-receptor complexes which dissociate only after exposure to an environmental pH of 7.4 during exocytosis in the basal membrane. The pH-gradient across the intestinal epithelium may determine this directional transport of antibodies and suggests that recycling of FcRn receptors may occur. These studies also have shown that the binding of IgG to receptors is a prerequisite for this specific transport. Following the intracellular selection, the mechanism of which is unknown to date, the membrane ligand-receptor complexes escape lysosomal degradation, in contrast to free molecules present in endocytotic vesicles (Rodewald et Kraehenbuhl, 1984). Absorption of antibodies by the small intestine of young rodents ceases when duodenal and jejunal epithelial cells are replaced by mature enterocytes originating from the crypts. These have no FcRn receptors on their surface (Borthistle et al., 1977).

1.2. Regulatory Role of Colostral Antibodyand Maternal Antibody Interference

It has been known for some time that maternal IgG downregulates neonatal Ig synthesis(Klobasa et al., 1981), (Klobasa et al., 1986), perhaps by removing environmental antigens in the same way as intravenous immunoglobulin treatment or through a direct effect on B cells. Inversely, maternal antibodies shape the repertoire of neonate (Fink et al., 2008).

Maternal immunization is commonly used in veterinary medicine to increase passive antibody transfer from the mother to the offspring. In current production conditions, sows are routinely immunized with vaccines against a number of bacterial and viral pathogens. These vaccines induce the production of high levels of IgG, IgM and IgA antibodies, which are passively transferred to the piglet via the colostrum, increasing the duration of protection against neonatal infections. However, these passively transferred maternal antibodies may interfere with the effects of active immunization on the neonatal immune system, delaying the development of active immune responses in the neonatal piglet. This interference involves a number of possible mechanisms, including the neutralization of attenuated live vaccines, the masking of B-cell epitopes and the inhibition of B-cell activation via Fc receptor-mediated signals (reviewed by (Siegrist, 2003)). Maternal antibodies (MatAbs) mostly interfere with humoral immune responses, as T-cell priming has been demonstrated in a number of models and species, the main determinant of interference being the Ag:MatAbs ratio (reviewed in (Siegrist, 2003)). It has to be stressed that the blocking of antibody formation after the vaccination of piglets with maternally acquired immunity did not mean the absence of immune response, since a memory anamnestic responses were induced in animals born to mothers subjected to immune suppression; such suppression is only apparent by the non-increase of antibody level in neonate after primary vaccination. Nevertheless, various strategies for overcoming interference of humoral response including mucosal vaccination have been evaluated. (Salmon et al., 2009) including some based on the use of recombinant adenoviral vectors. No data is currently available to support a possible regulatory role of sIgA of maternal origin, in the fetus by binding to the Fcα receptor of IgA-bearing cells.

1.3. Non Ig Factors

The transfer of cytokines from colostrum to the bloodstream of the piglet peaks two days after birth, coinciding with the gut permeability period (Nguyen et al., 2007a). Thus, colostrum may provide immune modulators (Salmon et al., 2009) during a critical window of development, facilitating the establishment of immune homeostasis. Telemo *et al.* (Telemo et al., 1991) showed that ovalbumin (Ova) fed to sows during pregnancy and lactation was transmitted to neonatal piglets in colostrum, together with the corresponding antibody, leading to a recall response in piglets weaned onto an egg-based diet. However, piglets from sows fed with Ova only during lactation displayed the rapid development of anti-Ova IgG, together with enteritis, suggesting that immunological exposure to dietary antigens in the mother may shape tolerance induction in the offspring. Furthermore, retinoic acid metabolites, which may serve as major maturation factors in the imprinting of gut lymphoidcells (Mora et von Andrian, 2004), (Saurer et al., 2007), have been found in bovine colostrums.

1.4. Cellular Factors: Colostral Cell Transfer in Neonatal Ungulata

The mammary secretions of all species contain polymorphonuclear cells, lymphocytes and macrophages, together with about $2x10^5$ to 10^7 epithelial cells/ml. These numbers in swine are consistent with those reported for mammary secretions in the alveolar and ductal lumina of other species, such as sheep(Tatarczuch et al., 2000). Cell types and numbers depend on the developmental stage of the MG and other individual conditions. For example, colostrum is enriched in phagocytic cells (PMN ~60%, macrophages 20%) (Magnusson et al., 1991), whereas milk is enriched in epithelial cells. A mean of 26% of the cells found in colostrum are lymphocytes (Evans et al., 1982).B lymphocytes are less numerous than T lymphocytes in mammary secretions, accounting for about ~30% of the lymphocyte population, with complement receptor expression observed in a smaller proportion of these B cells than of blood B cells (Schollenberger et al., 1986c). The intestinal absorption of colostral cells has been demonstrated by direct administration into the stomach of technetium-labeled cells (Tuboly et al., 1988).

Autoradiography showed that these lymphoid cells present in the colostrum were transported to the mesenteric lymph node via the lymphatic system. This absorption process is restricted to maternal colostral cells and does not apply to heat-treated colostral cells or blood lymphoid cells. However, when FITC-labeled colostral cells of maternal origin were fed to littermates (histocompatible SLA^{dd}), these cells were found in the blood and in the lumen of the gut, but were not found in the gut wall (Williams, 1993). These results suggested that colostral cells can cross the intestinal epithelium of the neonatal pig (Tuboly et Bernath, 2002). A study in newborn calves indicated that colostral leukocytes are taken up through the follicle-associated epithelium of the Peyer's patches (Liebler-Tenorio et al., 2002) and that maternal PBMC exposed to acellular colostrum entered the bloodstream of neonatal calves after ingestion(Reber et al., 2005), leading to phenotypic changes. These cells were clearly functional, as the transfer of colostral leukocytes from mother to offspring modified the blastogenic response to mitogenic stimuli in piglets (Williams, 1993) and stimulated the formation of antibodies against tetanus toxoid in newborn lambs (Tuboly et al., 1995). However, unlike gut cells, colostral cells (at least in humans) have tended to display weak cytotoxicity against virus-infected target cells and bacteria(Kohl et al., 1980). The compartment-specific adhesion molecules expressed on lymphocytes have yet to be determined.

2. Passive Mucosal Protection of the Neonate by Milk: Lactogenic Immunity

After gut closure, the colostrum is replaced by milk whichin addition to providing the neonatal intestinal environment with molecules capable of fulfilling defensive antimicrobial functions, milk like amniotic fluid, supplies growth factors, immunomodulatory and anti-inflammatory molecules promoting development and providing neonatal immune protectionand instruction (Labbok et al., 2004 ; Rumbo et Schiffrin, 2005) with Ig and cells whose level and isotype depend upon the species (Table 1).

2.1. Immunoglobulins

2.1.1. Gut absorption of colostrum and milk IgG in rodents

In mice, even though the milk contains predominantly IgA (see below), all immunoglobulin across the intestinal barrier is exclusively IgG (Appleby et Catty, 1983) and it was determined that 100% of milk IgG is serum-derived (Halsey et al., 1982a). A high level of maternally derived IgG is maintained in the circulation of youg mouse for 24d or more after gtut closure on the 16-20th day post-partum (Appleby et Catty, 1983).

2.1.2. Lacteal Ig Immunoactivity in Lumen of Ungulate Neonatal Gut

Gut closure occurs 24 to 36h after birth, barring the absorption of large macromolecules(Leece, 1973). However, the presence of the MHC class I-related molecule FcRn on gut epithelial cells may lead to IgG import or export (Stirling et al., 2005) and the epithelium lining mucous membranes is not a perfect barrier, so Ig may penetrate mucosal surfaces to some extent, as may IgA and antigens.

2.1.3. MG Translocated Lacteal IgG1 in Ruminant

In cow milk, 100% of Ig are derived by serum filtration through the MG (Newby (Newby et Bourne, 1977a) although some local synthesis of IgG1 persists in the MG during lactation (Sheldrake and Husband, 1985), which is consistent with immunohistochemical data regarding the number and location of IgG1 plasma cells in MG (Newby et Bourne, 1977b). During lactation, milk IgA are derived from the serum pool which are dimeric and primarily of intestinal origin (Lascelles et McDowell, 1974) since the secretory component (SC) molecules at the surface of mammary epithelial cells(Sheldrake et Husband, 1985) are available: in contrast, in the ewe, during the phase of involution of the mammary gland. secretory IgAs are produced locally by IgA plasma cells Sheldrake (Sheldrake et al., 1984).

2.1.4. Locally MG Produced Lacteal IgAand Immunoactivity in the Lumen of Neonatal Gut of Monogastric

Species in which IgA is present in milk and/or in colostrum (monogastric species) such as mouse (Weisz-Carrington et al., 1977 ; Tanneau et al., 1999a), rat (Parmely et Manning, 1983 ; Lee et al., 1978) and human(Brandtzaeg, 1983) have larger numbers of IgA plasma cells in mammary tissue than ruminants. Specially in human, owing to the same IgA1/IgA2 ratio in colostrum and milk as for the MG IgA plasma cells and due to the fact that 85% of IgA in serum is monomeric, it is generally accepted that all sIgA in human colostrum and milk is of local origin (Mestecky, 2001).

All immunoglobulin isotypes can protect the gut (see below) but, under natural conditions, IgM and IgG concentrations are high in colostrum and lower in milk. Thus, IgA is the dominant immunoglobulin in pig milk, as in gut secretion after weaning. This is why many studies have focused on means of eliciting high levels of IgA antibody in milk, particularly in diseases in which a continuous supply of milk antibody is required for protection. In sows, studies with a radiolabeled immunoglobulin isotype have shown that most milk antibodies (70% of IgG and more than 90% of IgM and IgA) are synthesized locally in the MG (Bourne et Curtis, 1973)which is consistent with immunohistochemical data regarding the number and location of IgA plasma cells in MG (Chabaudie et al., 1993 ; Salmon et Delouis, 1982 ; Abda et al., 1998).

2.1.5. Superimposition of Transudation and of Locally Produced Lacteal IgA

In mice, which resemble sows in terms of the predominance of IgA in milk, the IgA concentration in colostrum and milk until d5 remained to a plateau level of 0.05mg/ml and then increased to reach a five-fold higher level at day-13 of lactation(van der Feltz et al., 2001); in mice, it appears that the transport of circulating serum dimeric IgA from blood to milk is relatively important during early lactation: up to 60% of mouse milk IgA on the fourth day of gestation is blood derived, composed of 100% dimeric IgA, (monomeric/dimeric IgA~1 in blood) and by day 8 of lactation this transferred IgA was only 23% blood-derived, as diluted by the IgA produced by the MG plasma cells(Halsey et al., 1982a). The IgA transported from blood to milk and that secreted locally in the subepithelial connective tissue surrounding the plasma cells are mediated by the same receptor, the polymeric Ig receptor (pIgR) from which the secretory component is detached. Interestingly, in rat mammary gland, all the produced IgA are of local origin (Dahlgren et al., 1981) whereas the dimeric IgA are transported through the bile.

2.2. Anti-Infectious Antibody

sIgAs are more resistant than other immunoglobulins to proteolysis by secretory enzymes, such as trypsin, chymotrypsin and papain(Welsh et May, 1979). The primary function of these IgAs (and IgMs) at the mucosal surface is to block the first phase in the adhesion of enteropathogenic bacteria to enterocytes, by masking the surface receptors of these cells.These antibodies are also responsible for neutralizing bacterial enterotoxins and infectious enterotropic viruses. They induce the agglutination of many enteropathogenic microorganisms, favoring their elimination by intestinal peristalsis. However, for bactericidal action, the presence of complement and lysozyme is also required. IgAs activate complement through the alternative pathway rather than the classical pathway. They are therefore unable to induce lesions on the surface of enterocytes during the antigen-antibody reaction(Welsh et May, 1979). Moreover, in pigs, sIgA against enteropathogenic *E. coli* induce the irreversible loss of a plasmid encoding an adhesin (Porter et Chidlow, 1979). Finally, sIgAantibodies confer protection(Mazanec et al., 1995) against viruses present in the cytoplasm of epithelial cells, as shown in rotavirus infections in mice.

However, IgG antibodies in mammary secretions protect against TGE disease if they display a very high level of neutralizing activity (Bohl et Saif, 1975). This observation was later confirmed by directly testing(Stone et al., 1977) each isotype from immune colostrum. IgG was found to protect piglets if it was administered for a sufficiently long period to compensate for proteolytic degradation.

The predominance of *in situ* synthesized IgM in the guts of young piglets, up to one month of age, – and the properties of IgM antibodies (weaker adhesion to the mucous coat of epithelial surfaces than IgA and complement activation) render IgM more efficient that IgA for pathogen opsonization in the lumen of the gut. This may explain the strong correlation between levels of specific IgM in the colostrum and neonatal protection against colibacillosis (Porter et Chidlow, 1979). Polymorphonuclear cells, monocytes or macrophages and lymphocytes are present in both the gut and mammary secretions, so there may be a cooperative effect between these cells and immunoglobulins, including IgA (Honorio-Francca et al., 1997).

2.3. IgA against Dietary Antigens and Commensal Gut Microflora

IgAs antibodies are also involved in host defense mechanisms through a role in the immune exclusion of dietary antigens and gut commensal organisms. The formation of IgA and antigen complexes in the lumen of the intestine prevents the intestinal absorption of these molecules, thereby limiting their transfer to the bloodstream. In humans, IgA deficiency is often correlated with high concentrations of serum antibodies directed against antigens of the alimentary bolus or of enterotropic bacteria. The sIgAs therefore appear to play a major role in the control of allergen absorption and contribute to protection of the host against the development of allergies of dietary or environmental origin(Welsh et May, 1979). Human colostrum/milk contains IgA against food antigens (Rumbo et al., 1998).

It has also recently been shown that IgAs against commensal bacteria (Macpherson et al., 2000) are specifically induced and respond to antigenic changes within an established gut flora, this response being independent of T cell help and follicular lymphoid tissue organization. As expected, due to the enteromammary link in rodents, maternal milk contains IgA antibodies against the commensal flora of the mother/neonate and these polyreactive sIgA antibodies (Bouvet et Dighiero, 1998) limit the penetration of commensal intestinal bacteria through the neonatal intestinal epithelium in an apparently primitive process that does not require diversification of the primary natural Ab repertoire(Harris et al., 2006). The gradual decrease in the supply of maternal IgA antibodies during the suckling period allows bacterial colonization to occur (Inoue et Ushida, 2003 ; Inoue et al., 2005a ; Inoue et al., 2005b). These natural sIgA antibodies in milk (Brandtzaeg, 2003) were recently shown to decrease the spread of microbial pathogens through the population by reducing pathogen load in the feces, thereby protecting the entire herd, as well as the individual, against pathogens.

Milk may protect against both bacterial penetration and the development of allergies, by mediating antigen transfer. In mice, it has recently been shown that airborne antigens are efficiently transferred from mother to neonate through milk and that tolerance induction does not require the transfer of immunoglobulins.

The induction of this tolerance which is dependent on the presence of the transforming growth factor TGF-β during lactation was mediated by regulatory CD4+ T lymphocytes and depended on TGF-β signaling in T cells leading to antigen-specific protection against an allergy disease of the airways(Verhasselt et al., 2008). In conclusion, the transfer of an antigen to neonatal mice via breast milk resulted in the oral induction of tolerance. However, gut closure occurs much later in neonatal mice than in humans, at about 21 days post-partum, making it difficult to make a valid comparison unless the antigen gains access to the body through colostrum.

2.4. Non-Ig Factors

Mammary secretions from pig and other species contain substances with non specific antimicrobial and/or immunomodulating activities, including lactoferrin, transferrin, lysoszyme, lactoperoxidase, free cytokines, chemokines, complement, lipids, carbohydrates, non immunoglobulin macromolecules and factor binding proteins.

Lactoferrin and the derived peptide lactoferricin protect against neonatal infections by depleting the intestinal environment of iron, which is essential for bacterial growth. Lysozyme, sIgA, oligosaccharides and milk mucins, such as MUC-1, have bactericidal activity and may also protect against infection by inhibiting pathogen binding to epithelial cells.

Soluble pattern recognition receptors (PRRs), such as soluble CD14 (sCD14) and soluble Toll-like receptor 2 (sTLR2), have recently been identified in human milk(Labeta et al., 2000 ; LeBouder et al., 2003). The functional relevance of these molecules in the neonatal gut remains unclear (Rumbo et Schiffrin, 2005). Naive sows inoculated late in gestation with porcine reproductive and respiratory syndrome virus shed this virus into MG secretions. Prior vaccination seems to prevent shedding during subsequent lactation(Wagstrom et al., 2001).

2.5. Anti-Inflammatory Agents

Milk contains a number of anti-inflammatory agents, including protease inhibitors; enzymes (such as lysozyme, which binds elastin), anti-inflammatory cytokines (interleukin 10 (IL-10) and transforming growth factor-β (TGF-β), soluble receptors binding pro-inflammatory cytokines such as tumor-necrosis factor (TNF-α), enzymes degrading inflammatory mediators, epithelial growth factors and cytoprotective agents.

Epithelium-specific growth factors include lactoferrin, cortisol and polyamines, epidermal growth factor, insulin-like growth factor and transforming growth factor β (Donovan et al., 1994 ; Jaeger et al., 1987 ; Odle et al., 1996 ; Xu et al., 1999). These growth factors may provide the neonatal intestine with regulatory signals under both normal and pathophysiological conditions (Xu et al., 2000).

Other hormones and growth factors may also affect the growth and differentiation of epithelial cells, thereby limiting the penetration of antigens and pathogens and affecting other intestinal barrier functions. Thus, maturation of the biophysical and/or biochemical organization and function of mucosal barriers in neonates is accelerated by the ingestion of maternal milk in pigs.

2.6. Immunomodulatory Components

Sow colostrum/milk contains many immunomodulatory agents, including prolactin, anti-idiotypic antibodies, and nucleotides enhancing the activity of natural killer cells, macrophages, T helper (TH) cells and cytokines. These cytokines include pro-inflammatory factors (TNF-α, IL-1β and IL-6, present in milk at concentrations 10 to 20 times higher than those in serum), anti-inflammatory factors (TGF-β and IL-10), growth promoters (erythropoietin, granulocyte colony-stimulating factor and macrophage colony-stimulating factor), chemokines (IL-8, also known as CXCL8, and CCL5), Th1-type response-promoting agents (IFN-γ and IL-12) and Th2-type response-promoting agents (IL-6, IL-4 and IL-10) and TGF-β1 (Th-3) (Nguyen et al., 2007b). The biological significance of colostral/milk suppressor activity is unclear. These factors may be important for protection of the piglet's immune system against overstimulating due to sudden exposure to environmental antigens.

Maternal cytokine levels were correlated in the serum and MG, with the exception of TNF-α and TGF-β1 (Nguyen et al., 2007b), which were shown to be produced locally by MG cells. *In vitro* results have suggested that the two predominant lactogenic cytokines (TGF-β1 and IL-4) are involved in Th2 biased IgA responses and decreasing the immunologic responsiveness of neonates(Nguyen et al., 2007b). These cytokines have been detected in MG epithelial cells as well as CCL28 (MEC), a chemokine of the nasal and colonic mucosae, salivary and MG, which has been detected in sow's milk (Berri et al., 2008). This chemokine plays a dual role in mucosal immunity, as a chemoattractant for cells expressing CCR10 and/or CCR3, such as IgA plasma cells, and as a broad-spectrum antimicrobial protein secreted into low-salt body fluids (Hieshima et al., 2003). Antibacterial activity may be important for protection of both the MG itself and the neonatal gut. Two chemokines - interferon-γ-inducible protein and a monokine induced by interferon-γ, probably derived from milk cells and MG epithelial cells - have been found in human milk and are thought to enhance mucosal immunity during the early neonatal period (Takahata et al., 2003).

Intramammary inoculation with *E. coli* has been shown to lead to an increase in the local production of proinflammatory cytokines - IL1-β, IL-6, IL-8 and TNF-α - 24 h post-inoculation (Zhu et al., 2007). In the ruminant MG, IL-8 has been shown to be produced by mammary epithelial and myoepithelial cells (Barber et al., 1999 ; Zhu et al., 2007) and the increase in IL-8 mRNA levels at parturition, potentially accounts for the recruitment of a large number of PMN from colostrum at parturition (Magnusson et al., 1991). In addition to IL8/CXCL8, the ELR+CXC chemokine subfamily has six other known members. All known ELR+ CXC chemokines have been detected in human mammary epithelial cells and are present in larger amounts during lactation than during gestation(Maheshwari et al., 2003). These chemokines, like IL8, have protective effects on intestinal epithelial cells.

2.7. Cellular Transfer of Immunity by Milk Cells

Lymphocytes account for ~10% of the cells found in sow milk (Lee et al., 1983 ; Magnusson et al., 1991). Milk also contains epithelial cells (31% of all cells in sow's milk), nonnucleated cell fragments, granulocytes - mostly in the form of neutrophils (47%), a few eosinophils (1%), and macrophages (9%) (Schollenberger et al., 1986a ; Schollenberger et al.,

1986b). The phagocytes of sow milk take up heat-killed yeast, despite having a lower phagocytic index than autologous neutrophils and alveolar macrophages (Evans et al., 1982).

Porcine T-cell populations have some unique features(Piriou-Guzylack et Salmon, 2008), but little is currently known about the T-cell subsets in MG secretions. Milk T lymphocytes have a memory cell phenotype(Masopust et al., 2001, Campbell et Butcher 2002) as expected for lymphocytes in a peripheral tissue and are likely intra-epithelial lymphocytes. These cells are less reactive than those obtained from autologous blood lymphocytes, lymphocytes stimulated with certain mitogens or mixed lymphocyte cultures but they display a strong proliferative response to enteric antigens, whereas no such stimulation is observed with peripheral lymphocytes (Taylor et al., 1997). These differential reactivities suggest that the lymphocytes accumulating in mammary tissues correspond to a selected population different from that in the blood. This hypothesis has been confirmed by studies of adhesion molecules (see below). The mean ratios of CD4+ to CD8+ T lymphocytes in the peripheral blood and MG secretions were 1.53 and 0.85, respectively. Activated CD8+ T lymphocytes may play an important role in the regulation and expression of the local immune response to pathogens(Park et al., 1992). After gut closure, colostrum is replaced by milk, which contains all the immune components required for the passive transfer of specific cellular immunity. There has been some debate about the ability of milk cells to cross the neonatal intestinal epithelium to gain accesss to distant sites. For example, some studies in humans and rodents have demonstrated the transmission from mother to offspring, via mammary secretions, of such reactions as cell-mediated hypersensitivity(Mohr, 1972) and skin transplant rejection(Beer et al., 1974). However, transfer efficiency depends upon the ability of the cells to survive in the digestive tract of the young, unless passive acquisition of the immunity mediated by maternal cells results from the passage of soluble factors produced by lymphocytes, such as transfer factor, rather than from the actual transfer of lymphocytes themselves (Welsh et May, 1979). In newborn pigs, intestinal epithelium lymphocytes (IELs) are devoid of NK activity against TGEV-infected cells and the transfer of blood mononuclear cells from adult pigs has been shown to increase resistance to TGEV (Cepica et Derbyshire, 1984).

3. Induction of IgA Response in Colostrum/Milk: Humoral and Cellular Factors for Lymphocyte/Plasma Cells Recruitment to the Mammary Glandin Rodents and in Pig

In ruminants as IgG1 is main isotype present in MG secretion, by systemic immunization the higher IgG1 production is transported to MG and hence in the colostrum and milk. Similarly, the intestinal immunization enhances the IgA in the MG as resulting of selective transudation of IgA molecules release from the immunized gut. and conversly, immunization of the mammary gland may result in gut response(Sheldrake et al., 1985a).

In monogastric contrary to ruminants, the fact that colostral and milk IgA antibodies are produced locally in the MG by pIgA-secreting plasma cells prompted us to investigate the anatomical origin of such cells and the exploration of immunization routes effective in induction and excretion of sIgA antibodies in colostrum/milk.

3.1. Kinetic and Origin of Lymphocyte Subsets (Excluding Plasma Cells) Recruited to the MG during Pregnancy

In sows,all the cell types involved in the immune response were present in the mammary gland at the various stages of gestation and lactation, and moved closer to the alveolar epithelium as gestation proceeded: T lymphocytes, including CD4+ and CD8+, B lymphocytes and class II-bearing cells (epithelial cells and macrophages). T lymphocytes - specifically T helper cells - accumulated early in pregnancy(Salmon et Delouis, 1982; Salmon, 1986; Salmon, 1987). The local accumulation of immune cells and the increase in CD8+ cell numbers near the epithelium suggest a role in local immune defense(Salmon, 1987 ; Chabaudie et al., 1993).Only a few of the functions of mammary lymphocytes have been explored yet. These cells have levels of PHA/conA stimulation ratio similar to those of blood lymphocytes (Salmon, 1987), whereas milk lymphocytes tend to be less reactive to PHA than blood cells or to the antigen injected into the MG (Evans et al., 1982).

In bovine, the 3 components involved in production of an immune response, namely antigen-specific T cells, an immune recognition molecule (MHC class II molecule, Fitzpatrick et al, 1992) and antigen itself are present in the MG (see below, milk cells). In the MG of ewe, both pregnant and non-pregnant the great majority of the lymphocytes in the epithelium were agranulated CD8+ cells; B lymphocytes, were present in much lower concentrations and were located mainly in the connective tissues. Many of the lymphocyte transfer studies carried out in the 1980s were based on lymphocytes radiolabeled with ^{51}Cr (which labels lymphocyte and lymphoblasts) or ^{3}H uridine. and showed the presence of a pool of small virgin lymphocytes recirculating through the lymphoid organs (systemic or nodal pool): in addition, in sheep a further pool of gut lymphocytes (intestinal pool) (Cahill et al., 1977) was defined. Several mucosal compartments of the immune system were discovered only when it was realized that activation directed lymphocytes like lymphoblasts, to the peripheral non lymphoid tissues (Butcher et al., 1999a ; Butcher et Picker, 1996) and more specifically the precursor of IgA plasma cells.

It is now well known, that the migration pathways followed by lymphocytes and lymphoblasts in the organism, as reflected in the compartmentalization of the immune system into systemic and local immunity(Kunkel et Butcher, 2002), are determined by the expression of particular structures on the surface of endothelial cells (vascular addressin) and of complementary structures on the membranes of lymphocytes (homing receptor) (Springer, 1994 ; Butcher et Picker, 1996 ; Butcher et al., 1999b)and tissue chemokines interacting with their corresponding receptors on lymphocytes and lymphoblasts.As expected for a tertiary lymphoid organ, PNAd (Peripheral lymph node Addressin) is not present on the blood vessels of the MG at any stage of development, which is consistent with the absence in this organ of virgin lymphocytes as evidenced by the absence of L-selectin expression. This strengthens the above conclusion that only memory cells migrate into the MG and then in milk (see below). One of the ligands of L-selectin, GLYCAM-1 is not present on the endothelial cells of blood vessels, but is found in mouse epithelial cells and in milk. However, this protein lacks the sulfate-modified carbohydrate required for interaction with L-selectin

(Dowbenko et al., 1993). Unlike mammary lymph nodes, the endothelia of venules within the lamina propria of the gut and of the capillaries around MG acini express a common mucosal addressin cell adhesion molecule, MAdCAM-1 (the mucosal addressin cell adhesion molecule, specific to gut blood vessels) (Streeter et al., 1988). In mice, the MAdCAM-1 expression on MG blood vesselsincreases during pregnancy(Nishimura et al., 2000), (Tanneau et al., 1999a). We have also shown a correlated increase in the number of T cells expressing the α4β7 integrin (the corresponding homing receptor of MAdCAM-1) (Figure 1). Thus, α4β7 T lymphocytes (mostly CD8 T cells located in the epithelium), probably of gut origin (Bourges et al., 2004), accumulate in the MG during pregnancy depending on the proportion of endothelial cells expressing MAdCAM-1. The CCL25 was the only epithelial chemokine found to be increased during pregnancy (Bourges et al., 2008) (Figure 1). We hypothesized that CCL25 was responsible for T-cell recruitment in the MG during pregnancy as was demonstrated for T cells in the gut.

3.2. Plasma Cell Recruitment in Lactating MGand Local Ig Synthesis

In mice, during pregnancy and the early days of lactation, an increase in the number of plasma cells occurs, and by 1 week of lactation, there is a marked increase in the number, and most are synthesizing IgA. This increment parallels the development and proliferation of the glandular epithelium, in anatomical relation to which the plasma cells are observed (Weisz-Carrington et al., 1977). In both sows and mice, IgA plasma cells accumulate in lactating MG after delivery, whereas T cells do not (Chabaudie et al., 1993; Magnusson, 1999). This accumulation leads to an immediate increase in IgA synthesis in the MG and elsewhere in the mother after delivery, resulting in high postpartum serum IgA concentrations (Butler et al., 2006). Research on mice has demonstrated that IgA plasma cell precursors formed in the gut in response to antigenic stimulation home to both the lamina propria of the gut and to other secretory organs, including the MG (via the so-called gut-mammary axis), where they mature into plasma cells and secrete dimeric IgA (Roux et al., 1977). In sows, evidence for an immunological link between the gut (or the upper respiratory tract (URT)) and the MG is provided by observations that specific IgA antibodies are present in milk after stimulation of the gut or the URT with the corresponding antigens. This was particularly clear in experiments using very similar viruses replicating in the gut (TGEV) (Evans et al., 1980b). An IgA-based immunological link between the mammary gland and the upper respiratory tract is suggested by the presence of milk IgA in sows infected late in gestation with the porcine respiratory coronavirus (PRCV, a deletion mutant of TGEV replicating in the respiratory tract) (Lanza et al., 1995a).

3.2.1. Humoral and cellular factors responsible for recruiting distant plasma cells to the lactating MG: the enterobronchomammary pathwayRodents

We investigated the mechanisms of lymphocyte precursor homing to the MG, with a view to increasing the protective effects of maternal milk for the progeny and the MG itself (Salmon, 2000). The increase in α4β7/IgA plasma cell levels during lactation, at a time when MAdCAM-1 levels were beginning to decline (Figure 1) in both sows and mice(Tanneau et al., 1999a ; Czarneski et al., 2002) suggested a role for additional factors.

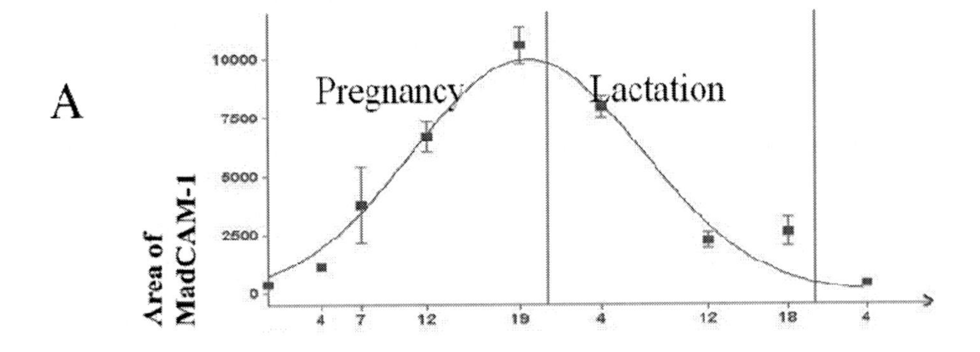

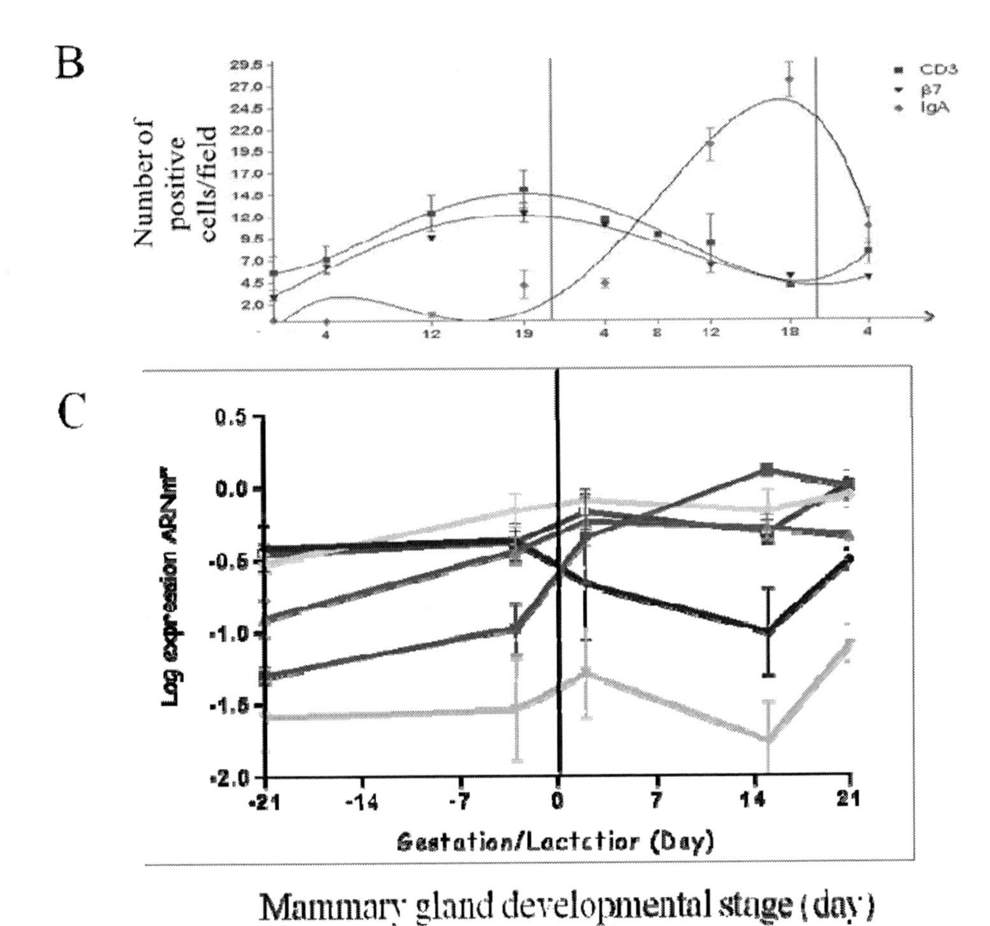

Figure 1. Kinetics of MAdCAM-1 (A) , T cells/ β7, IgA B cells (B) and epithelial chemokines (C) in mouse MG during pregnancy and lactation.

Thus, searching for substances in milk acting as chemoattractants (Czinn et Lamm, 1986), we have found a peptide derived from bovine β-caseinwhich ischemoattractant forpig lymphoblastsas strong as lactoserum itself(Fronteau et al., 1998), consistent with the similar levels of activity observed with mesenteric lymph node B-cell IgA and IgG in mice (Czinn et Lamm, 1986). More Recently, we have isolated, in a sow milk ultrafiltrate (10 kDa) (Fronteau et al., 1998) a small peptide of mammary-associated amyloïd SAA3 (McDonald et al., 2001) chemoattractant for the whole population of Ig-bearing B lymphoblast(Rodriguez B. et al., 2009). Furthermore, CCL28 (MEC) (Wang et al., 2000), a chemokine expressed by many different types of epithelial cells, including those of the gut and MG (Mickanin et al., 2001 ; Meurens et al., 2006), has been shown to be upregulated by lactation (Berri et al., 2008). As this molecule also acts as a gut IgA plasma cells chemoattractant (Lazarus et al., 2003 ; Hieshima et al., 2003), it was thought that it might play a major role in the accumulation of IgA plasma cells in the lactating MG. Such a role was demonstrated by blocking plasma cell immigration into MG with anti-CCL28 antibodies (Wilson et Butcher, 2004 ; Wilson et Butcher, 2004) or by knocking out CCR10 receptor (Morteau et al., 2008 ; Morteau et al., 2008 ; Morteau et al., 2008). These treatments resulted in an absence of MG plasma cells and correlatively (Wilson et Butcher, 2004) an absence of IgA in milk.In another hand, another vascular cell-adhesion molecule (VCAM-1) is detected to the large blood vessels of mouse mammary gland to the exclusion of small vessels, which represent the sites of lymphocytes extravasation(Tanneau et al., 1999b). At variance, in sow, we recently showed that VCAM-1 (vascular cell-adhesion molecule-1) was expressed on endothelial cells of small blood vessels as well as onto connective cells of the lactating mammary gland where it may contribute to α4β1/cIgA-cell retention (Bourges et al., 2008), those cells specially coming from the upper aerodigestive tract both in pig (Bourges et al., 2007) and in mice (Xu et al., 2003 ; Xu et al., 2003): indeed IgA plasma cell recruitment was blocked in mouse by anti-VCAM-1 whereas anti-MadCAM-1 was ineffective (Low et al., 2010 ; Low et al., 2010). Interestingly, in bovine MG, MAdCAM-1, was not detected at any of the four different physiological stages, (Hodgkinson et al., 2007) in agreement with the absence of β7+ cells in MG (Harp et al., 2005); furthermore, VCAM-1 was found onto large blood vessels as in lactating mice (Hodgkinson et al., 2007) supporting, together with other findings, that the immune responses of the ruminant MG are of a peripheral nature in contrast to the MG of monogastric mammal(Harp et al., 2005). These results show that MG recruit sIgA plasma cells cells from both nasal/bronchial and gut compartments whereas in ruminants sIgA is filtered from serum and in both case specificity of IgA is common to the enterobrocho-mammary axis (Figure2). In another hand, in ruminant the deficiency of endothelial cells membrane receptors preventing B cell traffic through the blood vesssels may be overcome by inflammation(Meeusen et al., 1991); 14 days after St. aureus antigen infusion, there was an accumulation of plasma cells, mostly of IgA isotype, suggesting that B cells and helper T cells interaction can take place at the local site of antigen stimulation in the MG (Lee et al., 1992), (Loving et Magnusson, 2002)

Figure 1-A: T-cell numbers (CD3, and β7, the lymphocyte homing receptor) increase during pregnancy, together with MAdCAM-1level, but decrease during lactation. By contrast, (Figure1B) IgA B cells, which are present in only very small numbers during pregnancy, increase in number during lactation, whereas MAdCAM-1 levels decrease (Modified from (Salmon, 2003).. In Figure1-C, kinetics of epithelial chemokines, (MEC/CCL28,

TECK/CCL25, BRAK/CXCL14, SDF-1/CXCL12, MIP3β/CCL19, MIP3α/CCL20) in mouse MG. Note that MEC (in red) is the only one chemokine which increase in lactation, corresponding to the sIgA cell increase.

3.2.2. Regulation of pIgR and Cell Trafficking

As one pIgR molecule transports only one dIgA molecule (1: 1 ratio) and as the pIgR is not recycled after each round of transportimplies that the amount of available receptor could be a rate-limiting factor for IgA into the milk.Therefore,(de Groot et al., 2000)transgenic mice over-expressing (60 to 270-fold above normal levels) the pIgR protein in mammary epithelial cells showed may express twice more IgA levels than conventional mice (de Groot et al., 1999). The expression of pIgR may be influenced by many biological effector molecules, including hormones, cytokines such as IL-4 and IFN-γ (Kaetzel, 2005), presumably derived from T cells found in the vicinity of epithelial cells (Phillips et al., 1990) and metabolic products (promoter regions with E-box elements (Martin et al., 1998).

Mammotropic hormones also seem to enhance the binding of antibodies to mammary epithelial cells and their transport across the epithelium(Weisz-Carrington et al., 1984). In other hand, cell trafficking itself may be also under hormonal control. The treatment of virgin mice with a combination of progesterone, estrogen and prolactin induces MG development and leads to an increase in the number of IgA plasma cells in this organand intra-epithelial IgA (Weisz-Carrington et al., 1977), (Weisz-Carrington et al., 1978). However, this increase in IgA plasma cell numbers may be a direct consequence of mammary tissue development, instead of concomitant increase in the density of receptors on epithelial cells to "trap" the circulating lymphoblast precursors of IgA plasma cells and/or to retain them in the MG. The presence of an ERE (estrogen-responsive element) in the promoter of MAdCAM-1 gene (Sampaio et al., 1995) is in agreement with the higher expression of MAdCAM-1 in pregnancy than in lactation.

Recently, it was shown that supplemental beta-carotene increases IgA-secreting cells in mammary gland and IgA transfer from milk to neonatal mice probably by increase of immigration of sIgA cells from the ileum to the MG (Nishiyama et al., 2010) and the observation of IgA increase in milk after supply of shark-liver oil to sows during gestation and lactation. (Mitre et al., 2005) may be relevant to this process. In sows, the density of prolactin receptors in mammary tissue has been shown to be correlated with the accumulation of lymphocytes in this organ (Salmon, 1987).

In both sowand bovine, the IgA plasma cells are originally produced and poised (they express α4β1 in URT, α4β7 in PP) in inductor sites such as tonsils for the upper respiratory tract (URT) and Peyer's patches for the gut, then they emigrated in effector sites belonging to the same compartment (respiratory or intestinal) according to the presence of the same vascular addressins and chemokines as those of the inductor site, such as VCAM-1 (recruiting the α4β1 cells) and MEC (plain green circles) for the URT and MadCAM-1 (recruiting the α4β7 cells) and TECK/MEC in the gut (2), respectively. As the endothelial cells of the sow MG express both VCAM-1 and MadCAM-1 and as epithelial cells secrete MEC, the IgA plasma cells originating from the URT and from the gut home to the MG thus constituting, respectively the broncho- (1) and entero- (2)-mammary axis. These plasma cells continue to secrete their IgA in the milk with the pIg receptor, such as to increase the original IgA production.

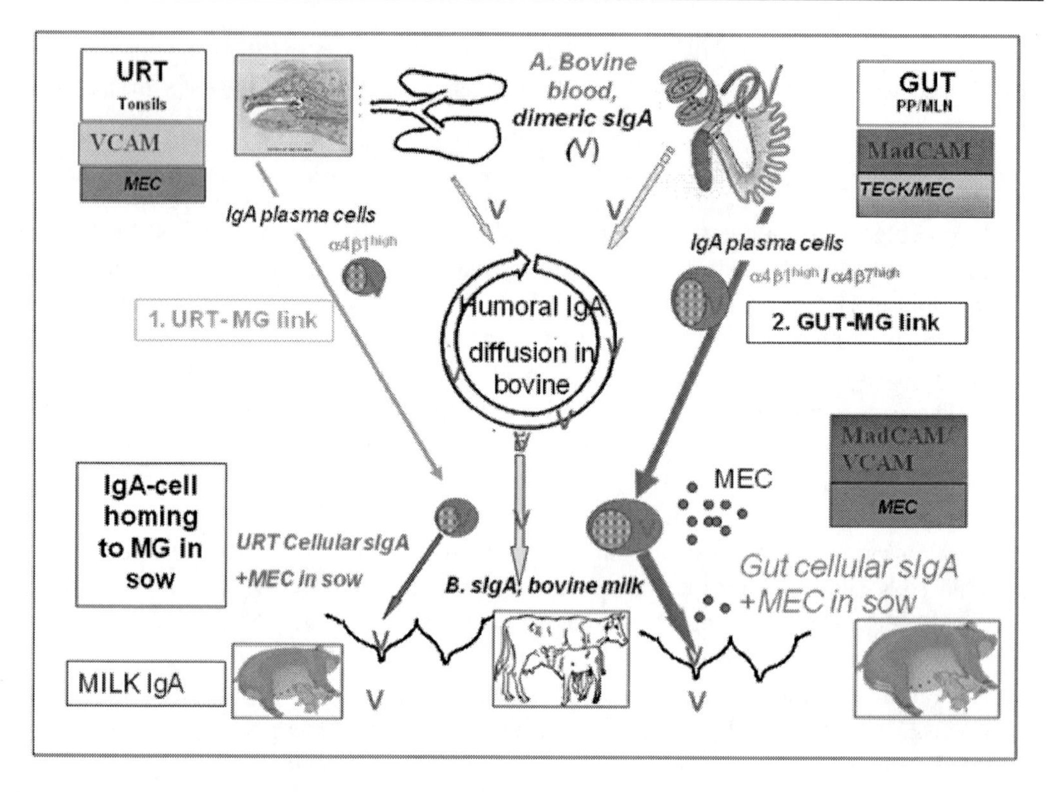

Figure 2. Broncho-enteromammary links, cellular in sow as compared to humoral in bovine.

In bovine, there is no vascular addressin to retain these cells, but as 100% of blood IgA are dimeric (V) and secreted from URT and/or the gut in the blood, they gain access to the milk.via the MG epithelium. Therefore there is much less sIgA in the bovine milk than in sow milk.

4. Intramammary Immunization

As there is a migratory flux of lymphoid cells from the mucosal inductor sites –distant to the MG- to the MG, one may envisage enhanced induction of these lymphoid cells. Several factors have been identified underlying the preferential expression of the IgA isotype in the B cells of Peyer's patches and that depend on the interaction of B cells with T cells and dendritic cells derived from these lymphoid follicles(Fagarasan et Honjo, 2003). The switch from sIgM B cells to sIgA B cells is controlled by TGF-β(Stavnezer, 1996); Th2 lymphocytes are involved in amplifying the pool of B cells expressing membrane IgA, through the production of IL-4. On the other hand, Th2 lymphocytes also act later on the effector sites of immune responses (intestinal mucosa) where they control the production of IgA antibodies, by secreting IL-5 and IL-6. IL-5 stimulates IgA production by IgA+ B cells stimulated with an antigen. It increases the number of IgA-secreting cells, whereas IL-6 increases the rate of antibody synthesis. This work requires extension to MG tissue to determine whether there is an MG-specific microenvironment. Meanwhile, there is only one recent study in cattle demonstrating the presence of cytokine mRNA in milk cells. No TNF-α, IL-2 or IL-4 was detected, whereas mRNAs for IFN-γ, IL-6, IL-10, IL-12 were found in milk (Taylor *et al.*, 1998). Inasmuch as MG of cattle is not a good IgA producing organ, further research on cytokine environment in MG of monogastric animals is needed.

The presence of lymphocytes capable of mounting an immune response in MG, and antigen-presenting cells (such as dendritic cells) raise the possibility that there is a genuine local immune response. Furthermore, the MG may represent a better route of immunization than the gut, which may require less antigen dose because of the absence of protein degradation in MG.

The use of this route would also simulate natural conditions, in which piglets may inoculate the MG during suckling. Intramammary immunization may mimic mucosal immunization if the Ag is infused in the teat canal, resulting in the introduction of antigens directly into the lumen. Thus, local infusions of antigen (ferritin) late in pregnancy increase the number of plasma cells in the immunized MG and the local IgA response (Bennell et Watson, 1979). Several experiments using intramammary inoculation of live virus, may be relevant to an Ag infusion inasmuch as the virus replicate into epithelial cells. Hence, intra-mammary inoculation with live attenuated TGEV in pregnant sows leads to the induction of persistently high levels of neutralizing IgG antibodies in milk, whereas inoculation of the MG during lactation results in the production of IgA antibodies (Saif et Bohl, 1983). However, injecting the virus by infiltration (to inject the live virus into the duct epithelial cells, those which are present in pregnant MG) into the MG of pregnant sows, we obtained an IgA immune response in the milk of each of the 18 tested glands (Salmon, 1995). Antibody activity was detected in the lactating glands that did not receive an injection (albeit at lower levels).

There are several possible explanations for this activity: the virus (or precursors of plasma cells) may be transferred from one gland to another or the virus (or dendritic cells conveying the virus) may gain access to the gut (although no signs of diarrhea were observed in this experiment) leading to the stimulation of gut lymphocytes, which may in turn have been trapped in the MG, with local restimulation in glands previously injected with the virus. Indeed, it was shown that antigens, although not formally required for the ultimate localization of IgA-cell precursors in secretory sites, are involved in the expansion of homed precursor populations. Thus, even if the immune response is initiated at a distant site, the sensitized lymphocytes may multiply further and some may undergo another cycle of homing (Husband et al., 1996).

In contrats to Ag infusion, intra-mammary injection in sows of ferritin alone induces an IgM response, whereas the injection of this molecule or ovalbumin together with complete Freund's adjuvant leads to local granuloma formation, the targeting of the antigen to the secretory site, a predominantly IgA response and the production of antigen-reactive lymphocytes in both peripheral blood and milk (Evans et al., 1982). The local inflammation may be responsible of the local IgA production, as exemplified in virgin ewes after intrammamary infusion of inactivate *Staphylococcus aureus*, by the modification of adhesion molecules on B and T cells and/or their ligands on the endothelium(Meeusen et al., 1991) and/or interactions between B cells and helper T cells at the local site of antigen stimulation in the MG (Lee et al., 1992).

Conclusion

The mechanism underlying the differences in the humoral and cellular components of colostrum and milk is beginning to be revealed by identifying involved adhesion molecules and chemokines (Figure 2). Gene expression profiling of MG development may reveal new important factors including immunological factors (Clarkson et al., 2004 ; Pfaffl et al., 2003). Besides the results in normal MG, chemokine receptors are involved in breast cancermatastasis (Muller et al., 2001). The findings about the adhesion molecules and chemokines in normal MG suggest that it should be possible (1) to increase protection of the MG itself by modifying MAdCAM-1 expression via its ERE (Sampaio et al., 1995) and hormonal stimulation or by inflammatory cytokines(Sikorski et al., 1993), resulting in the recruitment of more CD8 lymphocytes and (2) to enhance mucosal protection of the newborn through the production of milk richer in IgA. The recruitment of IgA plasma cells may be enhanced by increasing the levels of both MAdCAM-1 and chemoattractant factors, particularly MEC chemokine. Alternatively, T cell recruitment could be enhanced by increasing the secretion of TECK. Unfortunately, very few data are available concerning the control of chemokine or chemokine receptor levels in conditions other than inflammation(Janis et al., 2004 ; Fanti et al., 2003). The pathway of IgA lymphoblast circulation depends on the considered animal species; at least in absence of inflammation, ruminant MG is of a peripheral nature in contrast to the MG of mono-gastric mammal.

Lactogenic immunity is a potentially exciting area of research for the development of vaccines inducing mucosal/secretory immunity, because inactivated or attenuated antigens often fail to retain their immunogenicity for mucosal (i.e intestinal) IgA responses and, hence, for IgA-mediated milk immunity. We now need to improve our understanding by investigating further the mechanisms involved in the induction of IgA isotype antibody production by B-cell precursors, such as those in Peyer's patches, without neglecting the role of auxiliary T cells, antigen-presenting cells and the nature of the antigen. IgA lymphoblasts within the mucosa differentiate into plasma cells, which locally produce specific secretory antibodies. Little is known about the possible *in situ* regulation of this local humoral response by lymphokines and/or by tissue factors released by adjacent cells.

Finally, with improvements in our knowledge of the humoral and cellular factors governing the migration of lymphoblasts, it should be possible to induce the specific migration of immunocompetent cells to the MG, to initiate or enhance the specific humoral immune response and to improve the immune quality of the milk produced by the MG. Furthermore, similarly to gut lymphoblasts, dendritic cells may be involved in the transfer of some maternal lactobacillus strains to the neonatal gut (Martin et al., 2004). As porcine/bovine MG is a convenient bioreactor (Morcol et al., 1994) for complex proteins, milk IgA purified from sows immunized with human gut pathogens could be developed as a means of preventing human disease.

References

Abda, R., Chevaleyre, C., &Salmon, H. (1998). Effect of cryopreservation on chemotaxis of lymphocytes. *Cryobiology.*, 36, 184-193.

Ammann, A.J., & Stiehm, E. R.(1966). Immune globulin levels in colostrum and breast milk, and serum from formula- and breast-fed newborns. Proc Soc. *Exp Biol Med*, 122, 1098-1100.

Appleby, P., et Catty, D.(1983). Transmission of immunoglobulin to foetal and neonatal mice. *J Reprod Immunol*, 5, 203-213.

Baintner,K.(1994). Demonstration of acidity in intestinal vacuoles of the suckling rat and pig. *J Histochem Cytochem*, 42, 231-238.

Baintner K.(2007). Transmission of antibodies from mother to young: Evolutionary strategies in a proteolytic environment. *Veterinary Immunology and Immunopathology*, 117, 153-161.

Barber, M.R., Pantschenko, A.G., Hinckley, L.S., &Yang, T. J.(1999). Inducible and constitutive in vitro neutrophil chemokine expression by mammary epithelial and myoepithelial cells. *Clinical & Diagnostic Laboratory Immunology*, 6, 791-798.

Beer, A.E., Billingham, R.E., &Head, J.(1974). Proceedings: The immunologic significance of the mammary gland. *Journal of Investigative Dermatology*, 63, 65-74.

Bennell, M.A., et Watson, D. L.(1979). The local immune response in the mammary gland of the sow following infusion of a protein antigen. *Microbiology & Immunology*, 23, 1225-1231.

Berri, M., Meurens, F., Lefevre, F., Chevaleyre, C., Zanello, G., Gerdts, V., &Salmon, H.(2008). Molecular cloning and functional characterization of porcine CCL28: possible involvement in homing of IgA antibody secreting cells into the mammary gland. *Mol. Immunol.*, 45, 271-277.

Berthon,P., Tanneau,G., and Salmon,H. (2000). Immune factors of mammary secretions. p. 453-480. *In* Martinet J, and L.M. Houdebine (ed.) Biology of Lactation. INRA.

Bianchi, A.T., Scholten, J.W., Moonen-Leusen, B.H., &Boersma, W.J. (1999). Development of the natural response of immunoglobulin secreting cells in the pig as a function of organ, age and housing. *Dev. Comp Immunol*, 23, 511-520.

Binns, R. M.(1967). Bone marrow and lymphoid cell injection of the pig foetus resulting in transplantation tolerance or immunity, and immunoglobulin production. *Nature*, 214, 179-180.

Bohl, E.H., Gupta, R.K., Olquin, M.V., &Saif, L. J.(1972). Antibody responses in serum, colostrum, and milk of swine after infection or vaccination with transmissible gastroenteritis virus. *Infect. Immun.*, 6, 289-301.

Bohl, E.H., et Saif, L. J.(1975). Passive immunity in transmissible gastroenteritis of swine: immunoglobulin characteristics of antibodies in milk after inoculating virus by different routes. *Infect. Immun.*, 11, 23-32.

Borthistle, B.K., Kubo, R.T., Brown, W.R., Grey, H. M.(1977). Studies on receptors for IgG on epithelial cells of the rat intestine. *J Immunol*, 119, 471-476.

Bourges, D., Chevaleyre, C., Wang, C., Berri, M., Zhang, X., Nicaise, L., Meurens, F., &Salmon, H. (2007). Differential expression of adhesion molecules and chemokines between nasal and small intestinal mucosae: implications for T- and sIgA+ B-lymphocyte recruitment. *Immunology*, 122, 551-561.

Bourges, D., Meurens, F., Berri, M., Chevaleyre, C., Zanello, G., Levast B., Melo S., Gerdts V., Salmon H.(2008). New insights into the dual recruitment of IgA+ B cells in the developing mammary gland. *Mol Immunol*, 45, 3354-3362.

Bourges, D., Wang, C. H., Chevaleyre, C., and Salmon, H. (2004). T and IgA B lymphocytes of the pharyngeal and palatine tonsils: differntial expression of adhesion molecules and chemokines. *Scand. J Immunol.* Ref Type: Generic

Bourne F.J., et Curtis J.(1973). The transfer of immunoglobins IgG, IgA and IgM from serum to colostrum and milk in the sow. *Immunology*, 24, 157-162.

Bouvet J.P., et Dighiero G.(1998). From natural polyreactive autoantibodies to a la carte monoreactive antibodies to infectious agents: is it a small world after all? *Infect. Immun.*, 66, 1-4.

Bradley P.A., Bourne F.J., &Brown P.J. (1990). The respiratory tract immune system in the pig. I. Distribution of immunoglobulin-containing cells in the respiratory tract mucosa. II. Associated lymphoid tissues. *Veterinary Pathology*, 1976. 13, 2-89.

Brandtzaeg, P.(1983). The secretory immune system of lactating human mammary glands compared with other exocrine organs. *Ann. N. Y. Acad. Sci.*, 409, 353-382.

Brandtzaeg, P.(1988). Immunology of the human gastrointestinal tract in health and disease--a brief review. *Keio J Med*, 37, 168-176.

Brandtzaeg, P. (2003). Role of mucosal immunity in influenza. *Dev. Biol. (Basel)*, 115, 39-48.

Butcher, E. C. et Picker, L. J.(1996). Lymphocyte homing and homeostasis. *Science*, 272, 60-66.

Butcher E.C., Williams M., Youngman K., Rott L., Briskin M.(1999a). Lymphocyte trafficking and regional immunity. *Adv. Immunol.*, 72, 209-253.

Butcher, E.C., Williams, M., Youngman, K., Rott, L., &Briskin, M.(1999b). Lymphocyte trafficking and regional immunity. [Review] [178 refs]. *Advances in Immunology*, 72, 209-253.

Butler, J. E.(1998). Immunoglobulin diversity, B-cell and antibody repertoire development in large farm animals. *Rev. Sci. Tech.*, 17, 43-70.

Butler, J.E., Sun, J., Wertz, N., &Sinkora, M.(2006). Antibody repertoire development in swine. *Dev. Comp Immunol.*, 30, 199-221.

Cahill, R.N., Poskitt, D.C., Frost, D.C., Trnka, Z.(1977). Two distinct pools of recirculating T lymphocytes: migratory characteristics of nodal and intestinal T lymphocytes. *J Exp Med*, 145, 420-428.

Campbell, D.J., et Butcher, E. C.(2002). Intestinal attraction: CCL25 functions in effector lymphocyte recruitment to the small intestine.*J. Clin. Invest,* 110, 1079-1081.

Cepica, A., et Derbyshire, J. B.(1984). The effect of adoptive transfer of mononuclear leukocytes from an adult donor on spontaneous cell-mediated cytotoxicity and resistance to transmissible gastroenteritis in neonatal piglets. *Can. J. Comp. Med.*, 48, 360-364.

Chabaudie, N., Le Jan, C., Olivier, M., &Salmon, H.(1993). Lymphocyte subsets in the mammary gland of sows. *Res. Vet. Sci.*, 55, 351-355.

Clarkson, R.W., Wayland, M.T., Lee, J., Freeman, T., &Watson, C. J.(2004). Gene expression profiling of mammary gland development reveals putative roles for death receptors and immune mediators in post-lactational regression. *Breast Cancer Res.*, 6, R92-109.

Czarneski, J., Berguer, P., Bekinschtein, P., Kim, D.C., Hakimpour, P., Wagner, N., Nepomnaschy, I., Piazzon, I., &Ross, S. R.(2002). Neonatal infection with a milk-borne virus is independent of beta7 integrin- and L-selectin-expressing lymphocytes. Eur. *J Immunol*, 32, 945-956.

Czinn, S.J., et Lamm, M. E.(1986). Selective chemotaxis of subsets of B lymphocytes from gut- associated lymphoid tissue and its implications for the recruitment of mucosal plasma cells. *J. Immunol.*, 136, 3607-3611.

Dahlgren, U., Ahlstedt, S., Hedman, L., Wadsworth, C., &Hanson, L. A.(1981). Dimeric IgA in the rat is transferred from serum into bile but not into milk. *Scand J Immunol*, 14, 95-98.

Danielsen, M., Thymann, T., Jensen, B.B., Jensen, O.N., Sangild, P.T., &Bendixen, E.(2006). Proteome profiles of mucosal immunoglobulin uptake in inflamed porcine gut. *Proteomics.*, 6, 6588-6596.

de Groot, N., Kuik-Romeijn, P., Lee, S.H., de Boer, HA.(1999). Over-expression of the murine polymeric immunoglobulin receptor gene in the mammary gland of transgenic mice. [erratum appears in Transgenic Res 1999 Aug;8(4):319]. *Transgenic Research*, 8, 125-135.

de Groot, N., Kuik-Romeijn, P., Lee, S.H., de Boer, H. A.(2000). Increased immunoglobulin A levels in milk by over-expressing the murine polymeric immunoglobulin receptor gene in the mammary gland epithelial cells of transgenic mice. *Immunology*, 101, 218-224.

Defresne,M.P. (1998). The mouse model. p. 563-594. *In* P.P.Pastoret, P. Griebel, H. Bazin, and A. Govaerts (ed.) Handbook of vertebrate immunology. Academic Press.

Donovan, S.M., McNeil, L.K., Jimenez-Flores, R., Odle, J.(1994). Insulin-like growth factors and insulin-like growth factor binding proteins in porcine serum and milk throughout lactation. *Pediatr. Res.*, 36, 159-168.

Dowbenko, D., Kikuta, A., Fennie, C., Gillett, N., Lasky, L. A.(1993). Glycosylation-dependent cell adhesion molecule 1 (GlyCAM 1) mucin is expressed by lactating mammary gland epithelial cells and is present in milk. *J Clin. Invest*, 92, 952-960.

Evans, P.A., Newby, T.J., Stokes, C.R., &Bourne, F. J.(1982). A study of cells in the mammary secretions of sows. Vet. Immunol. *Immunopathol.*, 3, 515-527.

Evans, P.A., Newby, T.J., Stokes, C.R., Patel, D., &Bourne, F. J.(1980a). Antibody response of the lactating sow to oral immunization with Escherichia coli. *Scandinavian Journal of Immunology*, 11, 419-429.

Evans, P.A., Newby, T.J., Stokes, C.R., Patel, D., &Bourne, F.J. (1980b). Antibody response of the lactating sow to oral immunization with Escherichia coli. *Scand. J. Immunol.*, 11, 419-429.

Fagarasan, S., et Honjo, T., (2003). Intestinal IgA synthesis: regulation of front-line body defences. Nat. Rev. *Immunol.*, 3, 63-72.

Fanti, P., Nazareth, M., Bucelli, R., Mineo, M., Gibbs, K., Kumin, M., Grzybek, K., Hoeltke, J., Raiber, L., Poppenberg, K., Janis, K., Schwach, C., &Aronica, S. M.(2003). Estrogen decreases chemokine levels in murine mammary tissue: implications for the regulatory role of MIP-1 alpha and MCP-1/JE in mammary tumor formation. Endocrine., 22, 161-168.

Fink, K., Zellweger, R., Weber, J., Manjarrez-Orduno, N., Holdener, M., Senn, B.M., Hengartner, H., Zinkernagel, R.M., &Macpherson, A. J.(2008). Long-term maternal imprinting of the specific B cell repertoire by maternal antibodies. *Eur. J Immunol*, 38, 90-101.

Fishaut, M., Murphy, D., Neifert, M., McIntosh, K., &Ogra, P. L.(1981). Bronchomammary axis in the immune response to respiratory syncytial virus. *Journal of Pediatrics*, 99, 186-191.

Fronteau, D., Tanneau, G. M., Henry, G., Chevaleyre, C. C., Leonil, J., and Salmon, H.(1998). Activités chimiotactique de lait d'artiodactyle sur les lymphocytes porcins. Journées Rech. Porcine, [30], 363-367. Paris, Institut Technique du porc. 5-2-0098.

Gitlin, J.D., Gitlin, J.,& Gitlin D.(1976). Selective transfer of plasma proteins across mammary gland in lactating mouse. American. *Journal of. Physiology.* \{AM J PHYSIOL. }, 230, 1594-1602.

Goldblum, R.M., Ahlstedt, S., Carlsson, B., Hanson, L.A., Jodal, U., Lidin-Janson, G., &Sohl-Akerlund, A.(1975). Antibody-forming cells in human colostrum after oral immunisation. *Nature*, 257, 797-798.

Goldman, A.S., Chheda, S., &Garofalo, R.(1998). Evolution of immunologic functions of the mammary gland and the postnatal development of immunity. *Pediatric Research*, 43, 155-162.

Haelterman, E. O.(1975). Immunity fo tge transmissible gastroenteritis. Vet. Med. *Small Anim Clin.*, 70, 715-717.

Halsey, J.F., Mitchell, C., Meyer, R., &Cebra, J. J.(1982a). Metabolism of immunoglobulin A in lactating mice: origins of immunoglobulin A in milk. *European Journal of Immunology*, 12, 107-112.

Halsey, J.F., Mitchell, C., Meyer, R., &Cebra, J.J. (1982b). Metabolism of immunoglobulin A in lactating mice: Origins of immunoglobulin A in milk. *European. Journal of. Immunology* , 12, 107-112.

Hammerberg, C., Schurig, G.G., &Ochs, D. L.(1989). Immunodeficiency in young pigs. *Am. J. Vet. Res.*, 50, 868-874.

Hanson L.A., Carlsson B., Dahlgren U., Mellander L., Svanborg E.C., 1979. The secretory IgA system in the neonatal period. Ciba Foundation Symposium, 187-204.

Hanson, L.A., et Korotkova, M.(2002). The role of breastfeeding in prevention of neonatal infection. *Semin. Neonatol.*, 7, 275-281.

Harp, J.A., Waters, T.E., &Goff, J. P.(2005). Adhesion molecule and homing receptor expression on blood and milk polymorphonuclear leukocytes during the periparturient period of dairy cattle. Vet. Immunol. *Immunopathol.*, 104, 99-103.

Harris, N.L., Spoerri, I., Schopfer, J.F., Nembrini, C., Merky, P., Massacand, J., Urban, J.F., Jr., Lamarre, A., Burki, K., Odermatt, B., Zinkernagel, R.M., &Macpherson, A. J.(2006). Mechanisms of neonatal mucosal antibody protection. *J. Immunol.*, 177, 6256-6262.

He, W., Ladinsky, M.S., Huey-Tubman, K.E., Jensen, G.J., McIntosh, J.R., &Bjorkman, P. J.(2008). FcRn-mediated antibody transport across epithelial cells revealed by electron tomography. *Nature*, 455, 542-546.

Hennart, P.F., Brasseur, D.J., Delogne-Desnoeck, J.B., Dramaix, M.M., &Robyn, C.E. (1991). Lysozyme, lactoferrin, and secretory immunoglobulin A content in breast milk: influence of duration of lactation, nutrition status, prolactin status, and parity of mother. *Am J Clin Nutr*, 53, 32-39.

Hieshima, K., Ohtani, H., Shibano, M., Izawa, D., Nakayama, T., Kawasaki, Y., Shiba, F., Shiota, M., Katou, F., Saito, T., &Yoshie, O.(2003). CCL28 has dual roles in mucosal immunity as a chemokine with broad- spectrum antimicrobial activity. *JImmunol*, 170, 1452-1461.

Hodgkinson, A.J., Carpenter, E.A., Smith, C.S., Molan, P.C., &Prosser, C. G.(2007). Adhesion molecule expression in the bovine mammary gland. *Vet. Immunol. Immunopathol.*, 115, 205-215.

Honorio-Francca, A.C., Carvalho, M.P., Isaac, L., Trabulsi, L.R., &Carneiro-Sampaio, M. M.(1997). Colostral mononuclear phagocytes are able to kill enteropathogenic Escherichia coli opsonized with colostral IgA. Scand. *J. Immunol*, 46, 59-66.

Husband, A.J., Kramer, D.R., Bao, S., Sutherland, R.M., &Beagley, K. W.(1996). Regulation of mucosal IgA responses in vivo: cytokines and adjuvants. *Vet. ImmunolImmunopathol.*, 54, 179-186.

Inoue, R., Otsuka, M., &Ushida, K.(2005a). Development of intestinal microbiota in mice and its possible interaction with the evolution of luminal IgA in the intestine. *Exp. Anim*, 54, 437-445.

Inoue, R., Tsukahara, T., Nakanishi, N., &Ushida, K. (2005b). Development of the intestinal microbiota in the piglet. *J. Gen. Appl. Microbiol.*, 51, 257-265.

Inoue, R., et Ushida, K.(2003). Development of the intestinal microbiota in rats and its possible interactions with the evolution of the luminal IgA in the intestine. *Fems Microbiology Ecology*, 45, 147-153.

Israel, E.J., Patel, V.K., Taylor, S.F., Marshak-Rothstein, A., &Simister, N. E.(1995). Requirement for a beta 2-microglobulin-associated Fc receptor for acquisition of maternal IgG by fetal and neonatal mice. *J Immunol*, 154, 6246-6251.

Jaeger, L.A., Lamar, C.H., Bottoms, G.D., &Cline, T. R.(1987). Growth-stimulating substances in porcine milk. *Am. J. Vet. Res.*, 48, 1531-1533.

Janis, K., Hoeltke, J., Nazareth, M., Fanti, P., Poppenberg, K., &Aronica, S. M.(2004). Estrogen decreases expression of chemokine receptors, and suppresses chemokine bioactivity in murine monocytes. *Am. J. Reprod. Immunol.*, 51, 22-31.

Jensen, A.R., Elnif, J., Burrin, D.G., &Sangild, P. T.(2001). Development of intestinal immunoglobulin absorption and enzyme activities in neonatal pigs is diet dependent. *J. Nutr.*, 131, 3259-3265.

Kaetzel, C. S.(2005). The polymeric immunoglobulin receptor: bridging innate and adaptive immune responses at mucosal surfaces. *Immunol. Rev.*, 206, 83-99.

Kim, Y. B.(1975). Developmental immunity in the piglet. *Birth Defects Orig. Artic. Ser.*, 11, 549-557.

King,M.R., Kelly,D., Morel,P.C., and Pluske,J.R. (2003). Aspects of intestinal immunity in the pig around weaning. p. 219-257. *In* J.R.Pluske, J.Le. Dividich, and M.W.A. Verstegen (ed.) Weanning the pig:concepts and consequences. Wageningen Academic Publishers, Wageningen, Netherlands.

Kleinman, R.E., et Walker, W. A.(1979). The enteromammary immune system: an important new concept in breast milk host defense. *Dig. Dis. Sci*, 24, 876-882.

Klobasa, F., Butler, J.E., Werhahn, E., &Habe, F.(1986). Maternal-neonatal immunoregulation in swine. II. Influence of multiparity on de novo immunoglobulin synthesis by piglets. *Vet Immunol Immunopathol*, 11, 149-159.

Klobasa, F., Werhahn, E., &Butler, J. E.(1981). Regulation of humoral immunity in the piglet by immunoglobulins of maternal origin. *Res. Vet. Sci.*, 31, 195-206.

Kohl, S., Pickering, L.K., Cleary, T.G., Steinmetz, K.D., &Loo, L. S.(1980). Human colostral cytotoxicity. II. Relative defects in colostral leukocyte cytotoxicity and inhibition of peripheral blood leukocyte cytotoxicity by colostrum. *J. Infect. Dis.,* 142, 884-891.

Kumura, B.H., Sone, T., Shimazaki, K., &Kobayashi, E.(2000). Sequence analysis of porcine polymeric immunoglobulin receptor from mammary epithelial cells present in colostrum. *J. Dairy Res.*, 67, 631-636.

Kunkel, E., et Butcher, E.(2002). Homeostatic chemokines and the targeting of regional immunity.*Adv. Exp. Med. Biol.*, 512, 65-72.

Labbok, M.H., Clark, D., Goldman, A. S.(2004). Breastfeeding: maintaining an irreplaceable immunological resource. *Nat. Rev. Immunol.*, 4, 565-572.

Labeta, M.O., Vidal, K., Nores, J.E., Arias, M., Vita, N., Morgan, B.P., Guillemot J.C., Loyaux, D., Ferrara, P., Schmid, D., Affolter, M., Borysiewicz, L.K., Donnet-Hughes, A., &Schiffrin, E. J.(2000). Innate recognition of bacteria in human milk is mediated by a milk-derived highly expressed pattern recognition receptor, soluble CD14. *J. Exp.Med.*, 191, 1807-1812.

Lanza, I., Shoup, D., I, &Saif, L. J. (1995a). Lactogenic immunity and milk antibody isotypes to transmissible gastroenteritis virus in sows exposed to porcine respiratory coronavirus during pregnancy. *Am. J. Vet. Res.,* 56, 739-748.

Lanza, I., Shoup, D.I., &Saif, L. J.(1995b). Lactogenic immunity and milk antibody isotypes to transmissible gastroenteritis virus in sows exposed to porcine respiratory coronavirus during pregnancy. *Am. J. Vet. Res.*, 56, 739-748.

Lascelles, A.K., et McDowell, G. H.(1974). Localized humoral immunity with particular reference to ruminants. *Transplant Rev.*, 19, 170-208.

Lazarus, N.H., Kunkel, E.J., Johnston, B., Wilson, E., Youngman, K.R., &Butcher, E. C.(2003). A common mucosal chemokine (mucosae-associated epithelial chemokine/CCL28) selectively attracts IgA plasmablasts. *J. Immunol.*, 170, 3799-3805.

Le-Jan, C. (1993). Secretory component and IgA expression by epithelial cells in sow mammary gland and mammary secretions. *Res. Vet. Sci.*, 55, 265-270.

LeBouder, E., Rey-Nores, J.E., Rushmere, N.K., Grigorov, M., Lawn, S.D., Affolter, M., Griffin, G.E., Ferrara, P., Schiffrin, E.J., Morgan, B.P., &Labeta, M. O.(2003). Soluble forms of Toll-like receptor (TLR)2 capable of modulating TLR2 signaling are present in human plasma and breast milk. *J. Immunol.,* 171, 6680-6689.

Lee, C.G., Ladds, P.W., Watson, D. L.(1978). Immunocyte populations in the mammary gland of the rat at different stages of pregnancy and lactation. *Research In Veterinary Science*, 24, 322-327.

Lee, C.S., McCauley, I., &Hartmann, P. E.(1983). Light and electron microscopy of cells in pig colostrum, milk and involution secretion. *Acta Anat. (Basel),* 116, 126-135.

Lee, C.S., Meeusen, E., &Brandon, M. R.(1992). Local immunity in the mammary gland. *Vet. Immunol. Immunopathol.*, 32, 1-11.

Leece, J. G.(1973). Effect of dietary regimen on cessation of uptake of macromolecules by piglet intestinal epithelium (closure) and transport to the blood. *J. Nutr.*, 103, 751-756.

Lewis-Jones, D.I., Lewis-Jones, M.S., Connolly, R.C., Lloyd, D.C., &West, C.R. (1985). Sequential changes in the antimicrobial protein concentrations in human milk during lactation and its relevance to banked human milk. *Pediatr. Res*, 19, 561-565.

Liebler-Tenorio, E.M., Riedel-Caspari, G., &Pohlenz, J. F.(2002). Uptake of colostral leukocytes in the intestinal tract of newborn calves. *Vet. Immunol. Immunopathol.*, 85, 33-40.

Loving, M., et Magnusson, U.(2002). Sows intramammarily inoculated with Escherichia coli at parturition. II. Effects on the densities of MHC class II(+), CD4(+) and CD8(+) cells in the mammary gland. *Vet Immunol Immunopathol.*, 90, 45-54.

Low, E.N., Zagieboylo, L., Martino, B., &Wilson, E.(2010). IgA ASC accumulation to the lactating mammary gland is dependent on VCAM-1 and alpha4 integrins. *Mol Immunol*, 47, 1608-1612.

Macpherson, A.J., Gatto, D., Sainsbury, E., Harriman, G.R., Hengartner, H., &Zinkernagel, R. M.(2000). A primitive T cell-independent mechanism of intestinal mucosal IgA responses to commensal bacteria. *Science*, 288, 2222-2226.

Magnusson, U.(1999). Longitudinal study of lymphocyte subsets and major histocompatibility complex-class II expressing cells in mammary glands of sows. *Am. J Vet Res.*, 60, 546-548.

Magnusson, U., Rodriguez, Martinez, H., & Einarsson S.(1991). A simple, rapid method for differential cell counts in porcine mammary secretions. *Vet. Rec.*, 129, 485-490.

Maheshwari, A., Christensen, R.D., &Calhoun, D. A.(2003). ELR+ CXC chemokines in human milk. *Cytokine*, 24, 91-102.

Manz, R.A., Hauser, A.E., Hiepe, F., &Radbruch, A.(2005). Maintenance of serum antibody levels. *Annual Review of Immunology*, 23, 367-386.

Martin, M.G., Wang, J., Li, T.W., Lam, J.T., Gutierrez, E.M., Solorzano-Vargas, R.S., Tsai, A. H. (1998). Characterization of the 5'-flanking region of the murine polymeric IgA receptor gene. *American Journal of Physiology*, 275, G778-G788.

Martin, R., Langa, S., Reviriego, C., Jimenez, E., Marin, M.L., Olivares, M., Boza, J., Jimenez, J., Fernandez, L., Xaus, J., &Rodriguez, J. M.(2004). The commensal microflora of human milk: new perspectives for food bacteriotherapy and probiotics. *Trends in Food Science & Technology*, 15, 121-127.

Masopust, D., Vezys, V., Marzo, A.L., &Lefrancois, L.(2001). Preferential localization of effector memory cells in nonlymphoid tissue. *Science*, 291, 2413-2417.

Mayer, B., Zolnai, A., Freny, L.V., Jancsik, V., Szentirmay, Z., Hammarstrm, L., &Kacskovics, I.(2002). Localization of the sheep FcRn in the mammary gland. *Veterinary Immunology and Immunopathology*, 87, 327-330.

Mazanec, M.B., Coudret, C.L., &Fletcher, D. R.(1995). Intracellular neutralization of influenza virus by immunoglobulin A anti-hemagglutinin monoclonal antibodies. *J.Virol.*, 69, 1339-1343.

McDonald, T.L., Larson, M.A., Mack, D.R., &Weber, A.(2001). Elevated extrahepatic expression and secretion of mammary-associated serum amyloid A 3 (M-SAA3) into colostrum. *Vet. Immunol. Immunopathol.*, 83, 203-211.

McGhee, J.R., Michalek, S.M., &Ghanta, V. K.(1975). Rat immunoglobulins in serum and secretions: purification of rat IgM, IgA and IgG and their quantitation in serum, colostrum, milk and saliva. *Immunochemistry*, 12, 817-823.

Meeusen, E., Lee, C.S., &Brandon, M. (1991). Differential migration of T and B cells during an acute inflammatory response. *Eur. J. Immunol.*, 21, 2269-2272.

Mestecky, J.(2001). Homeostasis of the mucosal immune system - Human milk and lactation. *Bioactive Components of Human Milk*, 501, 197-205.

Meurens, F., Berri, M., Whale, J., Dybvig, T., Strom, S., Thompson, D., Brownlie, R., Townsend, H.G.G., Salmon, H., & Gerdts V.(2006). Expression of TECK/CCL25 and MEC/CCL28 chemokines and their respective receptors CCR9 and CCR10 in porcine mucosal tissues. *Vet. Immunol. Immunopathol.*, 113, 313-327.

Mickanin, C.S., Bhatia, U., &Labow, M.(2001). Identification of a novel beta-chemokine, MEC, down-regulated in primary breast tumors. *International Journal of Oncology*, 18, 939-944.

Mink, J.G., et Benner, R. (1979). Serum and secretory immunoglobulin levels in preleukaemic AKR mice and three other mouse strains. *Adv Exp Med Biol,* 114, 605-612.

Mitre, R., Etienne, M., Martinais, S., Salmon, H., Allaume, P., Legrand, P., &Legrand, A. B.(2005). Humoral defence improvement and haematopoiesis stimulation in sows and offspring by oral supply of shark-liver oil to mothers during gestation and lactation. *British Journal of Nutrition*, 94, 753-762.

Mohr, J. A.(1972). Lumphocyte sensitisation passed to the child from the mother. Lancet, 1, 688.

Mora, J. R., et von Andrian, U. H.(2004). Retinoic acid: an educational "vitamin elixir" for gut-seeking T cells. *Immunity*., 21, 458-460

Morcol, T., Akers, R.M., Johnson, J.L., Williams, B.L., Gwazdauskas, F.C., Knight, J.W., Lubon, H., Paleyanda, R.K., Drohan, W.N., &Velander, W. H.(1994). The porcine mammary gland as a bioreactor for complex proteins. *Ann. N. Y. Acad. Sci.*, 721, 218-233.

Morteau, O., Gerard, C., Lu, B., Ghiran, S., Rits, M., Fujiwara, Y., Law, Y., Distelhorst, K., Nielsen, E.M., Hill, E.D., Kwan, R., Lazarus, N.H., Butcher, E.C., & Wilson E.(2008). An indispensable role for the chemokine receptor CCR10 in IgA antibody-secreting cell accumulation. *J Immunol*, 181, 6309-6315.

Moughan, P.J., Birtles, M.J., Cranwell, P.D., Smith, W.C., &Pedraza, M. (1992). The piglet as a model animal for studying aspects of digestion and absorption in milk-fed human infants. *World Rev. Nutr. Diet.*, 67, 40-113.

Muller, A., Homey, B., Soto, H., Ge, N., Catron,D., Buchanan, M.E., McClanahan, T., Murphy, E., Yuan, W., Wagner, S.N., Barrera, J.L., Mohar, A., Verastegui, E., &Zlotnik, A.(2001). Involvement of chemokine receptors in breast cancermetastasis. *Nature*, 410, 50-56.

Newby, T.J., et Bourne, J.(1977a). The nature of the local immune system of the bovine mammary gland. *J. Immunol.*, 118, 461-465.

Newby, T.J., et Bourne, J.(1977b). The nature of the local immune system of the bovine mammary gland. *J. Immunol.*, 118, 461-465.

Nguyen, T.V., Yuan, L., Azevedo, M.S., Jeong, K.I., Gonzalez, A.M., &Saif, L. J. (2007a). Transfer of maternal cytokines to suckling piglets: in vivo and in vitro models with implications for immunomodulation of neonatal immunity. *Vet. Immunol. Immunopathol.*, 117, 236-248.

Nguyen, T.V., Yuan, L., Azevedo, M.S., Jeong, K.I., Gonzalez, A.M., &Saif, L. J.(2007b). Transfer of maternal cytokines to suckling piglets: in vivo and in vitro models with implications for immunomodulation of neonatal immunity. *Vet. Immunol. Immunopathol.*, 117, 236-248.

Nishimura, T., Koike, R., &Miyasaka, M.(2000). Mammary glands of Aly mice: developmental changes and lactation-related expression of specific proteins, alpha-casein, GLyCAM-1 and MAdCAM-1. *Am. J Reprod. Immunol*, 43, 351-358.

Nishiyama, Y., Sugimoto, M., Ikeda, S., &Kume, S.(2010). Supplemental beta-carotene increases IgA-secreting cells in mammary gland and IgA transfer from milk to neonatal mice. *Br J Nutr*, 1-6.

Odle, J., Zijlstra, R.T., &Donovan, S. M.(1996). Intestinal effects of milkborne growth factors in neonates of agricultural importance. *J. Anim Sci.*, 74, 2509-2522.

Ogra, P.L., Sinks, L.F., &Karzon, D. T.(1971). Poliovirus antibody response in patients with acute leukemia. *Journal of Pediatrics*, 79, 444-449.

Ogra, S.S., Weintraub, D.I., &Ogra, P. L.(1978). Immunologic aspects of human colostrum and milk: interaction with the intestinal immunity of the neonate. *Adv Exp Med Biol*, 107, 95-107.

Park, Y.H., Fox, L.K., Hamilton, M.J., &Davis, W. C.(1992). Bovine mononuclear leukocyte subpopulations in peripheral blood and mammary gland secretions during lactation. *J. Dairy. Sci.*, 75, 998-1006.

Parmely, M.J., et Manning, L. S.(1983). Cellular determinants of mammary cell-mediated immunity in the rat: kinetics of lymphocyte subset accumulation in the rat mammary gland during pregnancy and lactation. *Ann. N. Y. Acad. Sci.*, 409, 517-533.

Pfaffl, M.W., Wittmann, S.L., Meyer, H.H., &Bruckmaier, R. M.(2003). Gene expression of immunologically important factors in blood cells, milk cells, and mammary tissue of cows. *J. Dairy Sci.*, 86, 538-545.

Phillips, J.O., Everson, M.P., Moldoveanu, Z., Lue, C., &Mestecky, J. (1990). Synergistic effect of IL-4 and IFN-gamma on the expression of polymeric Ig receptor (secretory component) and IgA binding by human epithelial cells. *J. Immunol.*, 145, 1740-1744.

Piriou-Guzylack, L., et Salmon, H.(2008). Membrane markers of the immune cells in swine: an update. *Vet Res*, 39, 54.

Porter,P., and Chidlow,J.W. (1979). Response to E. coli antigens via local and parenteral routes linking intestinal and mammary immune mechanisms in passive protection against neonatal colibacillosis in the pig. p. 73-90. *In* P.L.Ogra, and D.H. Dayton (ed.) Immunology of the breast milk. Monograph of the National Institute of Child Health and Human Development. Raven Press, New York.

Porter,P.L. (1979). Structural and functionnal characteristics of immunoglobulins of the common domestic species. *Adv. Vet. Sc. Comp. Med,* (23):1-21.

Reber, A.J., Hippen, A.R., &Hurley, D. J.(2005). Effects of the ingestion of whole colostrum or cell-free colostrum on the capacity of leukocytes in newborn calves to stimulate or respond in one-way mixed leukocyte cultures. *Am. J. Vet. Res.*, 66, 1854-1860.

Redman, D.R., Bohl, E.H., &Cross, R. F.(1978). Intrafetal inoculation of swine with transmissible gastroenteritis virus. *Am. J. Vet. Res.*, 39, 907-911.

Rincheval-Arnold, A., Belair, L., Cencic, A., &Djiane, J.(2002). Up-regulation of polymeric immunoglobulin receptor mRNA in mammary epithelial cells by IFN-gamma. *Molecular and Cellular Endocrinology*, 194, 95-105.

Rodewald, R., et Kraehenbuhl, J. P.(1984). Receptor-mediated transport of IgG. *J Cell Biol*, 99, 159s-164s.

Rodriguez, B., Chevaleyre, C., Henry, G., Molle, D., Virlogeux-Payant, I., Berri, M., Boulay, F., Leonil, J., Meurens, F., &Salmon, H.(2009). Identification in milk of a serum amyloid A peptide chemoattractant for B lymphoblasts. *BMC. Immunol.*, 10, 4.

Roux, M.E., McWilliams, M., Phillips-Quagliata, J.M., Weisz, C.P., Lamm, M.E., (1977). Origin of IgA-secreting plasma cells in the mammary gland. *J. Exp. Med.*, 146, 1311-1322.

Rumbo, M., Chirdo, F.G., Anon, M.C., &Fossati, C. A.(1998). Detection and characterization of antibodies specific to food antigens (gliadin, ovalbumin and beta-lactoglobulin) in human serum, saliva, colostrum and milk. Clin. *Exp. Immunol.*, 112, 453-458.

Rumbo, M., et Schiffrin, E. J. (2005). Ontogeny of intestinal epithelium immune functions: developmental and environmental regulation. *Cell Mol. Life Sci.*, 62, 1288-1296.

Saif,L.J., and Bohl E.H. (1979). Role of secretory IgA in passive immunity of swine to enteric viral infections. p. 237-248. *In* P.Ogra, and D. Dayton (ed.) Immunology of breast milk. Raven Press, New York.

Saif, L.J., et Bohl, E. H.(1983). Passive immunity to transmissible gastroenteritis virus: intramammary viral inoculation of sows. *Ann. N. Y. Acad. Sci.*, 409, 708-723.

Saif, L.J., van Cott, J.L., Brim, T. A.(1994). Immunity to transmissible gastroenteritis virus and porcine respiratory coronavirus infections in swine. *Vet. Immunol. Immunopathol.*, 43, 89-97.

Salmon,H. (1986). Surface markers of swine lymphocytes: application to the study of local immune system in mammary gland and transplanted gut. p. 1855-1864. *In* M.E.Tumbleson (ed.) Swine in biomedical research. Plenum, New York.

Salmon, H.(1987). The intestinal and mammary immune system in pigs. *Vet. Immunol.Immunopathol.*, 17, 367-388.

Salmon, H.(1995). Immunité lactogène et protection vaccinale dans l'espèce porcine. *Vet. Res*, 26, 232-237.

Salmon, H.(1999). The mammary gland and neonate mucosal immunity. *VeterinaryImmunology & Immunopathology*, 72, 143-155.

Salmon, H.(2000). Mammary gland immunology and neonate protection in pigs. Homing of lymphocytes into the MG. *Adv. Exp. Med. Biol.*, 480:279-86., 279-286.

Salmon, H.(2003). Immunophysiology of the mammary gland and transmission of immunity to the young. *Reprod. Nutr. Dev.*, 43, 471-475.

Salmon, H., Berri, M., Gerdts, V., Meurens, F.(2009). Humoral and cellular factors of maternal immunity in swine. *Dev Comp Immunol*, 33, 384-393.

Salmon, H., et Delouis, C.(1982). (Kinetics of lymphocyte sub-populations and plasma cells in the mammary gland of primiparous sows in relation to gestation and lactation). *Ann. Rech. Vet.*, 13, 41-49.

Sampaio, S.O., Li, X., Takeuchi, M., Mei, C., Francke, U., Butcher, E.C., &Briskin, M. J.(1995). Organization, regulatory sequences, and alternatively spliced transcripts of the mucosal addressin cell adhesion molecule-1 (MAdCAM-1) gene. *J. Immunol.*, 155, 2477-2486.

Saurer, L., McCullough, K.C., &Summerfield, A.(2007). In vitro induction of mucosa-type dendritic cells by all-trans retinoic acid. *J. Immunol.*, 179, 3504-3514.

Sayed-Ahmed, A., Kassab, M., Abd-Elmaksoud, A., Elnasharty, M., &El-Kirdasy, A. (2010). Expression and immunohistochemical localization of the neonatal Fc receptor (FcRn) in the mammary glands of the Egyptian water buffalo. *Acta Histochemica,* 112, 383-391.

Schnulle, P.M., et Hurley, W. L.(2003). Sequence and expression of the FcRn in the porcine mammary gland. *Veterinary Immunology and Immunopathology*, 91, 227-231.

Schollenberger, A., Degorski, A., Frymus, T., &Schollenberger, A.(1986a). Cells of sow mammary secretions. I. Morphology and differential counts during lactation. *Zentralbl. Veterinarmed.* (A), 33, 31-38.

Schollenberger, A., Frymus, T., &Degorski, A.(1986b). Cells of sow mammary secretions. III. Some properties of phagocytic cells. *Zentralblatt.*, 33, 353-359.

Schollenberger, A., Frymus, T., Degorski, A., Schollenberger, A., (1986c). Cells of sow mammary secretions. II. Characterization of lymphocyte populations. *Zentralbl. Veterinarmed.* (A), 33, 39-46.

Sheldrake, R.F., et Husband, A. J.(1985). Immune defences at mucosal surfaces in ruminants. *J. Dairy. Res.*, 52, 599-613.

Sheldrake, R.F., Husband, A.J., Watson, D.L. (1985a). Specific antibody-containing cells in the mammary gland of non-lactating sheep after intraperitoneal and intramammary immunisation. *Research In Veterinary Science*, 38, 312-316.

Sheldrake, R.F., Husband, A.J., Watson, D.L., &Cripps, A. W.(1984). Selective transport of serum-derived IgA into mucosal secretions. *J Immunol*, 132, 363-368.

Sheldrake, R.F., Scicchitano, R., &Husband, A. J.(1985b). The effect of lactation on the transport of serum-derived IgA into bile of sheep. *Immunology*, 54, 471-477.

Siegrist, C. A. (2003). Mechanisms by which maternal antibodies influence infant vaccine responses: review of hypotheses and definition of main determinants. *Vaccine*, 21, 3406-3412.

Sikorski, E.E., Hallmann, R., Berg, E.L., Butcher, E. C.(1993). The Peyer's patch high endothelial receptor for lymphocytes, the mucosal vascular addressin, is induced on a murine endothelial cell line by tumor necrosis factor-alpha and IL-1. *J. Immunol.*, 151, 5239-5250.

Silim,A., Rekik,M.R., Roy,R.S., Salmon, H.and Pastoret, P.P. (1990). Immunité chez le foetus et le nouveau-né. p. 197-204. *In* P.P.Pastoret, A. Govaerts, and H. Bazin (ed.) Immunologie Animale. Medecines et Sciences. Flammarion, Paris.

Simister, N.E., et Mostov, K. E.(1989). An Fc receptor structurally related to MHC class I antigens. *Nature*, 337, 184-187.

Springer, T.A. (1994). Traffic signals for lymphocyte recirculation and leukocyte emigration: the multistep paradigm. *Cell*, 76, 301-314.

Stavnezer, J.(1996). Immunoglobulin class switching. *Curr. Opin. Immunol.*, 8, 199-205.

Stirling, C.M., Charleston, B., Takamatsu, H., Claypool, S., Lencer, W., Blumberg, R.S., Wileman, T. E.(2005). Characterization of the porcine neonatal Fc receptor--potential use for trans-epithelial protein delivery. *Immunology*, 114, 542-553.

Stone, S.S., Kemeny, L.J., Woods, R.D., &Jensen, M. T.(1977). Efficacy of isolated colostral IgA, IgG, and IgM(A) to protect neonatal pigs against the coronavirus of transmissible gastroenteritis. *Am. J. Vet. Res.*, 38, 1285-1288.

Streeter, P.R., Berg, E.L., Rouse, B.T., Bargatze, R.F., &Butcher, E. C.(1988). A tissue-specific endothelial cell molecule involved in lymphocyte homing. *Nature*, 331, 41-46.

Takahata, Y., Takada, H., Nomura, A., Nakayama, H., Ohshima, K., &Hara, T. (2003). Detection of interferon-gamma-inducible chemokines in human milk. *Acta Paediatr.*, 92, 659-665.

Tanneau, G.M., Hibrand-Saint, O.L., Chevaleyre, C.C., &Salmon, H. P.(1999a). Differential recruitment of T- and IgA B-lymphocytes in the developing mammary gland in relation to homing receptors and vascular addressins. *J. Histochem. Cytochem.*, 47, 1581-1592.

Tanneau, G.M., Hibrand-Saint, O.L., Chevaleyre, C.C., &Salmon, H. P.(1999b). Differential recruitment of T- and IgA B-lymphocytes in the developing mammary gland in relation to homing receptors and vascular addressins. *J Histochem Cytochem*, 47, 1581-1592.

Tatarczuch, L., Philip, C., Bischof, R., &Lee, C. S.(2000). Leucocyte phenotypes in involuting and fully involuted mammary glandular tissues and secretions of sheep. *J. Anat.*, 196 (Pt 3), 313-326.

Taylor, B.C., Keefe, R.G., Dellinger, J.D., Nakamura, Y., Cullor, J.S., &Stott, J. L.(1997). T cell populations and cytokine expression in milk derived from normal and bacteria-infected bovine mammary glands. *Cell Immunol.*, 182, 68-76.

Telemo, E., Bailey, M., Miller, B.G., Stokes, C.R., &Bourne, F. J.(1991). Dietary antigen handling by mother and offspring. Scand. *J. Immunol.*, 34, 689-696.

Tuboly,S., and Bernath, S. (2002). Intestinal absorption of colostral lymphoid cells in newborn animals. Journal???

Tuboly, S., Bernath, S., Glavits, R., Kovacs, A., &Megyeri, Z. (1995). Intestinal absorption of colostral lymphocytes in newborn lambs and their role in the development of immune status. *Acta Vet. Hung.*, 43, 105-115.

Tuboly, S., Bernath, S., Glavits, R., &Medveczky, I.(1988). Intestinal absorption of colostral lymphoid cells in newborn piglets. *Vet. Immunol. Immunopathol.*, 20, 75-85.

Vaerman, J.P., Langendries, A., Pabst, R., &Rothkotter, H. J.(1997). Contribution of serum IgA to intestinal lymph IgA, and vice versa, in minipigs. *Veterinary Immunology andImmunopathology*, 58, 301-308.

van der Feltz, M.J., de, G.N., Bayley, J.P., Lee, S.H., Verbeet, M.P., &de Boer, H. A.(2001). Lymphocyte homing and Ig secretion in the murine mammary gland. *Scand J Immunol*, 54, 292-300.

Verhasselt, V., Milcent, V., Cazareth, J., Kanda, A., Fleury, S., Dombrowicz, D., Glaichenhaus, N., &Julia, V.(2008). Breast milk-mediated transfer of an antigen induces tolerance and protection from allergic asthma. *Nat. Med.*, 14, 170-175.

Wagstrom, E.A., Chang, C.C., Yoon, K.J., &Zimmerman, J. J.(2001). Shedding of porcine reproductive and respiratory syndrome virus in mammary gland secretions of sows. *Am. J. Vet. Res.*, 62, 1876-1880.

Wang, W., Soto, H., Oldham, E.R., Buchanan, M.E., Homey, B., Catron, D., Jenkins, N., Copeland, N.G., Gilbert, D.J., Nguyen, N., Abrams, J., Kershenovich, D., Smith, K., McClanahan, T., Vicari, A.P., &Zlotnik, A.(2000). Identification of a novel chemokine (CCL28), which binds CCR10 (GPR2). *J. Biol. Chem.*, 275, 22313-22323.

Watson, D.L. (1980). Immunological functions of the mammary gland and its secretion--comparative review. *Austr. J. Biol. Sc.*, 403-422.

Weaver L. T.(1992). Breast and gut: the relationship between lactating mammary function and neonatal gastrointestinal function. *Proc Nutr Soc.*, 51, 155-163.

Weisz-Carrington, P., Emancipator, S., Lamm, M. E.(1984). Binding and uptake of immunoglobulins by mouse mammary gland epithelial cells in hormone-treated cultures. *J. Reprod. Immunol.*, 6, 63-75.

Weisz-Carrington, P., Roux, M.E., &Lamm, M. E.(1977). Plasma cells and epithelial immunoglobulins in the mouse mammary gland during pregnancy and lactation. *Journal of Immunology*, 119, 1306-1307.

Weisz-Carrington, P., Roux, M.E., McWilliams, M., Phillips-Quagliata, J.M., Lamm, M. E.(1978). Hormonal induction of the secretory immune system in the mammary gland. Proceedings of the National Academy of Sciences of the United States of America, 75, 2928-2932.

Welsh, J.K., et May, J. T.(1979). Anti-infective properties of breast milk. *J. Pediatr.*, 94, 1-9.

Williams P.P., (1993). Immunomodulating effects of intestinal absorbed maternal colostral leukocytes by neonatal pigs. *Can. J. Vet. Res.*, 57, 1-8.

Wilson, E., et Butcher, E.C., (2004). CCL28 controls immunoglobulin (Ig)A plasma cell accumulation in the lactating mammary gland and IgA antibody transfer to the neonate. *J. Exp. Med.*, 200, 805-809.

Xu, B., Wagner, N., Pham, L.N., Magno, V., Shan, Z., Butcher, E.C., &Michie, S. A.(2003). Lymphocyte homing to bronchus-associated lymphoid tissue (BALT) is mediated by L-selectin/PNAd, alpha4beta1 integrin/VCAM-1, and LFA-1 adhesion pathways. *J Exp Med*, 197, 1255-1267.

Xu, R., Doan, Q.C., &Regester, G. O.(1999). Detection and characterisation of transforming growth factor-beta in porcine colostrum. *Biol. Neonate*, 75, 59-64.

Xu, R.J., Wang, F., &Zhang, S. H.(2000). Postnatal adaptation of the gastrointestinal tract in neonatal pigs: a possible role of milk-borne growth factors. *Livestock Production Science*, 66, 95-107.

Zhu, Y., Berg, M., Fossum, C., &Magnusson, U. (2007). Proinflammatory cytokine mRNA expression in mammary tissue of sows following intramammary inoculation with Escherichia coli. *Vet. Immunol. Immunopathol.*, 116, 98-103.

In: Milk Production
Editor: Boulbaba Rekik, pp. 75-96
ISBN 978-1-62100-061-7

Chapter V

The Importance of Milk Fatty Acids in Human Nutrition

Rey Gutiérrez Tolentino[1*], Salvador Vega y León[1], Claudia Radilla Vázquez[2], María Radilla Vázquez[2], Samuel Coronel Nuñez[1] and Marta Coronado Herrera[1]

[1]Professor-Researcher of the Department of Agricultural and Animal Production. Universidad Autónoma Metropolitana- Xochimilco, Mexico City

[2]ConsultingMédica Sur Hospital, Puente de Piedra #150 Col. Toriello Guerra, Tlalpan, CP.14050, Mexico City

Abstract

Milk and dairy products are foods of great physiological functionality; however, this can be optimized by means of technological advances that make it possible to enhance their potentiality. Milk fat is perhaps the most complex of the edible fats, in which more than 400 different fatty acids (FA) have been detected, although 30 have been documented as majority, given that the rest are present in trace amounts, but which are nonetheless important from the functional viewpoint. Milk fat is constituted of approximately 25% monounsaturated fatty acids, 5 % polyunsaturated fatty acids and 70 % saturated fatty acids. Some of these FA have been shown to possess interesting properties for the health of individuals, especially butyric acid (C4:0), which intervenes in the elimination of cancer cells in the colon, and the vaccenic acids (trans-11. C18:1) and conjugated linoleic acid (CLA, C18:2) have anti-proliferative and anti-cancer actions, thus they have been added to foods as a preventive measure against cancer of the colon, rectum and prostate. There is sufficient scientific evidence that CLA participates in the inhibition of arteriosclerosis, increases bone mineralization and improves immunological functions. From the nutritional perspective, fatty acids naturally present in milk and dairy products are functional ingredients that play an important role in human health.

[*] E-mail: reygut@correo.xoc.uam.mx

Keywords: Fatty acids, milk, human nutrition, Mexico.

Introduction

Milk is one of the fundamental foods for humans. Unquestionably, it is an excellent food from the nutritional viewpoint, because it is an important source of proteins of high biological value, minerals, vitamins, and it contains an important number of polyunsaturated fatty acids. In recent years, it has been reported that milk fat contains nearly 400 different fatty acids,although 30 have been annotated as more realistic information, given that the rest are found in trace amounts. Milk fat is constituted of approximately 5% polyunsaturated fatty acids, 25% monounsaturated fatty acids and 70 % saturated fatty acids, and its composition can be influenced by diverse factors, such as lactation stage, feeding, genetic variation, seasonal variation, variations in the energy balance of the cows, among others. Some of these fatty acids such as omega 3 have potential benefits for human through their impact on certain illnesses such as cardiovascular diseases, inflammatory diseases, rheumatoid arthritis, cerebral functions, immunological functions, and depression, among others. Therefore, it is recommendable for the Mexican population to increase their consumption of milk and milk products. The objective of the present chapter is to provide up-to-date information on the importance of fatty acids of milk in human nutrition.

World Milk Production and Consumption

According to the FAO/WHO and to the Secretaría de Agricultura, Ganadería, Desarrollo Rural, Pesca y Alimentación (SAGARPA) in 2009, the consumption of dairy products was very heterogeneous across countries and regions, due to cultural patterns as well as income levels of populations. In general, the countries of Europe and North America (Canada and the United States) register the highest annual consumption levels, between 200 and 300 liters of milk or equivalent, although the relative importance of each type of product is different. For example, in Northern Europe, North America and Oceania, the consumption of pasteurized or ultra-pasteurized liquid milk (UTH) is very high, while the Mediterranean countries (France, Italy, and Spain) are distinguished by their high consumption of cheese. Latin America and the Caribbean, possibly as a consequence of European colonization, is a region with a relatively high consumption of milk and dairy products, which fluctuates between 40 and 60 liters per capita. In some countries, consumption is much higher, as much as 200 and 250 liters per capita, such as in Argentina and Uruguay, respectively. In general, the consumption in other countries of the southern region of Latin America is around 140 liters per capita (in Chile and Brazil). According to the FAO and SAGARPA in 2008, world cow's milk production was 692,7 million tons, and is concentrated in few blocks of nations, such as: the United States, which supplied 14.9 %, the European Union (with countries such as Germany, France, United Kingdom and Poland) with 13.7 %; developing countries, such as India with 7.6 %; China, with a contribution of 6.2 %, Russia with 5.5 % and Brazil with 4.8 %, while traditional milk producing countries, such as New Zealand, participated with 2.6 % of the world supply. In the case of Mexico, its contribution to world production in 2008 was 1.86 %, putting it in 16th place. The supply of the world market is heavily concentrated in few countries or blocks of countries, as in the case of the exports of the European Union, New Zealand and Australia, which represent nearly 80 % of the world total export (FAO, 2009; SAGARPA, 2010).

National Milk Production and Consumption

In Mexico, cow's milk is a priority product in the livestock subsector, due to its importance as a basic food for humans at all ages. In the first phases of urban growth, milk consumption depended on the proximity of stables to human settlements, given that a shorter distance would avoid the decomposition of milk. In this stage, the consumer knew the origin and treatment of milk, but as the cities grew, the security of the characteristics and qualities of the product became relative. Thus, during 1914 the government issued the decree "Expedition of patents for the sale of milk in the Federal District (Mexico) and conditions that must be satisfied by the distribution centers". During the 1930's, legislation was formulated for the processing and commercialization of dairy products. This motivated the producers between 1940 and 1950 to organize in order to comply with the new laws. Pasteurization industries were established to offer milk which satisfied health codes, although the preference of the consumers for raw milk persisted. Between 1950 and 1960, some companies began activities of pasteurization, transport, cooling and marketing in addition to participation in the input phase of the productive chain, such as the elaboration of balanced feed for animals, agricultural machinery centers, spare parts for transport, agrochemicals and seeds for pastures. However, most of the dairy farms of Mexico conserved their traditional structure of small production units, such as family or backyard dairies.

According to SAGARPA in 2009, national cow's milk production in Mexico was 10,549 million liters. In the past 10 years, the Mean Annual Growth Rate has presented an annual increment of 1.74 per cent. With data of the INEGI (Instituto Nacional de Estadística y Geografía), generated from the last livestock census of 2007, 63 % of national milk production is from dairy cattle, and the remaining 37 % from dual-purpose cattle. According to these data, the states with the highest milk production from specialized cattle are Jalisco, with 18.8 %, the Lagunera Region with 19.2 %, Chihuahua with 9.3 % and Veracruz with 6.9 %. Of the milk from dual-purpose cattle, the principal producing state is Veracruz with 15 %, followed by Jalisco with 10 %, Sinaloa with 8 %, Sonora with 6 % and Chiapas with 5.5 %. In Mexico, there are diverse production systems, the intensive system with Holstein and Swiss cattle fed continuously balanced rations regardless of the season to keep constant average production levels throughout the year, and the dual-purpose production system or family dairies where milk production depends on the availability of forage available essentially in the rainy season, and in recent years is linked to prices paid to the producer. Thus, higher production levels are observed particularly in the months of July to October (INEGI, 2007).

The apparent national consumption during 2009 was 13,323,846 thousand liters, which represents a higher volume than that reported in 2008. In the last 10 years, the apparent national consumption has shown an annual growth rate of 1.80 %, while the annual population growth rate has only been 1.1 %, this has permitted the per capita availability to rise from 116.14 liters to 124.23 liters per year (SAGARPA, 2010).

Industrial Milk Processing

SAGARPA alluded to the industrial processing of milk in Mexico looks over more than 300 formal companies, of which approximately 10 % are large, 30 % medium and 60 % small companies, within which 30 principal Industrial Groups are involved in more than 100 brands of dairy products, including milk, cheese, yoghurt, cream, butter, among others. These companies are mainly implemented in the North Central region of the country, with a national and regional distribution. The liquid milk in Mexico is distributed as follows: 30.9 % for the elaboration of pasteurized, homogenized and ultra-pasteurized milk; 17.6 % for whole milk and milk for infants; 15.7 % for industrial cheeses; 9 % for natural or fruit yoghurt; 6 % for the rehydration of milk; 4 % for cream, butter, margarine and butyric fats; and nearly 17 % for other products, including crafted cheese, sweets, and other regional dairy products (SAGARPA, 2010).

NAFTAand Dairy Products

In the North American Free Trade Agreement (NAFTA), a commercial liberation regimen was established for Mexico of up to 10 and 15 years for some agricultural products. For milk, a tariff-quota plan was established in substitution of the importation permit, which implied the elimination of the government importation permit and its substitution by an *ad valorem* (according to the value) tariff of 139 % or a specific tariff of 1160 dollars. Furthermore, for powdered milk imported from the United States, a tariff of 0 % was established in a capacity of less than 40 thousand metric tons in the first year to grow thereafter by 3 % each year. NAFTA has brought positive effects on milk production such as the reduction in prices of the inputs used for its production. With the elimination of tariffs on most agricultural products, the price of forage grains has fallen to historic levels. If it is considered that feed costs can represent as much as 70 % of the production costs in the intensive systems, this factor was the motor that has pushed the growth of national production. The examination of changes registered in the international price of milk, alfalfa and sorghum in the period 1994-2005 shows that NAFTA has had positive effects on milk production (Ramírez-Jaspeado *et al.*, 2010).

Milk Quality

The quality of any food destined for human consumption depends largely on its perception by consumers and its relation to health. Milk is one of the fundamental foods for humans; it is unquestionably an excellent food from the nutritional perspective. The increment of the indicators of its quality is fundamental, due to the importance of offering a more nutritious product to the consumer and with fewer risks to his health. Prerequisites of good quality milk begin at the dairy farm, thanks to demands for high quality dairy products. In addition to the nutritional requirements, hygiene parameters, the presence of residues of medicines and mastitis are decisive in the determination of milk quality as raw material (Ramírez *et al.*, 2007). The Comisión Federal para la Protección contra Riesgos Sanitarios (COFEPRIS, 2010) of the Secretaría de Salud (SSA) took one more step in the updating and simplification of the Mexican normative framework in the area of food, by harmonizing with international standards the regulations for dairy products, which will result in better prevention of risks for the population and greater competitiveness for the companies. In addition, the Norma Mexicana NOM-243-SSA1-2010, for Products and Services, milk, milk formula and combined dairy product and milk derivates contributes to the insurance of milk quality, given that it integrates five existing norms associated with the quality of milk and its products:

- NOM-035-SSA1-1991.-Whey cheeses. Sanitary specifications.
- NOM-036-SSA1-1993. - Ice cream, ice milk or made with vegetable fat, sherbets and bases or mixtures for ice cream. Sanitary specifications.
- NOM-121-SSA1-1994. - Cheeses: fresh, aged and processed. Sanitary specifications.
- NOM-184-SSA1-2002.- Milk, milk formula and combined dairy product Sanitary specifications, and
- NOM-185-SSA1-2002. - Butter, creams, sweetened condensed milk product, fermented and acidified milk products, milk based sweets. Sanitary specifications.

Functionality of Milk

Milk is probably the only food in nature that has been thought of, designed and evolved along with the species of the planet specifically, as a food. Whereas other foods originate from the capacity of adaptation of the species to their habitat, milk accompanies the most highly evolved animals on the zoological scale to insure the best nutrition possible in the first stages of life (Carmuega, 2004).

Many traditional dairy products possess physiological activity; this characteristic of going beyond the ordinary nutrimental effect could be attributed to a great variety of milk constituents, such as some proteins, vitamins and minerals, and especially its fat (Valenzuela*et al.*, 2003).

The interest in the composition of milk fat is not recent; in 1871 Fleischmann studied with details the diameter of the fat cell. However, the advent of new technologies such as liquid-gas chromatography has definitely contributed to the knowledge of the lipid composition of milk fat, making it possible to obtain the profiles of sterols, triacylglycerides and fatty acids. Diverse studies confirm that the variations in the composition of the lipids of the fatty material of cow's milk are the result of the feeding systems, lactation period, seasonal variation, and the geographic area, also including the degree of mastitis and race, among others (Pinto *et al.*, 2002; Gutiérrez *et al.*, 2009).

Fatty Acids in Human Nutrition

The importance of lipids in human nutrition and development has been recognized for decades. Lipids are important constituents of the structure of cell membranes, they fulfill energetic functions and of metabolic reserve, and form the basic structure of some hormones and bile salts. Furthermore, some lipids are essential, due to the fact that they cannot be synthesized from precursor structures. More still, recently the participation of some lipids has been identified in the regulation of the genetic expression in mammals. Within the great structural diversity that characterizes lipids, fatty acids are perhaps the structures of highest relevance.

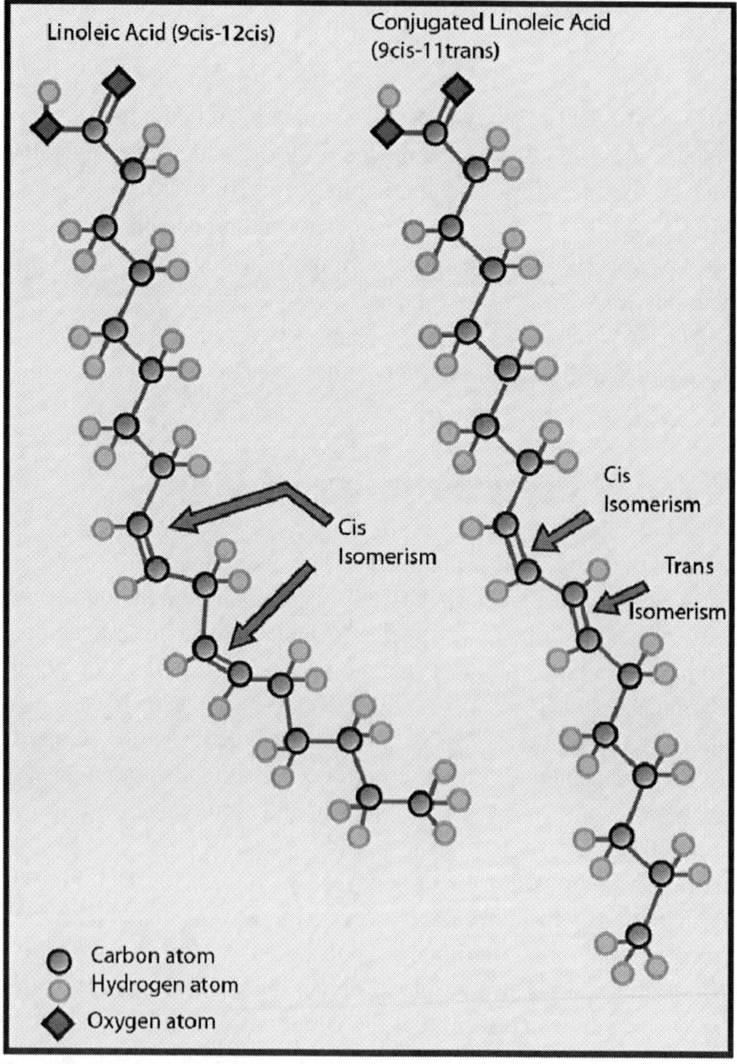

Source: Taken from Sanhueza *et al.*, 2002.

Figure 1. Chemical structure of linoleic acid (9 cis-12 cis) and of one of the isomers of conjugated linoleic acid (9 cis-11 trans).

The fatty acids are divided in two major groups according to their structural characteristics: saturated fatty acids (SFA) and unsaturated fatty acids (UFA). The latter, depending on their degree of unsaturation, can be classified as monounsaturated fatty acids (MUFA) and polyunsaturated fatty acids (PUFA). Depending on the position of the double bond, counting from the extreme carbon to the carboxylic functional group, the MUFA and the PUFA can be classified in three principal series: omega-9 fatty acids (first double bond in carbon 9), omega-6 fatty acids (first double bond in carbon 6) and omega-3 fatty acids (first double bond in carbon 3). The omega-9 fatty acids are not essential, given that humans can introduce an unsaturation to a SFA in this position (Valenzuela *et al.*, 2003).

The fat of cow's milk contains a great number of fatty acids, some of which have potential effects on human health. Milk fat is possibly the most complex of the edible fats. Nearly 400 different fatty acids (FA) have been detected in it, with chain lengths that go from C2 to C28, including even, uneven, saturated, unsaturated, cis and trans. Some of these fatty acids have been shown to have interesting properties for the health of the consumers, especially butyric acid (C4:0), vaccenic acid (trans-11, C18:1), and the conjugated linoleic acids (CLA), C18:2). The term conjugated linoleic (CLA) includes a group of isomers of linoleic acid with a conjugated arrangement in the double bonds. This means that the double bonds of the fatty acid are separated by one carbon atom, not two, as occurs in linoleic acid (Figure 1). Although these conjugated isomers of linoleic acid are widely distributed in nature, their concentrations are very low in most foods, except in the milk and meat of ruminants. Within the principal effects of linoleic acid are their strong anti-carcinogenic, anti-atherosclerotic, and anti-diabetic (type II diabetes) activity, increase in the immune response without muscular catalysis, increase in bone mineralization and lipolytic and anti-lipogenic effects (Rico *et al.*, 2007).

Benefits of Fatty Acids

Some years ago, the nutritional reputation of the lipids, fatty material derived from milk, was possibly one of the most deteriorated, not only with respect to milk components, but also in many other foods. Some cardiac diseases, colon cancer and other illnesses were attributed to these components. However, diverse investigations have revealed important functions of some lipids contained in foods. The potentiality of linoleic acid in the inhibition of cancer and atherosclerosis and the improvement of the immunological functions, the effects of attraction of the butyric acid for the elimination of cancer cells in the colon, and the cell regulatory function of the phospholipids (in the membrane) are some of the new discoveries of the positive functions of milk lipids (*Valenzuelaet al.,* 2003). Table 1 shows some ingredients derived from lipids to which some functionality has been attributed.

Some milk components have demonstrated beneficial effects on health, such as butyric acid on the reduction of colon cancer, participating in the elimination of cancer cells in the colon (Valenzuela*et al.*, 2003; Carmuega, 2004).In cell cultures derived from colon cancer, it has been observed that butyrate, at physiological concentrations, inhibits cell proliferation, blocking the cell in G1 phase of the cell cycle, inducing differentiation and apoptosis; and modulating the expression of multiple genes, including some of the oncogenes and suppressor genes implicated in colorectal carcinogenesis (Figure 2). Therefore; it is now believed that

C4:0 is one of the principal protector components in colorectal carcinogenesis, by inducing the detention of growth, differentiation and apoptosis of the colonocytes (Bravo *et al.*, 2000).

Table 1. Some functional ingredients derived from lipids

Funcional ingredient	*Function or objective*
Butyric gamma-amino acid	• Antihypertensive
Butyric acid	• Elimination of cancer cells in the colon
Omega-3 fatty acids (Only linoleic acid and a –linoleic acid are found in cow's milk)	• Development of the retina and the brain in the early infancy • Prevent coronary diseases and heart attacks • Prevention of autoimmune disorders • Prevention of Crohn's disease • Prevention of breast, colon and prostate cancer • Regulation of hypertension • Prevention of rheumatoid arthritis • Absence in the diet may cause neurological and/or visual deficiency
Conjugated linoleic acid	• Inhibition of cancer • Inhibition of atherosclerosis • Improvement of the immunological system • Anti-mutagenic
Sphingolipids of the membrane	• Related to the regulation of cell behavior • Control of colon cancer • Reduction of LDL colesterol (low density lipoproteins) • Increase of the HDLs (high density lipoproteins)
Metabolic products of triglycerides and phospholipids	• Antimicrobial and antiviral activities
Short chain fatty acids	• Prevention of colonization of enterogenic pathogens
Phospholipids	• Effect of protection against gastric ulceration
Phospholipids in butter wey	• Defense against listeria

Source: Modified from Valenzuela *et al.*, 2003.

Butyric Acid (C4:0)

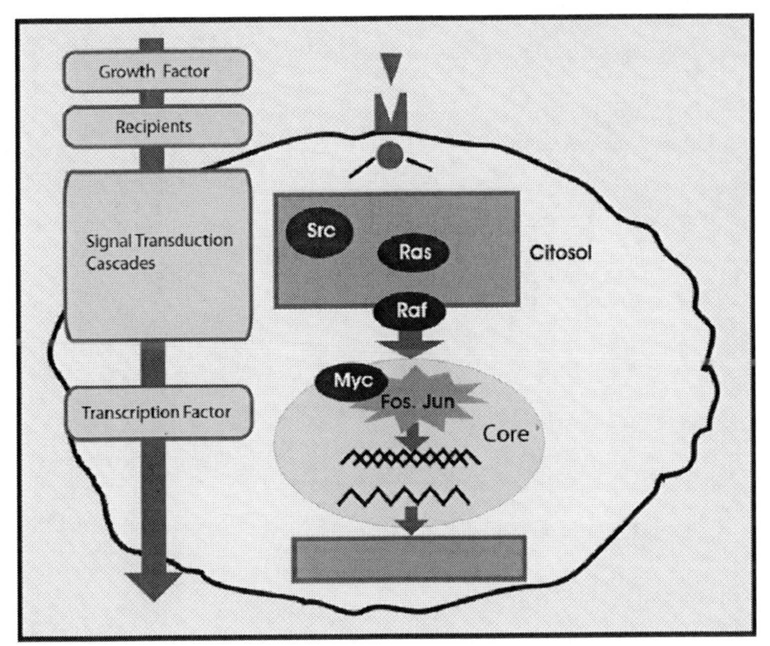

Source: Modified from Bravo *et al.*, 2000.

Figure 2. Regulation of the expression of oncogenes and suppressor genes by butyric acid.

Omega-3 Fatty Acids

There are numerous clinical and epidemiological investigations that have evidenced the essentiality in humans of the omega-3 fatty acids (n-3) and particularly the importance of eicosapentaenoic acid (EPA) and docosahexaenoic acid (DHA) in the prevention and management of diverse affections (Castro-González, 2002).

Benefits during Gestation

The n-3 FA are structural components of the brain and of the retina during the development of the fetus. It has been estimated that approximately 600 mg of the essential fatty acids (EFA) are transferred from the mother to the fetus during a gestation to term in a healthy mother; as shown in Figure 3. The diet of the mother prior to conception is of great importance, given that it determines in part the type of fats that will accumulate in the tissues of the fetus.

The placenta selectively transports arachidonic acid (AA) and DHA from the mother to the fetus. This produces an enrichment of these FA in the circulating lipids of the fetus, which is vital during the third trimester of gestation when the development of the nervous system is highest. A notable increase has been observed in the content of DHA in the cerebral tissue during the third trimester and after birth (Castro-González, 2002; Valenzuela *et al.*, 2003).

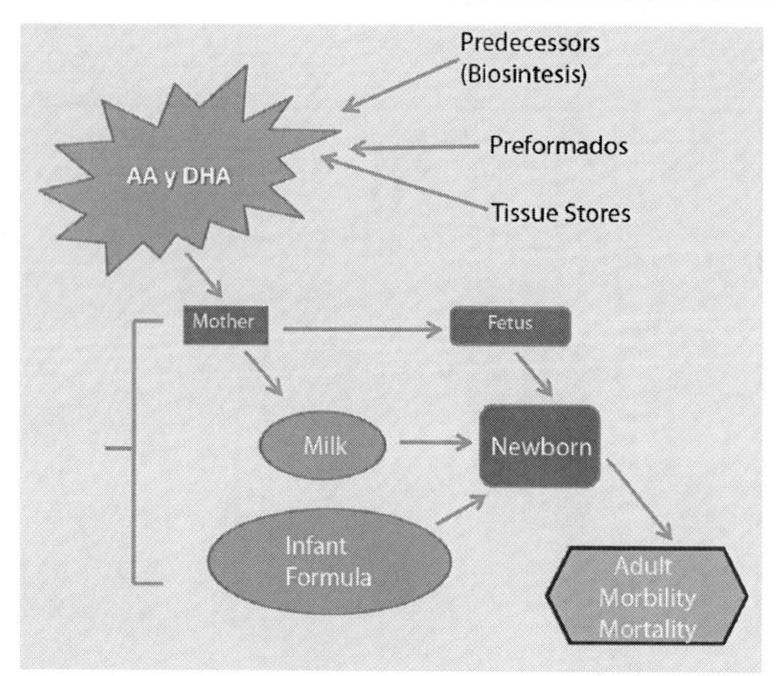

Source: Modified from Valenzuela *et al.*, 2003.

Figure 3. Contribution of omega-3 fatty acids during the gestation and lactation periods.

An association has been reported between a lower ingestion of vitamins and PUFA and a higher incidence of low birth weight. Other studies have informed a correlation between maternal nutrition during the third trimester and the serum lipids in newborns. These results underline the need for an adequate nutritional state of fatty acids from the early stages of pregnancy and during the lactation period, in order to achieve a good transference of FA to the fetus, through the placenta, and to the newborn through human milk (Rodríguez-Cruz *et al.*, 2005). The AA and the DHA perform their metabolic functions forming part of the structure of the phospholipids of the cell membranes, particularly of the phosphatidylcholine, phosphatidylethanolamine and phosphatidylserine. Because of their high degree of polyunsaturation, these FA provide great fluidity to the membranes in whose formation these phospholipids participate. Fluidity is essential so that the proteins of the membrane (ion channels, receptors, junctions, catalytic receptors, enzymes, vesicle forming structures, etc.) can have the mobility required by their functions, whether on the surface of the membranes or in the interior of the lipid bilayer. In the formation of the nerve tissue, and particularly of the brain, the fluidity of the membranes is very important. The most critical stages in the formation of the structure of the brain occur during the last gestational trimester in humans and continue until two years after birth. This morphogenic process that begins in the neural crest is characterized by successive stages of neurogenesis, neuronal migration, selective apoptosis, synaptogenesis and myelination stages, which in relatively sequential form give form and functionality to the cerebral tissue. These cell processes require the active participation of the glial cells, particularly of the astrocytes, which provide the neurons with the metabolites and particularly with the astrocytes, which provide the neurons with the metabolites and the physical support required for their mobilization within the brain.

Morphogenesis, intimately associated with the function of the brain, requires an extraordinary supply of PUFA, particularly AA and DHA. These FA are particularly concentrated in the cones of axonal growth and in the synaptic vesicles, thus they have great relevance in the formation and propagation of the electric impulse and in the mobilization of the vesicles that contain the neurotransmitters. Figure 4 sums the effects of the LCPUFA in the development of the brain (Valenzuela A. *et al.*, 2003).

Benefits during Growth

The ingestion of lipids during gestation and the first year of life of humans is fundamental, not only to satisfy the energy requirements, but also as vehicle of the liposoluble vitamins to favor their absorption and as an important source of UFA. The n-3 and n-6 PUFA are basic for the cerebral development of the fetus and cognitive development of the newborn (Figure 4), given that the phospholipids that integrate the cell membranes of the nervous system contain large amounts of this type of fatty acids (Rodríguez-Cruz *et al.*, 2005). It has been observed that babies that are breast-fed or fed with formulas that contain DHA present greater visual acuity and a greater capacity to respond to light, which is associated with the cognitive ability to integrate information. A better intellectual coefficient has also been observed (Castro-González, 2002). On the other hand, the studies with children born at term indicate that the levels of DHA in the cerebral cortex are higher when they are breast-fed, with respect to those fed with milk formula. In addition, some reports suggest that the groups of infants that are breast-fed have better results in psychometric tests than those that are fed with milk formula (Rodríguez-Cruz *et al.*, 2005).

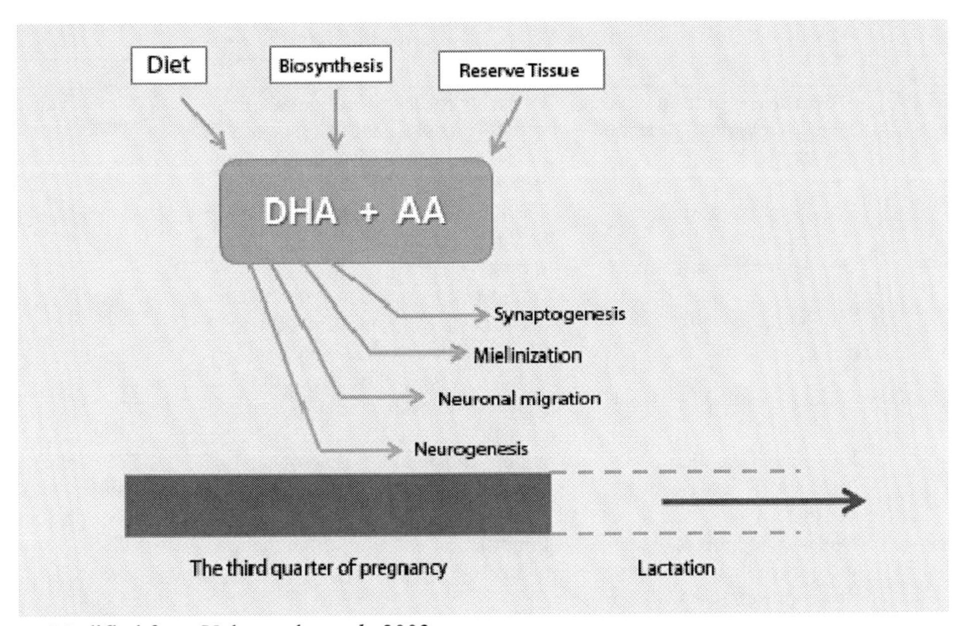

Source: Modified from Valenzuela *et al.*, 2003.

Figure 4. Participation of the LCPUFA, omega-6 and omega-3 in the development of the central nervous system.

Polyunsaturated Fatty Acids (PUFA) in Disease

There are a variety of studies carried out in humans that show the possible effects of the PUFA in different diseases (Rodríguez-Cruz *et al.*, 2005). Table 2 presents a summary of information associated with the consumption of polyunsaturated fatty acids and their effect on human health.

Table 2. Effects of the consumption of the PUFA on different diseases in humans

Fatty acid	Consumption	Effect	Population
		Diabetes	
ADH AEP	Provided in capsules; 4g/d of ADH or AEP during 6 weeks	Reduction of the serum lipids and short-term adverse effects in glucose levels	30 men 12 women
PUFA	Content in the diet; 6.0± 1.9 g/d, men; 5.9± 1.9 g/d, women	Higher consumption of PUFA is associated with a lower concentration of glycated hemoglobin	Diabetics: 2,759 men, 3,464 women
PUFA	Content in the diet; 2 to 7 % of energy	Higher consumption of PUFA is associated with a lower risk of developing type 2 diabetes	84,204 women
EPA DHA	Provided in capsules 0.69/0.20 g/d to 5.40/2.30 g/d During 2 a 36 weeks	No changes are observed in the glucose nor in the concentration of glycated hemoglobin Reduction of triacylglycerols (-30%).	Meta-analysis of 26 studies
DHA EPA	Provided as fish oil 3 to 18 g/d during 12 weeks	No effect on the concentration of glucose Reduction of triacylglycerols by 0.56mmol/l Increase of LDL cholesterol by 0.21 mmol/l	823 women with type 2 diabetes
		Cancer	
n-3 PUFA	Provided as fish from <once/week, once/week and ≥ twice/week	Reduction of the risk of developing various types of cancer, especially of the digestive tract (esophagus, stoma and rectum)	7990 individuals
n-3 PUFA	Fish from once/week, 1 to 2 days/week 3-4 days/week and almost daily	The higher the consumption of fish, the higher the content of n-3 PUFA in serum. An inverse association is observed between the concentration of n-3 PUFA in the serum and prostate cancer. This is not observed in other cancers (stomach, colon, rectum and breast).	297 men
ALN	Provided in the diet (0.22 a 4.78 de ALN g/d)	Higher consumption is associated with a reduction in the formation of platelets in the arteries	1,575 adults (698 men, 877 women)
ALN	Dietary manipulation; increase in the consumption of foods rich in ALN such as nuts and nut oil (4.8 g of ALN + 2/1 for 6 weeks)	Inhibition of vascular inflammation and vascular activity	Adults 20 men y 3 women

Fatty acid	Consumption	Effect	Population
		Cardiovascular diseases	
n-3 PUFA	Fish (once a week)	The consumption of fish is associated with a reduction in primary cardiac arrest	Cases of primary cardiac arrest 334 Controls 493
n-3 PUFA	Fish (<once a month to ≥ 6 times a week	No reduction was observed in the risk of coronary disease by increasing the consumption of fish in individuals without antecedents of vascular disease	44,805 men
n-3 PUFA	Fish (<once a month to ≥6 times a week	Higher consumption of fish is associated with a low risk of coronary disease	84,688 women
n-3 PUFA	Fish (<once a month to ≥5 times a week)	Higher consumption of fish is associated with a lower incidence of cardiac disease and mortality from cardiac disease in diabetic women	5,103 women with type 2 diabetes
n-3 PUFA	n-3 PUFA of intermediate chain ≥ 1.08 g/d of sea or plant origin	Reduction by 50 % in risk of cardiac disease	45,722 individuals

Source: Modified from Rodríguez-Cruz *et al.*, 2005.

Table 3. Effects of polyunsaturated fatty acids on several genes encoding enzyme proteins involved in lipogenesis, glycosis, and glucose transport[1]

Function/gene	Linoleic acid	α-linolenic acid	Araquidonic acid	Eicosapentaenoic acid	Docosahexaenoic acid
Hepatic cells					
Lipogenesis					
FAS	↓	↓	↓	↓	↓
S14	↓	↓	↓	↓	↓
SCD1	↓		↓	↓	↓
SCD2	↓	↓	↓	↓	↓
ACC	↓	↓	↓	↓	↓
ME	↓	↓	↓	↓	↓
Glicolisis					
G6PD	↓				
GK	↓	↓	↓	↓	↓
PK	-	↓	↓	↓	↓
Mature adipocytes					
Glucose transport					
GLUT4	-	-	↓	↓	-
GLUT1	-	-	↑	↑	-

[1]FAS: fatty acid sintetasa; SCD: esteril-CoA desaturasa; ACC: acetil CoA carboxilasa; ME: malic enzime; G6PD: glucosa 6-fosfato deshidrogenasa; GK: glucocinasa; PK: piruvato cinasa; GLUT: glucose transporter. ↓ Supress or decrease; ↑ Induce or increase.

Source: Modified from Simopoulos, 2000.

Benefits for the Cardiovascular System

The n-3 FA, particularly those of long chain, may affect numerous factors implied in the development of atherosclerosis, which initially could result in slower progression of the disease. It has been described that the n-3 FA reduce the concentrations of chemoattractants, growth factors and products of molecular adhesion, which can favor a reduction in the migration of leucocytes and vascular smooth muscle cells on the inside of the vascular wall, and thus delay the atherosclerotic process (López*et al.*, 2006). Three principal mechanisms seem to be involved in the cardiovascular protector affect of the n-3 FA: their anti-inflammatory effect, their anti-thrombotic effect and their anti-arrhythmic action; they increase the time of bleeding, thus avoiding the adherence of platelets in the arteries. They prevent atherosclerosis by reducing the concentrations of cholesterol in plasma, and are useful in hyper tense patients, as they contribute to lowering blood pressure and reduce the concentration of triacylglycerides (TG) in plasma. They also reduce total cholesterol and the VLDL-C (Very Low Density Lipoprotein Cholesterol); which are shown in Table 3 (Castro-González, 2002; Simopoulos, 1999; Benatti *et al.*, 2004; López*et al.*, 2006).

Benefits for the Nervous System

In continuation information is given by tabulation, of the relationship of the consumption of n-3 FA with the central nervous system:

1) The n-3 FA are essential for an adequate development and functioning of the brain and the central nervous system. They concentrate in the retina and cerebral cortex, and have the capacity to correct visual and cerebral problems in patients with a proven deficiency. Many aspects of location, anxiety, learning ability, memory, and retinal function are favored by the consumption of the n-3 FA (Castro-González, 2002).
2) The DHA and the AA are the principal components of the brain, given that they are found in more than 30 % of the fatty acids that form the phospholipids of the membranes.
3) On the other hand, they are precursors of hormonal substances such as the prostanoides (prostaglandines and thromboxanes) that facilitate the transmission of messages in the central nervous system; furthermore, when there are adequate levels of DHA in the brain, cerebral activity improves (Castro-González, 2002; Simopoulos, 1999).
4) Two thirds of the fatty acids of the membranes of the photoreceptors of the retina are n-3, principally DHA (Castro-González, 2002). In the formation of visual tissue, the external membranes of the cones and of the rods of the retina accumulate a great quantity of LCPUFA,particularly of DHA. The rods of the retina have more than 50 % of the fatty acids of the n-3 family, principally DHA. The fluidity of these membranes is essential for the process of transduction of the light signal and its conversion into an electric signal, which is later processed by the brain. The photoreceptors are concentrated in the external membranes of the cones and the rods, and when they receive a light signal, in the form of photons, they move through the

membrane, modifying the concentration of cyclic Guanosin monophosphate (GMPc) (a second messenger). The reduction of the concentration of cyclic GMP stimulates the closing of the sodium channels, producing a hyperpolarization of the membrane, which generates the electric impulse that is sent to the brain. Figure 4 summarizes the effects of the LCPUFA in the visual system (Valenzuela *et al.*, 2003; Rodríguez-Cruz *et al.*, 2005).

5) Another relationship between the DHA and the cerebral function has been found in the organization pattern of sleep in children. A low consumption of DHA results in fewer slow waves of sleep, which serve as an indicator of the maturation and development of the central nervous system (SNC) and of the brain (Castro-González, 2002).
6) The n-3 are related with problems of depression and violence. It has been demonstrated that the dietary DHA has protective effects against an increase in hostility of students under stress conditions (Castro-González, 2002).
7) Low concentrations of DHA are a useful indicator for predicting greater problems of behavior in children who have been diagnosed with the syndrome of attention deficit with hyperactivity (TDAH). These problems may be a reflection in part of the problems in the serotoninergic neurotransmission (Castro-González, 2002).

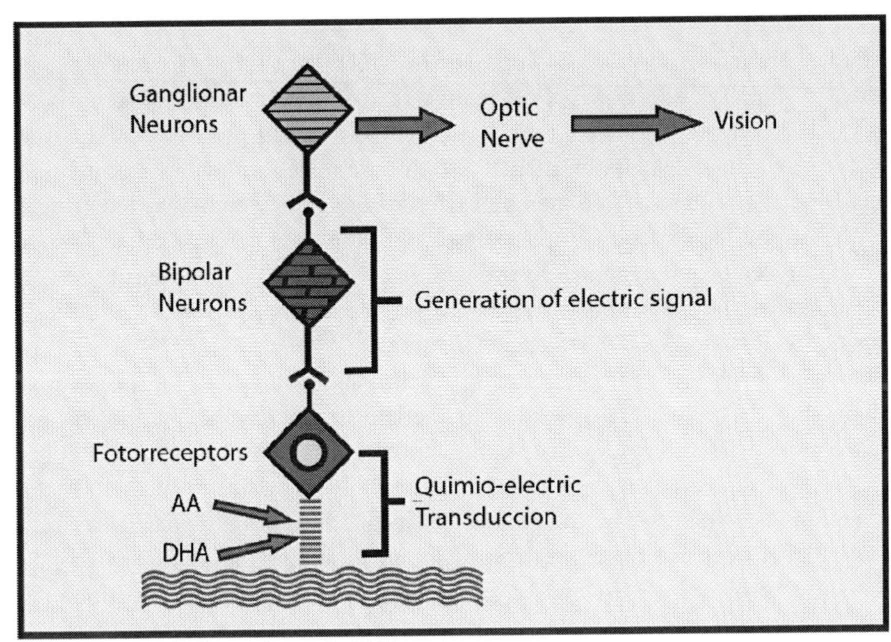

Source: Modified from Valenzuela *et al.*, 2003.

Figure 5. Participation of the LCPUFA, omega-6 and omega-3 in the development of the visual system.

Benefits in Diabetes

It has been proposed that the PUFA may have a beneficial effect on the development or control of diabetes through different mechanisms. One of these mechanisms refers to the capacity of the PUFA to act as activating ligands of the receptor activated by peroxisomic proliferators (Peroxisome proliferator-activated receptor gamma) (PPARγ)- The active PPARγ stimulates the differentiation of the preadipocytes to adipocytes, which generates in this cell an increase in the receptors for insulin, thus reducing insulin resistance. Another mechanism of action of the PUFA is the protection of the pancreatic beta cells from the damage caused by the increase in free radicals produced during diabetes (Rodríguez-Cruz *et al.*, 2005). However, there is still controversy with respect to the beneficial effects of the PUFA in diabetes in humans. A study entitled "Nurses' Health Study" of the United States reported that the risk of developing diabetes is inversely associated with the consumption of polyunsaturated fat. Another study carried out on 35,988 women who initially did not have diabetes, showed that the consumption of PUFA has a protective effect, given that a decrease of 40 % was observed in the incidence of type II diabetes. In contrast, other investigations do not show beneficial effects in the consumption of PUFA, as reported by Woodman *et al.* (2001), who evaluated the effect of DHA and EPA on the control of glucose in 59 subjects with type II diabetes, who consumed 4g/d of DHA or EPA during six weeks. Their results showed adverse effects, given that an increase was observed in the concentration of glucose in fasting (Rodríguez-Cruz *et al.*, 2005).

Benefits in Obesity

Different mechanisms have been proposed for which the PUFA can delay or control the development of obesity. The PUFA are negative regulators of hepatic lipogenesis, which is mediated by the repression of Sterol regulatory element-binding protein-1 (SREBP-1). It has been observed that the consumption of PUFA by obese mice diminishes the mature form of the protein SREBP-1 and therefore the expression of lipogenic genes is reduced, such as the synthase of FA, and the esteroil CoA desaturase-1 in the liver of these mice. Consequently, both the hyperglycemia and hyperinsulinemia are improved with the administration of PUFA. The PUFA improve the biochemical and metabolic alterations associated with obesity, such as the hepatic steatosis and insulin resistance in mice.

In humans, the PUFA of the n-3 family oxidize more rapidly than the saturated fatty acids, given that they have the special characteristic of increasing thermogenesis and in consequence reduce the efficiency of depositing body fat (Rodríguez-Cruz *et al.*, 2005).

It is also well known that the serum concentration of leptine in patients with type 1 diabetes mellitus is influenced by the type of fat in the diet. In particular it has been found that the n-3 PUFA decrease the expression of the gene for leptine both *in vitro*and *in vivo*. The direct effects of the PUFA in the promotion of the activity of leptine, indicate a regulation of their specific action in the regulation of their expression (Benatti *et al.*, 2004).

Benefits in Cancer

The causes of cancer are not clearly defined, but it is known that both the internal and external factors such as type of diet play an important role for initiating and promoting carcinogenesis. It is estimated that approximately 35 % of all deaths by cancer are related to diet. The amount and type of fat of the diet consumed can be important in the development of human cancer.

Epidemiological evidence in humans has suggested that the ingestion of PUFA-CL of the n-3 family, have beneficial effects in cancer. The biochemical mechanisms by which the n-3 PUFA-CLs inhibit cell development in some tumors are not exactly known. However, studies carried out in rodents and in human tumor cells have proposed some hypotheses that involve different ways of signaling (Rodríguez-Cruz *et al.*, 2005):

1) DHA induces the arrest of the cellular cycle due to the dephosphorylation of the protein pRBI, which is found in its active form and detains the cell cycle. Furthermore, DHA promotes the apoptosis of human cell lines of cancer of the pancreas and of leukemia.
2) The PUFA are cytotoxins for certain tumor cells*in vitro*. This cytotoxic action may be related to the peroxidation of their double bonds, which generates a persistent oxidative stressdue to the increase in the production of free radicals, which damage the DNA.
3) Another hypothesis suggests that the n-3 PUFA of the diet suppress the cell growth of the tumor by means of the inhibition of the cyclooxygenase, enzyme that regulates the synthesis of the prostaglandins (PG). This proposal arose given that it exists in breast tumors and in metastasis among other neoplasias, compared with normal tissues, a higher amount of PG that are metabolic products of the AA. However, this hypothesis is under controversy given that it in mice a suppression of the cell growth was demonstrated independently of the cyclooxygenase of implanted tumor cells of the colon (Rodríguez-Cruz *etal.*, 2005).

Benefits in the Treatment of Inflammatory Processes

Arachidonic acid (AA) does not exist freely within the cells; it is found forming part of the structure of the phospholipids of membrane, esterified to the position sn-2. The concentration of free arachidonic acid is very low, lower than 10^{-6} Molar, thus the first step in its metabolism consists of its liberation of the membrane phospholipids, catalyzed by the phospholipase A2.

This enzymatic reaction is stimulated in states that may be physiological or pathological (Pérez-Ruiz *et al.*, 1998; Companioni, 1995). The eicosanoids derived from the AA and EPA, present a very similar molecular structure, but different marked biological effects. These eicosanoids are considered to diminish the inflammatory process, given that at a reduction in the amount of inflammatory products derived from the AA (PGE2 and LTB4) have been implied as base anti-inflammatory mechanisms. Table 4 shows the effects of the n-3 PUFA as factors involved in the inflammatory processes and in atherosclerosis (Benatti *et al.*, 2004; Simopoulos, 2000).

There now is growing evidence from studies in animals and humans in which it is shown that the n-3 PUFA have an important immune-modulator effect. This has incited researchers to investigate its usefulness for the treatment of diseases such as: ulcerative colitis and intestinal inflammatory disease, rheumatoid arthritis, psoriasis, asthma, atopic dermatitis, lupus, and cystic fibrosis, among others (Ruxton *et al.*, 2004; Benatti *et al.*, 2004).

Table 4. Effects of n-3 PUFA on the factors involved in the physiopathology of atherosclerosis and inflammation[2]

Factor	Function	Effect of the n-3 PUFA on factor concentrations
Arachidonic acid	Precursor of eicosanoids; platelet aggregation; stimulation of white blood cells	↑
Thromboxane A_2	Platelet aggregation; vasoconstriction; increment of intracellular Ca^{2+}	↓
Prostacyclins	Prevents platelet aggregation; vasodilator, increments the AMPc	↑
Leukotriene B_4	Chemoattractant of Neutrophils; increment of the intracellular Ca^{2+}	↓
Activator of tissue plasminogen	Increment of endogenous fibrinolysis	↑
Fibrinogen	Blood coagulation factor	↓
Platelet activation Factor	Activation of platelets and white blood cells	↓
Growth factor derived from platelets	Chemoattractant and mitogen for muscle cells and macrophages	↓
Free radicals	Cause cell damage; improves LDL levels, stimulates the metabolism of the arachidonic acid	↓
Lipid hydroperoxydases	Stimulate the formation of eicosanoids	↓
Interleukin 1 and tumor necrosis factor	Stimulation of the neutrophils for the formation of free radicals, lymphocyte proliferation, and the platelet activation factor. Expression of molecular adhesion in the endothelial cells, and activation of the plasminogen	↓
Platelet relaxing factor	Reduces the vasoconstriction response	↑
VLDL	Relationship with the concentration of LDL HDL	↓
HDL	Reduction of the risk of coronary disease	↓
Lipoprotein (a)	Atherogenic and thrombogenic	↑
Triacylglycerols and chylomicrons	Contribute to postprandial lipemia	↓

Source: Modified from Simopoulos, 2000. ↑ Increases; ↓ Decreases

Conjugated Linoleic Acid (CLA)

There are now studies that have demonstrated the role of conjugated linoleic acid in the regulation of the immune function, thus reducing the risk of certain forms of cancer and atherosclerosis (Nagedra, 2000; Carmuega, 2004). Milk products contain numerous components that potentially may play a role in cancer prevention. One which has been the focus of investigation in recent years is cis linoleic acid (CLA). Although there are various isomers of CLA, practically all those that are normally found in milk products exert anti-carcinogenic effects both in studies *in vitro*and experiments in colon and breast cancer (Carmuega, 2004). From its purification and synthesis, its efficacy was demonstrated in the suppression of distinct tumors (stomach, prostate, colon and breast) in different animal models, at concentrations as low as 0.05 % or equivalent consumptions of 2g/d in the human diet (De Blas, 2004).

Although it is likely that the amounts habitually found in milk products may not be sufficient to reproduce the effects found in experimental animals, they can be increases by improving their preventative profile (Carmuega, 2004).

TheCLA also seems to have antilipogenic and lipolitic effects in a wide variety of animal species, both of ruminants and monogastric species, which is of commercial interest for animal feeding and for the human species. However, it should be considered that the doses of CLA required for this effect seem to be higher (around 0.5 % of the diet or equivalent consumptions equivalent of 15-20 g/d in the human species) and more difficult to reach through the diet. In consequence, the results of the addition of CLA to the diet of persons with problems of obesity are unequal as a function of the dose employed and the duration of the treatment; and seem to have more effect by reducing body weight gain than the total live weight. Supplementation with CLA also seems to protect against the accumulation of lipids in the arteries in rabbits and hamsters, but at even higher doses (De Blas, 2004).

The mechanisms of the mode of action of CLA are not yet known but relationships have been suggested with its high antioxidant activity, or with its effects on the metabolism of the eicosanoids, the genetic expression or the increment of the lipolitic activity of the adipocytes. Some effects could be specific of the different types of isomers, thus the cis-9, trans-11 (the most abundant in milk) would be the most active as anticarcinogenic agent, while the trans-10 cis-12 would be the most active for its effects against adipocytes (De Blas, 2004).

Vaccenic Acid

The secretion of vaccenic acid in milk is parallel to that of CLA and superior in a proportion between 1.5:1 and 3:1. The presence of vaccenic acid in foods could be considered unfavorable, as it is an unsaturated FA in trans configuration. However, the desaturase Δ-9 is also found in human tissues, having been demonstrated their capacity to convert vaccenic acid in CLA. Therefore, an increase in the consumption of this fatty acid could have the same beneficial effects for health associated with the ingestion of CLA (De Blas, 2004).

Sphingolipids of the Membrane

In diverse investigations at the experimental level, it has been demonstrated that the sphingolipids of milk are effective in reducing the risk of colon cancer in a dose very close to what is now consumed when the calcium requirements are satisfied (Carmuega, 2004).

Conclusion

The information included in this chapter demonstrates that the consumption of milk in some Latin American countries is below the amount recommended by the WHO and the Comité de la Secretaría de Salud particularly in the case of Mexico. The mean consumption of milk in Mexico is approximately 250 ml of milk, which corresponds to less than 50 % of the required amount. The functional substances present in milk, such as butyric and conjugated linoleic acidwhich have anticancergenic properties, are a solid argument for the governments with deficient milk production and consumption, to implement integral measures that will give a solution to this economic problem and enhance milk consumtion, and of course Public Health.

References

Benatti P., Peluso G., Nicolai R., Calvani M. (2004). Polyunsaturated Fatty Acids: Biochemical, Nutrtitional and Epigenetic Properties. *Journal of the American College of Nutrition.* 23(4), 281-302.

Bravo A., Alfonso JJ., Medina V., Lorenzo N., Fernández MV., Gonzpalez F. (2000). Butirato y carcinogénesis colorrectal. *Cir Esp.* 68, 57-64.

Carmuega E. (2004). Los beneficios de la leche para la dieta del ser humano. *Serie: Industria Lechera (Argentina).* 65(736), 17-25.

Castilla-Pinedo Y., Alvis-Estrada L., Alvis-Guzman N. (2010). Exposición a órganoclorados por ingesta de leche pasteurizada comercializada en Cartagena, Colombia. *Rev. Salud pública.* 12(1), 14-26.

Castro-González M. (2002). Ácidos grasos omega 3: beneficios y fuentes. *INCI.* 27(3), 128-136.

COFEPRIS. (2010). COFEPRIS armoniza con estándares internacionales marco regulatorio para productos lácteos. 19 Octubre 2010. http://www.cofepris.gob.mx/work/sites/cfp/resourses/Local/Content/1918/21/NOM_243_19102010.pdf

Companioni M. (1995). Ácido araquidónico y radicales libres: su relación con el proceso inflamatorio. *Rev Cubana Invest Bioméd.* 14(1), 13-17.

De Blas C. (2004). Cambios en el perfil de ácidos grasos en productos animales en relación con la alimentación animal y humana. Importancia del ácido linoleico conjugado.1.Rumiantes. *XX CURSO DE ESPECIALIZACION FEDNA.* 79-100. http://vaca.agro.uncor.edu/~pleche/material/04CAP_5.pdf

Espinoza Y., Rodríguez Y. (2008). Estudio bacteriológico de leche cruda por el sistema Diralec en un municipio de la región oriental del país. *REDVET. Revista Electrónica de Veterinaria.* 9(7), 1-7. http://www.veterinaria.org/revistas/redvet/n070708/070804.pdf

FAO.Ganadería Bovina en América Latina: Escenario 2008-2009 y tendencias del Sector. http://www.rlc.fao.org/es/ganaderia/pdf/ganbov.pdf

Gutiérrez, R., Vega, S., Díaz-González, G., Sánchez, J., Coronado, M., Ramírez, A., Pérez, J., González, M. y Schettino, B. (2009). *Detection of non-milk in milk fat by gas chromatography and linear discriminat analysis. J. of Dairy Sci.* 92, 1846 -1855.

INEGI. (2007). Censo Agrpecuario 2007. Síntesis Metodológica. 1-75. http://www.inegi.org.mx/est/contenidos/espanol/proyectos/censos/agropecuario2007/

López A., Macaya C. (2006). Efectos antitrombóticos y antiinflamatorios de los ácidos grasos omega-3. *Rev Esp Cardiol Supl.* 6, 31D-37D.

Máttar S., Calderón A., Sotelo D., Sierra M., Tordecilla G..(2009). Detección de Antibióticos en Leches: Un Problema de Salud Pública. *Rev Salud pública.* 11(4), 579-590.

Miller-Pérez C., Sánchez-Islas E., Mucio-Ramírez S., Mendoza-Sotelo J., León-Olea M. (2009). Los contaminantes bifenilos policlorinados (PCB) y sus efectos sobre el Sistema Nervioso y la salud. *Salud Ment.* 32(4), 335-346.

Morales S. (1999). Factores que afectan la composición de la leche. *Tecno Vet.* 5(1). http://www.tecnovet.uchile.cl/CDA/tecnovet_articulo/0,1409,SCID%253D9670%2526IS ID%253D459,00.html

Nagendra P. (2000). Effects of milk-derived bioactives: an overview. *British Journal of Nutrition.* 84(Suppl 1), S3-S10.

Pérez A., Cartaya L., Valencia V., Sanjurjo V., Ilisástigui T. (1998). Biosíntesis de los productos del ácido araquidónico y su repercusión sobre la inflamación. *Rev Cubana Estomatol.* 35(2); 56-61.

Pérez N., Díaz G., Gutierrez R., Vega S., Urbán G., Prado M., González M., Ramírez A., Pinto M. (1998). Composición en ácidos grasos de la grasa de leches pasteurizadas mexicanas. *Veterinaria México.* 29(4),329-335.

Pinto M., Rubilar A., Carrasco E., Shun K., Brito C., Molina L. (2002). Efecto estacional y del área geográfica en la composición de ácidos grasos en la leche de bovinos. *Agro sur.*30(2), 75-90.

Prado G., Diaz G., Gutiérrez R., Vega S., Noa M., Chávez E.(2007). Residuos plaguicidas organoclorados en leche de cabra de Querétaro, Querétaro, México. *Veterinaria México.* 38(003), 291-301.

Ramírez A., Vega S., Prado G., Gutierrez R., Pérez C. (2008). Detección de suero de quesería en leches ultrapasteurizadas mexicanas mediante la cuarta derivada del espectro de absorción. *Veterinaria México.* 39(001), 17-27.

Ramírez-Jaspeado R., García-Salazar J.A., Mora-Flores JS., García-Mata R. (2010). Efectos del Tratado de Libre Comercio de Amércia del Norte sobre la producción de leche en México. *Universidad y ciencia.* 26(3), 283-292. www.ujat.mx/publicaciones/uciencia

Ramírez N., Álvarez J., Ponce P., Suárez E., Hernández J. (2007). Versión avanzada del sistema Diralec: una tecnología para el análisis de la calidad de la leche. *Biotecnología Aplicada.* 24, 290-293.

Rico J., Moreno B., Pabón M., Carulla J. (2007). Composición de la grasa láctea en la sabana de Bogotá con énfasis en ácido ruménico – CLA cis 9, trans -11. *Rev Col Cienc Pec.* 20(1), 30-39.

Rivera A., Muñoz-Hernández O., Rosas-.Peralta M., Aguilar-Salinas C., Popkin B., Willett W. (2008). Consumo de bebidas para una vida saludable; recomendaciones para la población mexicana. *Salud Pública Mex.* 50(2), 173-195.

Rodríguez-Cruz M., Tovar A., Del Prado M., Torres N. (2005). Mecanismos moleculares de acción de los ácidos grasos poliinsaturados y sus beneficios en la salud. *Revista de Investigación Clínica.* 57(3), 457-472.

Ruxton C., Reed S., Simpson M., Millington K. (2004). The health benefits of omega-3 polyunsaturated fatty acids: a review of the evidence. *Hum Nutr Dietet.* 17, 449-459.

SAGARPA (Secretaría de Agricultura, Ganadería, Desarrollo Rural, Pesca y Alimentación). (2008). *Revista Claridades Agropecuarias.* 207(Especial),1-56. ISSN:01889974. http://www.infoaserca.gob.mx/claridades/revistas/207/ca207.pdf

SAGARPA (Secretaría de Agricultura, Ganadería, Desarrollo Rural, Pesca y Alimentación). (2009). *Revista Claridades Agropecuarias.* 207(Especial),1-56. ISSN:01889974. http://www.infoaserca.gob.mx/claridades/revistas/207/ca207.pdf

SAGARPA (Secretaría de Agricultura, Ganadería, Desarrollo Rural, Pesca y Alimentación). (2010). *Revista Claridades Agropecuarias.* 207(Especial),1-56. ISSN:01889974. http://www.infoaserca.gob.mx/claridades/revistas/207/ca207.pdf

Silva E., Verdalet I. (2003). Revisión: alimentos e ingredientes funcionales derivados de la leche. *ALAN.* 53(4), 333-347.

Simopoulos A. (1999). Essential fatty acids in health and chronic disease. *Am J Clin Nutr.* 70(Suppl), 560S-569S.

Valencia E., Ramírez M. (2009). La industria de la Leche y la contaminación del Agua. *Elementos.* 73, 27-31.

Valenzuela A., Nieto S. (2003). Ácidos grasos omega-6 y omega-3 en la nutrición perinatal: su importancia en el desarrollo del sistema nervioso y visual. *Revista Chilena de Pediatría.* 74(2), 149-157.

Part III: Milk Production by Different Species in Varied Geographical Areas

In: Milk Production
Editor: Boulbaba Rekik, pp. 99-124
ISBN 978-1-62100-061-7

Chapter VI

Investigations on the Geographical Origin of Cow Milk and Comparison of the Lipid Composition of Cow and Buffalo Milk by Means of Traditional and Innovative Physico-Chemical Analyses

A. Sacco*[*], *D. Sacco, G. Casiello, V. Mazzilli, A. Ventrella and F. Longobardi
Dipartimento di Chimica, Università di Bari "A. Moro",
Via Orabona 4, 70126 Bari (Italy)

Abstract

The quality of food products depends upon several factors such as pedoclimatic conditions, humidity level, and average temperature that characterize the geographic area where the foodstuff is obtained or produced. Since the geographic origin is widely considered as a valid criterion for quality assessment, the European Community set rules to confer the Protected Designation of Origin (PDO) label to peculiar local foodstuffs on the basis of their origin. In particular, the PDO label can be assigned only if the original raw materials are harvested, produced, and processed in the specific geographic region that gives the final food product its name. In the last decades, efforts have been directed towards exploring innovative and competitive methods to certificate the geographical origin, i.e. for the assessment of food quality and for the establishment of authenticity and traceability of food products. Several food products from the South Italian region of Apulia, are widely appreciated and considered as typical; among these, one important example is represented by the Apulian cow mozzarella, that is a fresh and stringy-

*Email: antonio.sacco@chimica.uniba.it

textured cheese, whose authenticity could be proved by determining the geographical origin of milk.

In this chapter, traditional techniques, High Performance Ion Chromatography (HPIC), Inductively Coupled Plasma Atomic Emission Spectroscopy (ICP-AES), Nuclear Magnetic Resonance (NMR) spectroscopy and Isotope Ratio Mass Spectrometry (IRMS) were used in combination with Multivariate Statistical Analysis (MSA) to determine a number of components and, hence, for discriminating the geographical origin of Apulian from foreign milk samples. Furthermore, the lipid fractions (FA) of raw Apulian cow and buffalo milk samples were analysed by using an innovative and powerful technique, i.e. ^{1}H NMR spectroscopy in combination with MSA, in order to search for a method to discriminate cow from buffalo milk, on the basis of the species. For the attribution of whole signals observed by ^{1}H NMR, 2D-NMR techniques, such as Correlated SpectroscopY (COSY), Heteronuclear Multiple Quantum Coherence (HMQC), and Heteronuclear Single Quantum Coerence (HSQC) were usefully employed; the quantitative determinations were achieved by integration of selected peaks.

Keywords: Cow milk, Buffalo milk, Lipid composition, Geographical origin, Routine, analyses, NMR, IRMS.

1. Introduction

Product authenticity and its authentication are emerging topics within the food sector. They are of great importance, not only for consumers but also for producers and distributors. The EC regulation n° 509/2006 [1] protects consumers through a system of effective and impartial controls that define, within the Common Market, the safeguard of the 'Protected Designation of Origin' (PDO). PDO is a term used to describe foodstuffs with a strong regional identity that are produced, processed and prepared in a specific geographical area using prescribed techniques that may be unique to that region. This regulation meets the producers' requirements because only objective and clear controls can protect the authenticity of products that are to be promoted on the market.

In Apulia, a region of Southern Italy, the milk obtained from autochthonous cows bred on local farms is very characteristic. Cow Apulian mozzarella is a fresh and stringy-textured cheese made from this local milk.It isgreatly appreciated by consumers and is well known on the national and international scales.It is thusvery important that a PDO certification be obtained for this particular local foodstuff, in order to protect its authenticity. To do this, the most distinctive constituents of the product have to be determined. The typical characteristics of a PDO dairy product are the production technology, the origin of the milk and materials associated with the animal feeding. It is therefore necessary to define and analyze milk parameters that are influenced by these factors.

The use of IRMS to characterize origin and animal feeding has been widely investigated [2-5]. The ratio of carbon isotopes in plants is influenced mainly by the pathway of CO_2 fixation used by the plant species, be it the C_3 Calvin cycle, as for wheat, or the C_4 Hatch-Slack cycle, as for corn and maize. The $\delta^{18}O$ content in milk water reflects the isotope composition of the ground water drunk by animals, which, in turn, is influenced by geographical factors such as altitude, latitude, and distance from the sea [6]. Seasonal effects are also of great importance: in summer, milk water contains higher $\delta^{18}O$ content because animals eat fresh plants containing $\delta^{18}O$ –enriched water due to evapotranspiration phenomena in leaves. The $\delta^{15}N$ values in dairy products reflect, through the plants consumed by animals, the isotope composition of the original soil [7-10] which is influenced by many factors such as agricultural practices (intensive or extensive cultivation) [11], climatic and geographical conditions [12]. The use of massive quantities of organic fertilizers and other factors such as aridity, salinity, and closeness to the sea, tend to increase the $^{15}N/^{14}N$ isotope ratio in soil, plants and animal products. Moreover, for animals' products, the presence of nitrogen-fixing plants in the diet can lead to lower values, because these plants use both atmospheric and soil nitrogen as a nitrogen source, which results in lower ^{15}N content than in plants relying only on soil nitrogen (enriched in ^{15}N relative to the atmospheric nitrogen).

Recently, associations of producers of Italian PDO cheeses became interested in the study of the characteristic amino acid contents of cheese products in order to add these values to other parameters used in production control. Amino acid determination requires extraction and other selective steps in sample preparation. Nuclear magnetic resonance (NMR) spectroscopy, although limited in sensitivity, enables the free amino acid, sugar and organic acid composition of milk and dairy products to be characterized rapidly [5,13,14]. The importance of NMR application in food science is underlined in the third volume of the handbook Modern Magnetic Resonance [15], dedicated to the application of this innovative technique to various food products.

The principal aim of the first partof this chapter was to establish that a combination of NMR and IRMS analyses allows more precise identification of the geographical origin of milk produced in two neighbouring Italian regions (Apulia and Basilicata) of Southern Italy and in some countries located in Central-Eastern Europe. The data obtained by classical analytical methods and those obtained by ^{1}H-NMR and IRMS were treated separately by multivariate analyses in order to compare the discriminating potential of each methodological approach. In addition, we have assessed and compared the use of chemical and spectroscopic techniquesfor determining various compounds in milk samples.

In Italy buffalo's milk mozzarella is considered a very high quality cheese; in fact it has obtained Denomination of Origin (DOP) mark in a well defined geographical area in the Campania region, in southern Italy. There buffaloes' breeding is favored by the presence of wide grazing grounds and stagnant water during all the year. Many producers in other areas of southern Italy are activating for requesting DOP, therefore the problem of analytical authentication will represent a major challenge in the near future. The production disciplinary for Buffalo's milk mozzarella provides the exclusive use of buffalo milk;in this way even a partial employment of cow milk has to be excluded. Otherwise, it should be declared in the ingredients list. According to this premise, in order to protect the consumer, it is necessary to set out analytical methods to detect the addition of cow milk in buffalo mozzarella.

The methods that gave the most promising results are based on the analysis of the proteic fraction, and, in particular, on the research of homologous caseinic or whey protein fractions. Various electrophoretic methods based on the identification of caseinαs1 [16], β-casein [17] and γ2-casein [18] were proposed. HPLC of whey cheese proteins soluble at pH 4.6 permitted to detect and quantify low quantities of cow milk added to buffalo milk in buffalo milk mozzarella. Indeed, since cow milk whey contains β-lactoglobulin A, that is absent in buffalo milk, it is possible to distinguish up to 1% of cow milk in buffalo mozzarella [19].

The official analytical method of the European Community [20] for the detection of cow milk in cheeses made from goat or buffalo milk is based on the isolation of caseins from cheeses and submission to plasmin cleavage. Plasmin-treated caseins are subjected to isoelectric focusing in the presence of urea and staining of proteins. Afterwards, the stained casein patterns are evaluated by comparison of the pattern obtained from the sample with those obtained in the same gel from the reference standards, containing 0% and 1% cows' milk. The use of capillary zone electrophoresis to determine cow milk in buffalo milk and mozzarella cheese has also recently been evaluated [21].

These methods, although precise, require various stages of preparation and analysis and/or large amounts of reagents and solvents. For this reason a rapid and simple method that could be used as a screening test could be interesting. Recently methods based on innovative techniques have been proposed [22-24]. Among these the potential of ^{13}C NMR spectroscopy has also been investigated succeeding in the identification of different milk species, through the determination of the triacylglycerolic composition [25]. A similar approach has never been applied using ^{1}H-NMR spectroscopy. However it could be more interesting as a screening method because it has the advantage of a higher sensibility compared to ^{13}C-NMR and therefore requires shorter acquisition time.The aim of the second part of this chapter was to evaluate the possibility of using ^{1}H-NMR spectroscopy in combination with chemometric methods for the classification of cow and buffalo milk samples according to species.

2. Materials and Methods

2.1. Milk

2.1.1. Milk Samples

Thirty nine samples of Southern Italy and Central-Eastern Europe cow milk were analysed. Twenty Italian samples were supplied by different farms in Apulia (1-18) and Basilicata (19-20), two contiguousadministrative regions of Southern Italy. All Italian samples from morning milking were collected in polypropylene boxes and stored in a freezer at –80°C until being analyzed. The nineteen milk samples (21-39) were purchased from supermarkets located in some countries located in Central-Eastern Europe since samples of local dairies perished easily. From now on these samples are denoted as “foreign”. The origin of the samples is reported in table 1.

Table 1.Origin of cow milk samples

Standard number	Origin
1	Corato
2	Corato
3	Corato
4	Gioia del Colle
5	Gioia del Colle
6	Gioia del Colle

Standard number	Origin
7	Gioia del Colle
8	Gioia del Colle
9	Gioia del Colle
10	Gioia del Colle
11	Gravina in Puglia
12	Gravina in Puglia
13	Gravina in Puglia
14	Ruvo di Puglia
15	Martina Franca
16	Martina Franca
17	Murgia Barese
18	Murgia Tarantina
19	Irsina
20	Scanzano
21	Leibnitz - Austria
22	Canton Ticino - Switzerland
23	Bayernland - Germany
24	Kutina - Croatia
25	Krapina - Croatia
26	Bjelovar - Croatia
27	Veszprem - Hungary
28	Tiszafured - Hungary
29	Szekesfehervàr - Hungary
30	Vac' - Hungary
31	Szged - Hungary
32	Vasvàr - Hungary
33	Ceglèd - Hungary
34	Jàszberèny - Hungary
35	Ljubiana - Slovenia
36	Vipava - Slovenia
37	Murska Sobota - Slovenia
38	Kobarid - Slovenia
39	Celje - Slovenia

2.1.2. Routine Analyses

Contents of protein, fat, and lactose of milk samples were determined using a Milko Scan FT 120 (Foss Italia, Padova, Italy). Samples were heated to 102°C to give dry extracts whose ash content was obtained by heating to 600°C for 6 hours.

2.1.3. Chromatographic Analyses

Lithium, sodium, potassium, magnesium and calcium contents were measured by means of high performance ion chromatography (HPIC, DX 120 EX, Dionex, Sunnyvale CA, USA, equipped with a conductometric detector), following the procedure described by [26]. For these determinations, milk ash samples were dissolved in 50 ml of 0.1 N HCl.

2.1.4. Atomic Emission Spectrometric Measurements

The total concentrations of Al, Cr, Cu, Fe, Mn, Ni, Zn, Pb, Se and Ba were measured by inductively coupled plasma atomic emission spectrometry (ICP-AES) on a Varian instrument model Liberty 110 (Varian Inc. Palo Alto, USA) equipped with an ultrasonic nebulizer Cetac model U-5000AT$^+$ (Cetac Tecnologies Inc., Omaha, Nebraska, USA). Before analysis, the samples were freeze-dried and subjected to a mineralization process: 0.5g of sample was dissolved in 7 ml of 70% HNO_3 Ultrapure Reagent (J.T. Baker, Phillipsburg, USA) and 2 ml of 30% H_2O_2 (J. T. Baker, Phillisburg, USA) and digested in a microwave system Milestone model MLS-1200 MEGA (Milestone, Bergamo, Italy). For each digestion step, a blank sample was prepared with the same amount of reagents. Normal precautions for trace element analysis were observed throughout. The reaction vessels for microwave oven were cleaned before each digestion with 5 ml of INSTRA-Analyzed 70% HNO_3 (J.T. Baker, Phillipsburg, USA), heated for fifteen minutes at 600 W and then rinsed with ultra pure water. Glassware was cleaned by soaking overnight in 10% nitric acid solution and then rinsed with ultra pure water obtained from a water deionization system Millipore model MILLI-Q RG (Millipore, MA, USA). Calibration was carried out using an external standard solution obtainedby diluting a 1000 mgl^{-1} standard solution for inductively coupled plasma (J.T. Baker, Phillipsburg, USA). For the freeze-dry process of the milk, a Heto-Holten A/S LyoLab 3000 instrument was used, connected with a vacuum rotary pump. The freeze-dry process was performed at −55°C.

2.1.5. Fatty Acid Composition

The fatty acid composition of the Apulian and foreign milk was determined by gas chromatography (GC) as fatty acid methyl esters (FAME). This FAME were prepared using saponification/methylation with potassium methylate. Fresh samples of milk were centrifugated at 15000 rpm for 40 min, so an aliquot of extracted milk fat was saponified for 30 minutes in the water bath at 60-70°C with the methanolic solution of KOH (2mol/l). Methyl esters of fatty acids were separated and quantified by using a HR-GC, Carlo Erba gas chromatograph, equipped with a DB Wax-type analytical column, (30 m length x 0.32 mm i.d) coated with 0.5 μm film thickness (100% PEG, polyethylene glycols) (J&W Scientific Inc. Rancho Cordova, CA, USA), and coupled with an FID detector (Varian, Middelburg, The Netherlands). Helium was used as the carrier gas. The injection is performed with a split ratio 1:50 and constant flow operating mode at 50m/sec. The injector temperature and the heated block temperature are both set at 220°C. The injected volume was 1μL. Fatty acids were identified by comparison with retention times of fatty acids in standard samples.

2.1.6. Stable Isotope Ratio Analysis

The determination of $^{13}C/^{12}C$, $^{15}N/^{14}N$ ratios of milk samples were carried out on the freeze-dried samples. About 1.5 mg of sample was directly weighed into tin capsules (Santis

Analytical Italia, Bareggio, Italy) for $^{13}C/^{12}C$ and $^{15}N/^{14}N$ determination. The analyses were performed using an isotopic ratio mass spectrometer (IRMS, Finnigan Delta V Advantage, Thermo Fisher Scientific, Bremen, Germany) coupled with an Elemental Analyser (EA, FlashEA 1112 HT, Thermo Fisher Scientific, Bremen, Germany). The EA was equipped with a combustion reactor for determination of $^{13}C/^{12}C$, $^{15}N/^{14}N$. The EA was connected with an autosampler (MAS 200R, Thermo Fisher Scientific, Bremen, Germany) and interfaced with the IRMS through a dilutor (Finningam, Conflo III, Thermo Fisher Scientific, Bremen, Germany) dosing the samples and reference gases. Variations in stable isotope ratios were reported as parts per thousand (‰) deviations from internationally accepted standards: Pee Dee Belemnite (PDB) for carbon, Atmospheric nitrogen (AIR) for nitrogen. The isotopic values were expressed using the formula:

$$\delta(‰) = [(R_{sample} - R_{standard})/R_{standard}] \times 1000,$$

where R is the ratio between the heavy and light isotopes, R_{sample} is the isotopic ratio of the sample and $R_{standard}$ is that of the reference material. Each sample was analyzed twice and values were averaged. The analysis was repeated if the difference between the two values was higher than 0.2‰ for $\delta^{15}N$ and $\delta^{13}C$, respectively. The values were referenced against reference gases (N_2, CO_2) previously calibrated against International Standards from International Atomic Energy Agency (IAEA). Moreover, for each run at least one in-house standard (casein for carbon and nitrogen) was analyzed to check the accuracy of the analysis.

2.1.7. NMR Spectra

^1H NMR spectra were obtained on freeze dried milk samples. 0.024 g of samples was dissolved in 3.6 ml of D_2O containing 0.75% of 3-(trimethylsilyl)propionic-2,2,3,3-d_4 acid, sodium salt (TSP), as reference. The mixture was placed in a ultrasonic bath to ensure solubilization. The aqueous extract obtained was filtered to remove insoluble particles and placed in a 5mm NMR sample tube. Spectra were obtained on a Varian Inova 400 MHz spectrometer, using a presaturation sequence for water suppression. The following conditions were used: 32768 data points, 512 scans, and spectral width of 7002.801 Hz. ^1H 2D-NMR spectra were recorded on a Varian 800 MHz NMR spectrometer. Double quantum filtered COSY spectrum was recorded using the following parameters: spectral width = 8000 Hz, number of time domain data points = 2546 x 256, zero filled to 4096 x 4096. FIDs were multiplied, in both dimensions, by a sine function before Fourier transformation. The TOCSY spectrum was acquired using the same parameters of COSY. A mixing time of 80 ms was used and acosine function multiplication was applied to FIDs, in both dimensions, before Fourier transformation.

2.1.8. Multivariate Statistical Analysis

The experimental data obtained by all the performed determinations were subjected to multivariate statistical analysis to evaluate the possibility of differentiating milk samples according to their origin. Results were grouped into two data sets. The first contained the results of routine, chromatographic analyses, ICP-OES and fatty acid composition; the second contained results of IRMS and NMR determinations. Multivariate statistical analysis was applied to all data sets using Statistica Software (StatSoft Inc., Tulsa, OK, USA). The chemometric methods used were principal components analysis (PCA) and discriminant analysis (DA).

2.2. Fatty Acids in Cow and Buffalo Milk

2.2.1. Milk Samples

Fourteen samples of raw buffalo milk and twenty-twosamples of raw cow milk were supplied by local farms in the Apulia region. Cows and buffaloes were in the same phase of lactation. All samples were collected in polypropylene bottles and stored at -80°C, until the moment of analysis.

2.2.2. ^{1}H-NMR Determinations

Triacylglycerols were obtained from milk by extraction with chloroform and methanol (ratio 2:1 by volume) according to an already described procedure [27]. Spectra were acquired at 500 MHz, using a DRX500 Advance Bruker spectrometer.

The following experimental conditions were applied: spectral width = 3500 Hz (~7 ppm); time domain = 32K points; number of transients = 256, pulse width = 7.2 μs(90° pulse), recycle delay = 5 sec. The sample temperature in the probehead was kept to 300 K for all the samples.Spectra were processed by applying an enhancement multiplication with a 0.6 Hz line broadening factor.

Fourier-transformed spectra were phased and then baseline corrected by spline interpolation of 14 baseline selected points. Signals integration was carried out by the xwinnmr integration routine,defining manually the regions of integration and setting to 1000 the intensity of the signal resonating at 1.27 ppm due to $-(CH_2)_n-$ groups of fatty acids. The integration regions were the same for all the spectra.

2.2.3. 2D-NMR Spectra

2D gradient-COSY spectrum (cosygspulse sequence in Bruker software) was obtained using 1024 x 256 time domain data points; 7 ppm of spectral width; 16 dummy scans; 8 transients for each increment (F1 dimension).Gradient-HSQC 2D spectrum (invieagssi pulse sequence in Bruker software) was acquired by using with 2048 x 512 time domain data points; 7 ppm (F2 dimension) and 140 ppm (F1 dimension) of spectral width; 32 transients for each increment (F1 dimension).

It was processed by using 2048 x 1024 data points and no shifted square sine bell functions in both dimensions. Gradient-HMBC 2D spectrum (inv4gplplrndpulse sequence in Bruker software) was acquired by using the same parameters employed for gradient-HSQC, apart an additional delay for evolution of long range couplings of 0.06 s.

2.2.4. Statistical Analysis

Principal Component Analysis (PCA), Hierarchical Clustering Analysis (HCA) and Discriminant analysis (DA) and Analysis of Variance (ANOVA) were carried out on the obtained data using STATISTICA software (StatSoft Inc., Tulsa, OK, USA).

3. Results and Discussion

3.1. Milk

In table 2are reported the contents in protein, fat, and lactose of milk samples whileLi, Na, K, Mg and Ca concentrations are given in Table 3. Those of Al, Ba, Cr, Cu, Fe, Mn and Zn, in the freeze-dried milk samples, are summarized in Table 4. It should be pointed out that results given in table 4 were obtained on "freeze-dried" milk samples. Therefore, these values cannot be immediately compared with those reported in the literature for whole cow milk. However, if we consider that the water content in cow milk is on average 86-87%, it is possible to recalculate the data on the basis of water loss due to freeze-drying process. The results herein obtained are in agreement with data found in the literature. In tables 5 and 6 are reported the composition of fatty acids for Apulian and foreign milk samples, respectively.

Table 2. Routine analyses of milk samples

Standard number	Protein %	Fat %	Lactose %
1	3.61	3.86	4.84
2	3.49	3.38	4.75
3	3.29	3.38	4.81
4	3.38	3.29	4.74
5	3.78	3.45	4.58
6	3.27	3.96	4.89
7	3.67	3.49	4.86
8	3.45	3.41	4.74
9	3.48	3.76	4.86
10	3.43	3.92	4.83
11	3.24	3.35	4.70
12	3.18	3.61	4.57
13	3.24	3.71	4.80
14	3.45	3.60	4.91
15	3.15	3.50	4.76
16	3.30	3.50	4.88
17	3.26	3.26	4.56
18	3.01	3.12	4.65
19	3.82	3.50	4.84
20	3.35	3.88	4.76
21	3.26	1.50	4.55
22	3.65	3.80	4.76
23	3.31	2.80	4.91
24	3.40	2.80	4.82
25	3.82	1.50	4.73
26	3.56	2.80	4.51
27	3.82	3.20	4.39
28	4.63	1.50	4.84

Table 2. (Continued)

Standard number	Protein %	Fat %	Lactose %
29	4.61	1.50	4.75
30	3.60	3.50	4.81
31	3.24	4.20	4.78
32	3.52	2.80	4.91
33	3.70	3.50	4.70
34	3.85	4.20	4.79
35	3.40	3.50	4.76
36	3.90	3.50	4.79
37	3.54	3.50	4.50
38	3.79	4.20	4.80
39	3.27	3.50	4.86

Table 3. Li, Na, K, Mg and Ca content (mg/100g) of milk samples

Standard number	Li	Na	K	Mg	Ca
1	0.08	66.9	46.5	8.11	85.7
2	0	35.7	80.9	9.11	93.9
3	0.02	43.2	134.9	11.1	118.4
4	0.10	49.6	96.8	10.2	106.1
5	0.22	44.2	120.0	10.5	120.1
6	0.01	37.1	102.6	9.48	109.4
7	0.28	46.0	89.5	11.3	121.8
8	0.15	49.5	135.6	10.8	108.9
9	0.24	38.0	70.8	8.13	79.8
10	0.01	41.3	140.5	10.2	93.7
11	0.22	64.6	64.1	11.5	124.8
12	0.13	73.6	62.2	10.3	109.5
13	0.12	48.8	129.1	13.1	142.0
14	0.31	38.3	131.0	13.5	111.0
15	0.09	54.4	116.5	10.5	108.2
16	0.40	54.6	115.8	11.5	114.5
17	0.19	46.4	112.4	12.2	123.3
18	0.19	57.5	11.5	10.8	110.8
19	0	37.7	119.3	12.0	135.6
20	0.12	25.4	61.8	6.49	106.0
21	0.26	47.8	122.5	12.2	120.4
22	0.29	50.3	128.0	13.0	122.9
23	0.55	51.4	104.4	11.6	118.0
24	0.10	36.5	116.9	11.4	112.3
25	0	30.1	124.0	11.2	107.6
26	0.08	34.3	112.8	10.5	106.6
27	0.28	44.0	115.9	11.1	112.8
28	0.12	61.6	117.2	11.2	110.6

Standard number	Li	Na	K	Mg	Ca
29	0.11	44.1	131.0	12.0	111.7
30	0	40.5	133.3	11.5	101.1
31	0	38.1	131.5	11.6	107.6
32	0	38.4	138.0	12.0	102.0
33	0.24	48.6	105.4	11.4	110.2
34	0.14	42.8	130.7	6.36	113.9
35	1.08	52.1	85.1	6.32	118.6
36	0.18	42.8	132.1	6.92	114.4
37	0.16	31.6	106.4	4.29	110.2
38	0.01	41.0	138.7	8.27	117.3
39	1.09	53.3	86.8	8.54	120.6

Table 4. Atomic emission spectrometric measurements of the freeze-dried milk samples (mg/1000g)

Standard Number	Al	Ba	Cr	Cu	Fe	Mn	Zn
1	< 0.06	0.35	0.02	0.41	1.27	0.14	19.9
2	0.07	0.33	0.17	0.33	2.40	0.16	25.6
3	2.05	0.49	0.02	0.43	2.24	0.15	17.8
4	3.76	0.49	0.04	0.34	3.23	0.27	17.2
5	0.43	0.48	0.09	0.43	2.09	0.13	17.4
6	< 0.06	0.31	0.07	0.34	1.06	0.10	21.2
7	< 0.06	0.50	0.13	0.71	1.11	0.15	20.7
8	< 0.06	0.30	0.10	0.42	0.97	0.15	22.0
9	0.35	0.26	0.08	0.42	1.12	0.10	16.8
10	< 0.06	0.31	0.07	0.46	1.06	0.12	18.3
11	10.30	0.41	0.28	< 0.30	3.63	0.12	17.9
12	< 0.06	0.43	< 0.02	< 0.30	1.46	0.10	18.0
13	1.78	0.37	< 0.02	0.51	1.36	0.12	18.6
14	< 0.06	0.26	< 0.02	< 0.30	0.84	0.08	13.2
15	< 0.06	0.81	0.12	0.57	1.28	0.11	18.7
16	< 0.06	0.58	< 0.02	0.34	1.21	0.11	13.8
17	< 0.06	0.55	0.07	< 0.30	3.47	0.12	15.3
18	< 0.06	0.25	< 0.02	0.40	0.95	0.09	15.6
19	3.08	< 0.15	0.05	0.37	1.21	0.14	17.5
20	< 0.06	0.26	< 0.02	< 0.30	1.12	0.10	12.0
21	6.18	1.03	< 0.02	0.56	2.78	0.23	29.3
22	8.64	1.56	< 0.02	0.62	1.79	0.18	26.2
23	4.86	0.78	< 0.02	0.55	2.12	0.21	30.2
24	< 0.06	0.99	< 0.02	0.46	1.67	0.18	25.5
25	< 0.06	0.96	< 0.02	< 0.30	1.44	0.14	18.5
26	< 0.06	0.88	< 0.02	< 0.30	1.44	0.14	18.6
27	< 0.06	0.48	< 0.02	0.56	1.35	0.15	22.7
28	< 0.06	0.45	< 0.02	< 0.30	1.41	0.13	21.7
29	< 0.06	0.80	< 0.02	< 0.30	1.67	0.26	22.2

Table 4. (Continued)

Standard Number	Al	Ba	Cr	Cu	Fe	Mn	Zn
30	< 0.06	0.48	< 0.02	0.32	1.51	0.12	21.3
31	< 0.06	0.52	< 0.02	0.37	1.85	0.15	28.7
32	< 0.06	0.47	< 0.02	0.39	2.04	0.16	26.8
33	< 0.06	0.41	< 0.02	0.41	1.72	0.14	25.2
34	4.45	0.47	< 0.02	0.31	2.03	0.15	26.6
35	< 0.06	0.60	< 0.02	< 0.30	1.72	0.15	24.9
36	< 0.06	0.60	< 0.02	< 0.30	1.57	0.15	26.3
37	< 0.06	0.77	< 0.02	< 0.30	1.86	0.23	22.5
38	< 0.06	0.60	< 0.02	< 0.30	1.58	0.15	32.2
39	< 0.06	0.68	< 0.02	< 0.30	1.52	0.15	28.9
detection limit	0.06	0.15	0.02	0.30	0.22	0.04	0.04

Multivariate statistical method was applied to the previously obtained results. PCA was performed on a matrix of 15 analytical parameters for 39 samples. The first eight Principal Components (PC) explained 83% of the total variance. A scatter plot of PC2 vs. PC1, describing 45% of the total variance, is reported in Figure1.

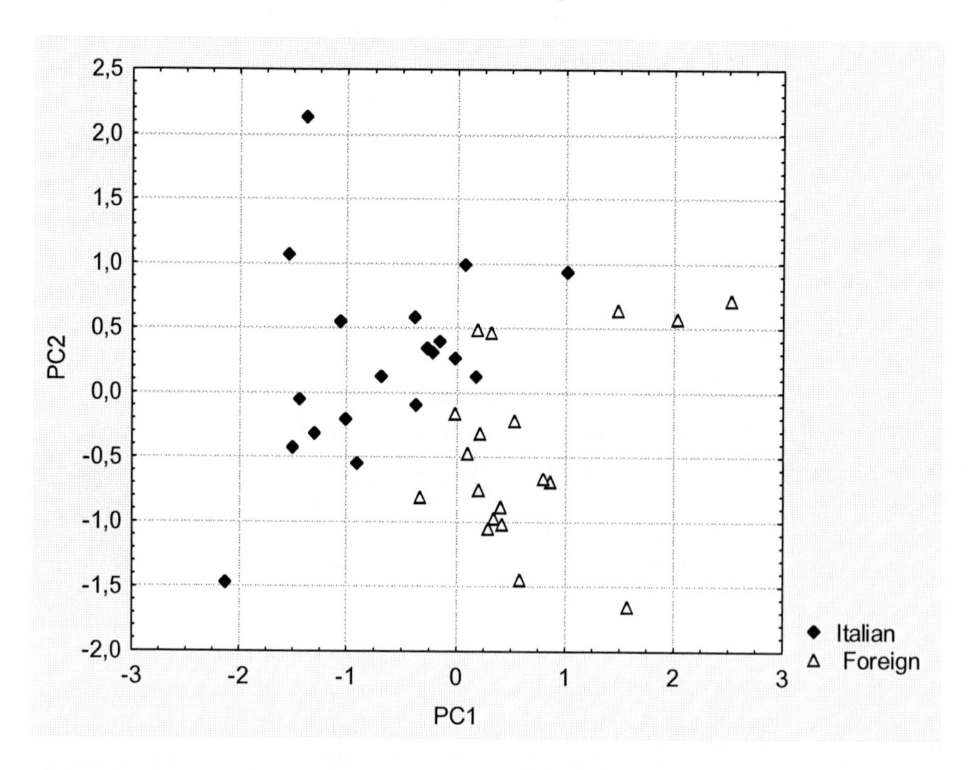

Figure 1. Scatter plot of the scores of the milk samples from the first two principal components PC1and PC2 obtained using analytical data.

Table 5. Fatty acids in Apulian milk samples

Standard Number	C4%	C6%	C8%	C10%	C:12%	C14%	C14:1%	C16%	C16:1%	C18%	C18:1%	C18:2%
1	0.15	0.21	0.46	2.10	3.56	12.7	0.94	34.7	1.74	11.1	22.0	1.90
2	0.68	0.41	0.89	0.13	1.08	4.08	0.40	55.0	0.54	3.60	21.2	0.72
3	0.90	0.65	1.53	0.15	1.77	5.48	0.43	67.5	0.71	4.80	9.01	0.84
4	1.98	1.84	1.36	3.33	3.83	12.5	0.87	30.5	1.08	9.58	17.2	2.09
5	2.87	2.32	1.59	3.66	4.11	12.2	0.62	27.2	1.25	10.2	19.4	3.21
6	0.00	0.13	0.30	1.44	2.63	10.9	0.70	31.5	0.70	12.7	23.8	3.77
7	2.66	2.08	1.34	2.78	3.11	11.2	1.07	30.7	0.81	10.8	21.9	2.73
8	2.17	1.80	1.11	2.80	3.21	10.8	0.10	28.5	1.23	11.8	21.5	4.15
9	2.69	2.35	1.60	3.57	3.86	11.8	0.84	29.0	1.19	10.9	19.3	2.64
10	0.07	0.11	0.26	1.44	2.73	11.3	0.64	30.9	1.03	13.8	21.5	3.77
11	0.55	0.47	1.43	0.17	1.79	6.07	0.55	68.9	0.76	4.31	9.20	1.16
12	0.94	0.58	1.30	0.16	1.52	5.24	0.50	71.2	0.67	4.73	8.64	0.83
13	0.66	0.47	1.19	0.09	1.43	10.2	0.21	49.8	0.37	5.50	25.5	1.66
14	3.00	2.49	5.78	0.55	6.15	15.7	1.24	30.5	1.57	7.65	14.9	2.28
15	1.74	1.82	1.36	3.31	4.01	12.4	0.78	29.1	1.43	8.88	20.4	2.75
16	2.41	2.35	1.79	4.65	5.27	12.9	1.21	25.3	1.52	8.08	17.3	2.27
17	2.17	1.74	1.17	2.69	3.06	10.1	0.30	28.4	1.16	12.7	21.1	3.24
18	0.00	1.43	0.99	2.89	3.41	11.5	0.31	31.8	0.97	13.3	21.9	3.14
19	0.93	0.64	1.49	0.12	1.67	5.05	0.29	67.1	0.48	7.85	8.65	1.51
20	1.84	1.12	2.52	0.25	3.03	10.3	0.73	33.7	1.22	12.9	19.5	2.67

Table 6. Fatty acids in foreign milk samples

Standard Number	C4 %	C6 %	C8 %	C10 %	C:12 %	C14 %	C14:1 %	C16 %	C16:1 %	C18%	C18:1 %	C18:2 %
22	0.36	0.33	1.42	0.18	2.58	10.9	0.82	32.7	1.33	12.2	23.6	2.19
23	0.17	0.15	0.57	0.07	0.96	4.17	0.33	71.9	0.62	5.14	9.17	1.34
24	0	0.31	1.11	0.15	1.9	7.7	0.56	51.4	1.15	9.23	16.7	1.69
25	3.72	1.97	3.07	0.93	3.3	0.41	10.6	0.96	26.7	1.36	9.73	1.61
26	1.05	0.72	1.14	0.31	2.04	8.9	0.58	30.8	1.38	13.2	24.4	2.83
27	0.78	0.57	1.38	0.17	1.71	5.54	0.46	71.1	0.8	3.71	7.93	0.99
28	1.29	0.76	1.44	0.23	1.78	6.44	0.5	57.3	0.79	6.89	11.8	1.15
29	2.57	1.54	2.67	0.47	2.94	9.76	0.73	28.2	1.3	12.3	20.9	2.23
30	2.83	1.59	2.99	0.54	3.34	10.7	0.87	31.6	1.43	10	19.8	2.93
31	2.36	1.34	2.8	0.41	3.2	9.87	0.85	25.2	1.13	6.66	12.7	1.83
32	2.52	1.44	2.61	0.61	3.24	10.4	0.92	30.5	1.53	9.35	20.4	2.91
33	1.09	0.66	1.51	0.21	1.86	6.2	0.52	61.3	1	5.6	11.7	1.83
34	1.21	0.79	1.84	0.23	2.16	6.59	0.53	59.8	1.01	5.82	11.6	1.81
35	2.04	1.27	2.65	0.45	3.2	11.2	0.98	28.8	1.52	10.3	20	1.93
36	2.82	1.62	3.15	0.54	3.41	11.1	0.28	30	1.46	9.76	19.3	1.88
37	2.65	1.54	3.04	0.57	3.47	11.2	0.98	29.4	1.55	9.6	20.7	1.85
38	2.43	1.43	2.73	0.51	3.11	10.6	0.3	27.2	1.35	10.9	20.7	1.97
39	2.59	1.54	2.93	0.58	3.43	11.5	0.27	29.1	1.48	9.62	20.4	1.94

From this plot, a separation of samples according to geographical origin was realized on PC1: samples from southern Italy have negative scores on this component, whereas foreign samples have positive scores. Discriminant analysis (DA) was applied to the first eight PCs in order to classify milk samples into two groups according to their geographical origin. 94% of samples were correctly classified while 88% of the samples were correctly assigned when the test set method was applied. In Figure 2 the ^{1}H NMR spectrum of a southern and a foreign milk sample is reported.

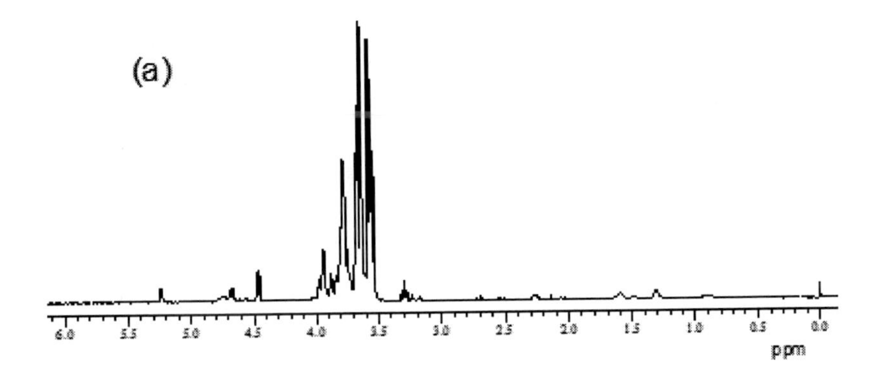

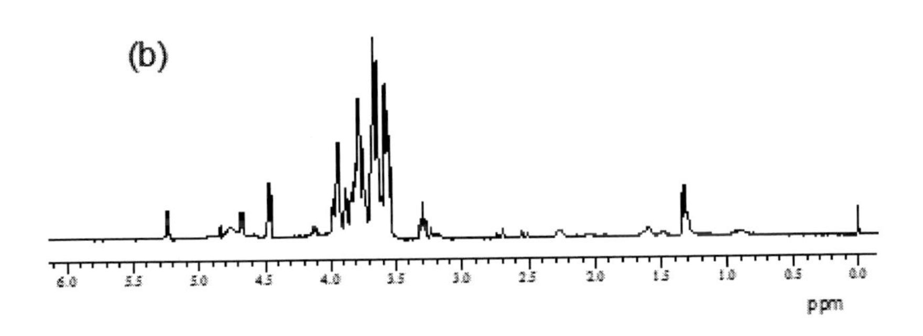

Figure2. ^{1}H NMR spectrum of the aqueous extract of foreign (a) and southern Italy (b) milk.

Peaks of some organic acids, sugars and amino acids can be distinguished in these spectra. They were assigned on the basis of correlations observed in TOCSY and ^{1}H–^{13}C hetero–correlated 2D–NMR spectra (table 7). Bi-dimensional analysis enables signals to be assigned to particular metabolites by examining the existing correlation. The assignments are based on the comparison of chemical shifts and spin multiplicities with data reported in the literature [28].

The comparison between ^{1}H NMR spectra of the aqueous extracts of southern Italyand foreign milk samples showedhigher sugar content in foreign milk than Southern Italy. This could be due to different feeding regime of animals. Peaks at 1.32 ppm and 4.11 ppm were observed in the ^{1}H NMR spectrum of a southern Italy milk sample, due respectively to CH and CH_3 of lactic acid.

Table 7. ^{1}H and ^{13}C chemical shift assignments of compounds detected in milk samples

$\delta^{1}H$	COSY	TOCSY	$\delta^{13}C$	Assignment
1.32 d	4.11		20.82	CH_3 lactic acid
2.11 s	2.35			CH glutamate
2.52 d	2.70		45.45	CH_2 citric acid
2.70 d	2.52		45.45	CH_2 citric acid
3.02 m	1.69			CH_2 lysine
3.04 m	3.15			CH tyrosine
3.15m	3.04			CH tyrosine
3.27 t	4.65		74.65	CH β-glucose
3.54 dd			63.26	CH_2 glycerol
3.58 m			75.58	CH α-glucose
3.63 dd			63.26	CH_2 glycerol
3.65 m	3.77		73.51	CH β-galactose
3.70 m		4.07	61.87	CH α-galactose
3.75 m	1.88			CH lysine
3.75 m	2.05 - 2.11			CH glutamate
3.76 m			72.80	CH glycerol
3.77 m	3.65		63.26	CH β-galactose
3.84 m			72.09	CH α-glucose
3.85 m			60.70	CH β-glucose
3.91 m	3.65		69.36	CH β-galactose
3.95 m	3.80	4.07 - 3.70	70.88	CH α-galactose
4.07 t	3.71 - 3.95	3.70	71.22	CH α-galactose
4.11 q	1.32		69.31	CH lactic acid
4.56 d	3.47	3.91 - 3.65	97.22	CH anomeric β-galactose
4.65 d	3.27		96.65	CH anomeric β-glucose
5.21 d	3.58		92.51	CH anomeric α-glucose
5.24 d	3.70 - 3.95 - 4.07		93.09	CH anomeric α-galactose
8.45 s				CH formic acid

These peaks are absent in foreign milk spectra. Moreover, in the latter samples, the peak at 3.75 ppm of glycerol is more intense than in Southern Italymilk. The heights of signals showing neither overlapping nor correlation with other signals were considered and normalized to the resonance of internal standard.The results of IRMS measurements ($\delta^{13}C$ and $\delta^{15}N$) are reported in table 8.

$\delta^{13}C$ values were significantly different among the analysed samples. In fact, the average value for Italian milk samples (−23.6) is lower than the value for the foreign samples (−21.6). This probably reflects the different amounts of C3 and C4 plants employed in feeding regimes[29], and particularly on the higher amount of maize, a C4 plant in the diet of cows from foreign countries. It is well known that the $\delta^{13}C$ content is determined by the botanical origin of the plant, on the basis of different photosynthetic pathways: C3 plants have more negative $\delta^{13}C$ values than C4 plants.

Table 8. Values of $\delta^{13}C$ and $\delta^{15}N$ for milk samples

Standard number	$\delta^{13}C$	$\delta^{15}N$
1	-23.0	5.9
2	-23.2	4.8
3	-27.8	5.4
4	-25.8	4.7
5	-22.6	5.2
6	-24.6	5.9
7	-24.5	4.5
8	-22.9	5.6
9	-25.0	5.2
10	-24.8	5.2
11	-21.6	5.5
12	-23.4	6.2
13	-22.4	6.0
14	-23.4	4.7
15	-25.2	4.6
16	-24.9	4.1
17	-22.9	5.8
18	-23.0	5.6
19	-21.5	5.8
20	-19.7	6.0
Average value	-23.6	4.9
Std. Dev.	±1.8	±0.7
21	-24.8	5.2
22	-19.3	5.6
23	-19.2	4.8
24	-21.4	6.2
25	-22.8	4.2
26	-23.4	5.4
27	-19.9	4.7
28	-18.7	5.6
29	-22.5	5.7
30	-23.7	6.2
31	-19.2	5.8
32	-22.5	5.5
33	-19.2	6.0
34	-18.8	5.5
35	-25.6	4.5
36	-23.1	5.8
37	-23.1	5.3
38	-21.1	6.8
39	-22.3	4.3
Average value	-21.6	5.0
Std. Dev.	±2.1	±0.6

The result concerning $\delta^{15}N$ values showed instead that there were no significant differences betweenSouthern Italyand foreign samples. NMR signals from different metabolites in milk samples were submitted to analysis of variance to select the most discriminating peaks. The selected peaks and the isotope ratios were used as variables for the statistical analysis.The first two PCs obtained by PCA, explained 75% of the total variance. From the scatter plot (Figure 3), it can be seen that the separations of samples according to their geographical origin was realized on PC1: Italian samples have positive scores whereas foreign samples have negative scores.

In the scatter plot of the PCA scores based on NMR and IRMS data (Figure 3) samples from all over the central Europe are very close, while samples from a small Apulian area are more spread. This behaviour can be explained by the main existing difference between central Europe and Apulian samples; the first were purchased in local markets, while the latter are raw milk samples. On the commercial samples some technological processes (i.e. pasteurization) were carried out, determining modifications on the concentrations of some components, like sieroproteins. These processes have probably reduced some of the existing compositional differences between the central Europe samples. The statistical analysis of the data setindicated a better geographical separation between the two groups of samples than the first data set which included results obtained from routine, chromatographic analyses and the emissionspectroscopy measurements.

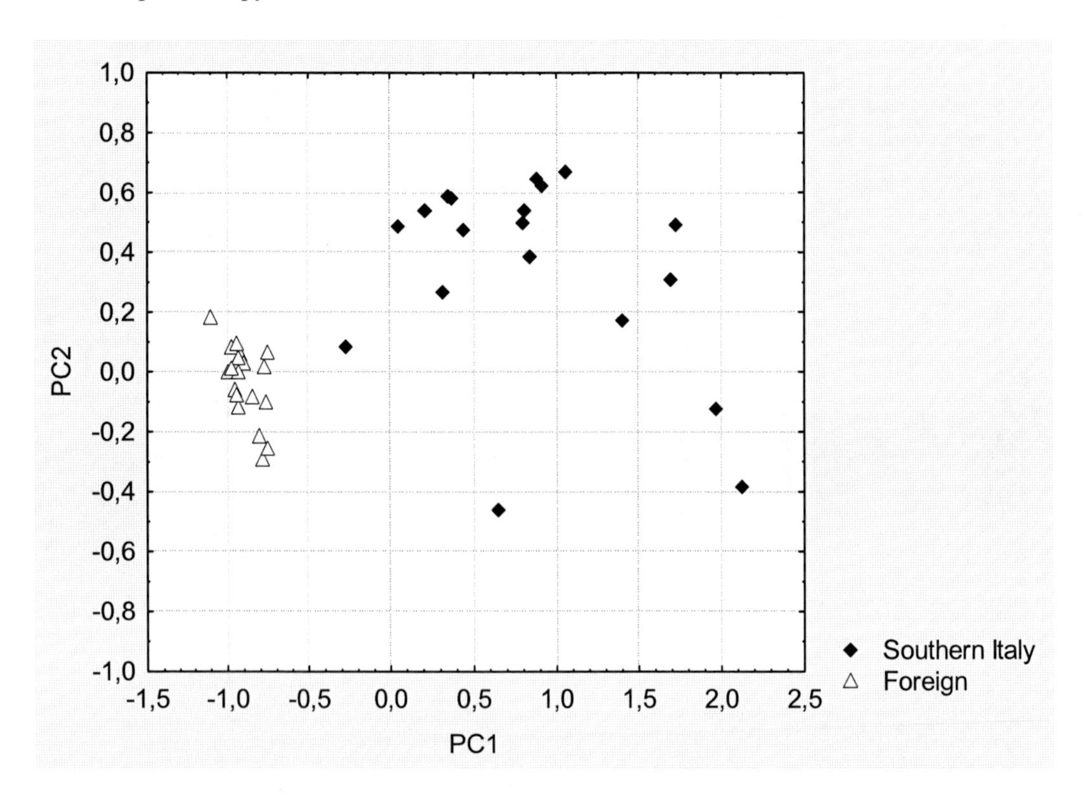

Figure 3. Scatter plot of the scores of milk samples from the first two principal components PC1 and PC2 obtained using NMR and IRMS data.

^1H NMR spectra of the aqueous extractsindicated differences in amino acidcomposition between the foreign and Italian samples. The smaller proportionof amino acidsin the latter milk samples was probably due to the different feeding regime of Italian animals. The classification ability and the prediction ability were shown by discriminant analysis to be 95% and 90%, respectively, using samples of known geographical origin for assignment.

3.2. Fatty Acids in Cow and Buffalo Milk

The ^{1}H-NMR spectrum of the triglyceridic fraction of milk (Figure 4) consists of two series of signals having very different intensities. It is possible to distinguish principal and secondary compounds. The assignment of the main resonances, whose pattern is typical of the ^{1}H-NMR spectrum of a fat sample [30], was performed with the help of COSY (Figure 5), HMBC, and heteronuclear multiple quantum coherence(HMQC) experiments (Table 9).

FAcomposition of milk has been determined through the calculation of the areas of the NMR signals due to triacylglycerols. In order to evaluate the linoleic and linolenic acid concentrations the integrals of the signals due to the diallylic moieties, resonating respectively at 2.79 and 2.83 ppm [31], were calculated. The following equations were derived using the E and F signals in figure 4:

$$\text{linoleic acid}(\%) = 3B/2(E+F)$$

$$\text{linolenic acid } (\%) = 3A/4(E+F)$$

The percentage of monounsaturated fatty acids (MUFA) can therefore be obtained by subtracting the percentages of these two acids from the total content of unsaturated acids, determined as follows:

$$\text{unsaturated acids } (\%) = \text{MUFA} + \text{linoleic acid} + \text{linolenic acid} = D/2C$$

The signal resonating at 0.96 ppm (signal E in Figure 4) has been attributed to the CH_3 of linolenic acid and to the CH_3 of butyric acid, as confirmed by COSY and HSQC and comparing the chemical shifts of the $\square_1$ carbon with those present in literature [32]. The percentage of this acid can be determined with the following equation:

$$\text{butyric acid } (\%) = E/(E+F) - 3A/2(E+F)$$

Saturated fatty acids content is obtained subtracting the unsaturated fatty acids content from the total fatty acids:

$$\text{saturated fatty acids } (\%) = F/E+F + E/(E+F) - D/2C$$

The fatty acid composition was calculated for all the samples and a data set was built with all these values. Multivariate statistical analysis was conducted on the determined fatty acid percentages to evaluate if they can be used to discriminate the milk species. The height of the peak resonating at 0.697 ppm, due to cholesterol, was added to the data set for each sample.

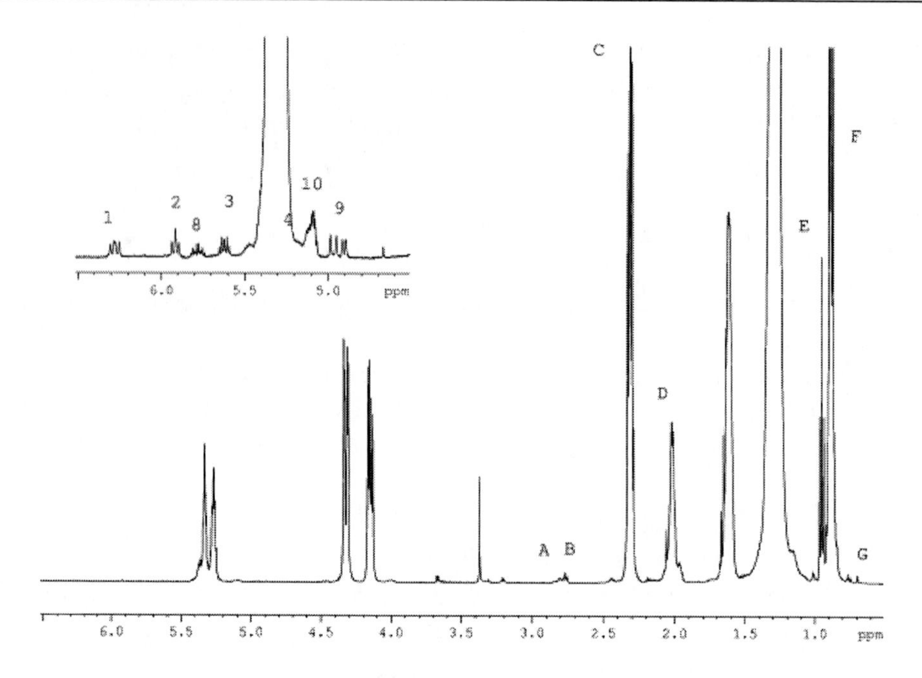

Figure 4. 500 MHz ^{1}H-NMR spectrum of the triacylglycerols of milk fat TAG. The insert shows the signals due to secondary components.

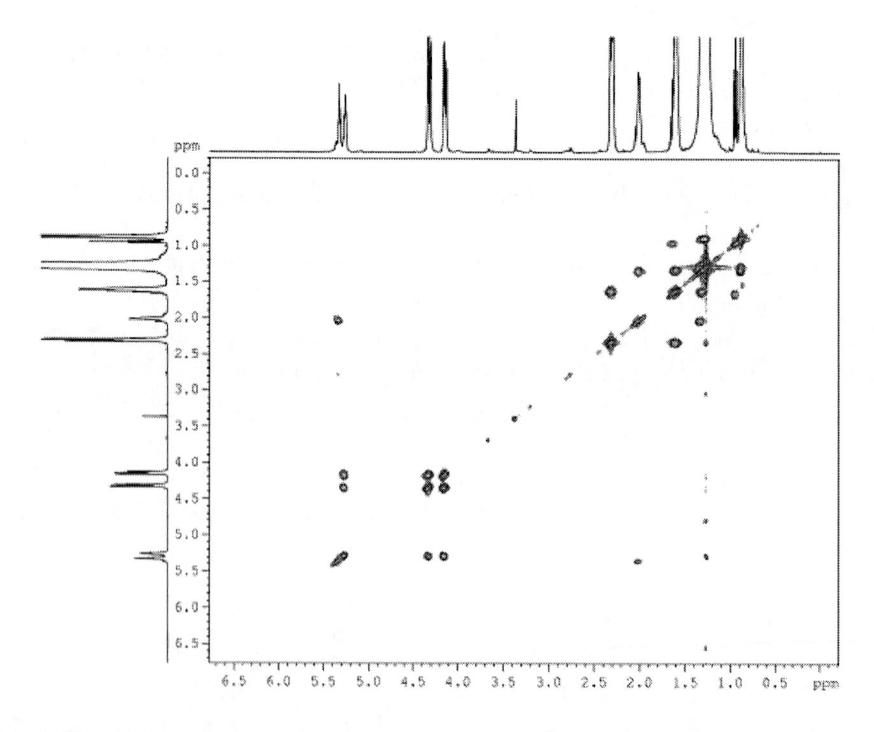

Figure 5. 500 MHz COSY spectrum of the triacylglycerols of milk fat TAG.

Table 9. Assignment of triacylglycerol resonances in 500 MHz ^{1}H-NMR spectrum of the milk fat

$\delta^{1}H$ (ppm)	$\delta^{13}C$ ^{1}H ^{13}C HMQC	$\delta^{13}C$ ^{1}H ^{13}C HMBC	Assignment
0.88 t	14.23	39.0; 34.6-34.0; 32.3; 30.0-29.0; 27.34; 27.30; 23.00	CH_3 fatty acids
0.96 t	13.65	18.44; 36.2	CH_3 linolenic and butyric acid
	22.84		
1.27 m	29.9	35.0-20.0	$-(CH_2)_n-$
	32.12		
	32.10		
1.32 m	29.9 – 29.2		$-(CH_2)_n-$
1.62 m	25.02	174; 36.2; 34.6-34.0; 32.12; 30.0-29.0	$-\underline{CH_2}-CH_2-COO-$
1.65 m	18.44		
2.02 m	27.30	129.7-130.1	$-\underline{CH_2}-CH=CH-$
	27.34		
2.31 t	36.2	173.6-173.0	$-\underline{CH_2}-COO-$ butyric acid
	34.12		
2.33 t	34.28	173.6-173.0	$-\underline{CH_2}-COO-$
	35.96		
2.79 t	25.96	129.7-130.1	$-CH=CH-\underline{CH_2}-CH=CH-$ linoleic acid
2.83 t	25.96	129.7-130.1	$-CH=CH-\underline{CH_2}-CH=CH-$ linolenic acid
4.15 dd	62.28		CH_2 glycerol
4.33 dd	62.58		CH_2 glycerol
5.26 m	69.25		CH glycerol
5.33 m	129.76		-CH=CH- fatty acids

The means and standard deviations of the determined parameters for cow and buffalo milks are summarized in Table 10. ANOVA was applied to test if the differences in the average values relative to the considered species are significant. The F (test Fihser) values represent the ratio of the between-groups variance over the within-groups variance. The F-critical valuewas calculated by the software at 1.34. The calculated values for the considered variables are listed in Table 10 and the relative p-level tended to zero in all cases, indicating that all the NMR data are significantly different for the considered species.

Table 10. F-statistics, means (%), and standard deviations of determined parameters for cow and buffalo milks

Variables	F (1.34)	Cow mean	SD	Buffalo mean	SD
Linolenic acid	15.45	0.78	0.12	0.58	0.19
MUFA	17.45	25.00	4.60	19.68	1.43
Linoleic acid	10.89	2.12	0.69	1.46	3.56
Unsaturated acids	19.28	27.92	5.10	21.73	1.51
Saturated acids	20.21	72.08	4.98	78.27	1.51
Butyric acid	32.27	10.39	1.07	12.57	1.26
Cholesterol	20.07	1.34	0.23	1.03	0.15

Compositions of saturated and unsaturated acids were eliminated from the data set being highly correlated with MUFA concentration. Butyric acid was eliminated because it is highly correlated with linolenic acid concentration. PCA was applied on a data set consisting of thirty-six samples. In examining the plot of the samples in the space defined by the two first principal components that accounted for 70% of the total variance (Figure 6), a separation between milk samples according to their species was found on PC1.

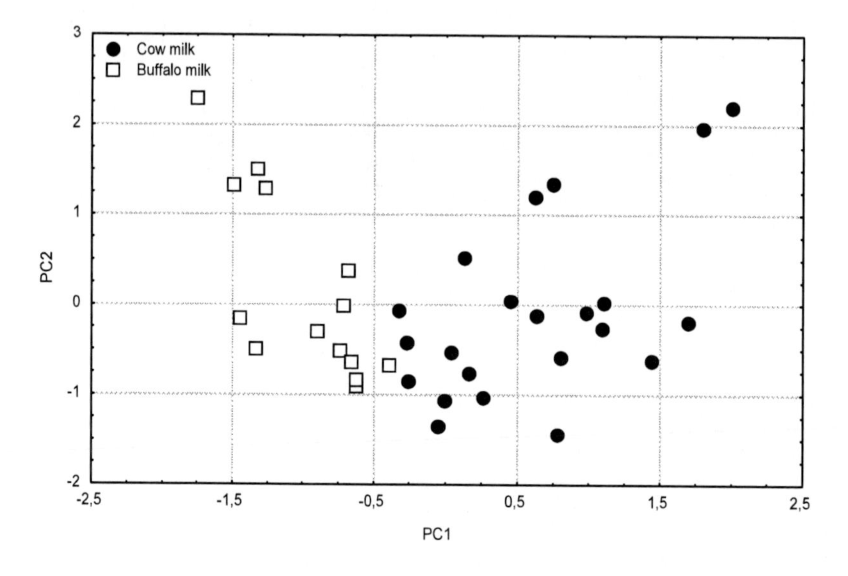

Figure 6. Score plot of milk samples on the first two principal components obtained from ^{1}H-NMR data.

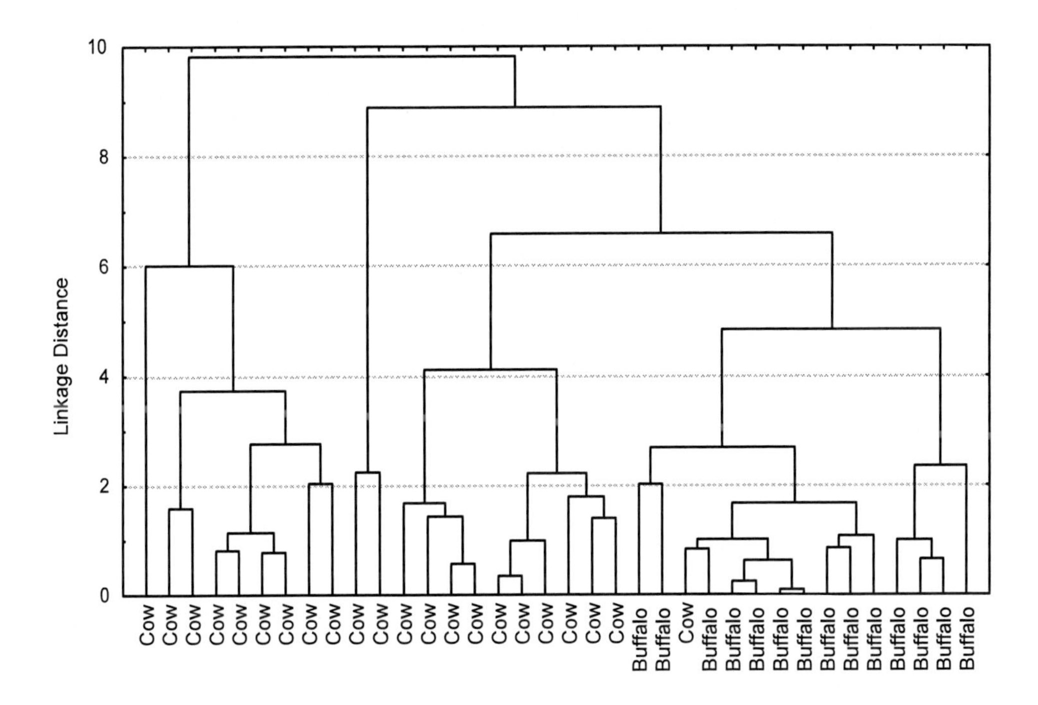

Figure 7. Dendrogramof milk samples obtained from ^{1}H-NMR data.

All the analyzed variables have high loadings on the first principal component. The numerical value of the loading of each variable on a given principal component shows how much the variable has in common with that component. Hence, the loadings can be interpreted as correlation between the variables and the components. In the present case, since PC1 loadings are high and positive for all the determined parameters. This reflects the relevant influence of the fatty acid composition for achieving a discrimination of samples according to species.HCA was carried out on the same variables. In this case, a matrix consisting of the Manhattan distances was used as similarity matrix. A hierarchical agglomerative method was employed to obtain clusters: the complete linkage method.In this method, the distances between clusters are determined by the greatest distance between any two objects in the different clusters.

The dendrogram in Figure 7 shows the results of this analysis. At a similarity level of 6.5, four clusters were found: the first three clusters consisted of 21 samples of cow milk; the fourth cluster was composed of all buffalo milks and one cow milk.

Both PCA and HCA are unsupervised statistical methods and give complementary information on the similarities and groupings of the considered samples. If such a trend exists it is interesting to evaluate the possibility of classifying the samples.

DA was applied on the first five PCs using the species of the samples as classes for assignment. The recognition ability for the two classes was highly satisfactory since all samples were correctly classified.

Afterwards, in order to verify the prediction ability of the obtained model, the data set was split into a training set, for developing a discriminant model, and a test set (made of four samples of buffalo milk and eight samples of cow milk). The obtained model assigned 97% of training set correctly and 94.5% of the test set samples.

Conclusion

Statistical comparison of the constituents of milk sampled from a variety of sources has contributedto the geographical characterization of milk. The most important parameters for discrimination were fatty acids, identified bychemicalanalysis.Thesewere influenced by the feeding diet. The occurrence and levels of sugars, amino acids and organic acids,obtained from NMR spectra,provide a further useful contribution to the discrimination. The $\delta^{13}C$ results, obtained from IRMS measurements, were found to be the most discriminating parameter, and are also influenced by the cows' feeding regime.This method could be used when investigating other dairy products which have been awarded European Community Protected Designation of Origin (e.g. Blue Stilton cheese).The investigation carried out showed that the composition of the fatty acids in cows and buffalo milk triacylglycerols, determined by NMR, can be used to distinguish the two milks. Moreover, the ^{1}H-NMR spectroscopy was used to study the triacylglycerolic fraction of milk and a quantitative measure of the most important classes of fatty acids was obtained. The assignment of the minor compounds will need further study but appears to be interesting because they could carry information for milk species discrimination.The obtained results constitute a good starting point for the extension of this study to processed products, like mozzarella, thus yielding a potential suitable method for cheese authenticity issue. Furthermore, proton NMR spectroscopy appears to be an advantageous technique in this context because of the rapidity with which information can be obtained about a large number of compounds and because of the small amount of sample preparation required for analysis.

Acknowledgments

Tables 1 and 2, figures 1, 2 and 3 were reused with permission of Elsevier. Tables 9 and 10, figures 4, 5, 6 and 7 were reused with permission of Springer.

References

[1] Council Regulation (EC) n. 509/2006 of 20 March 2006 on agricultural products and foodstuff astraditional specialities guaranteed. Official Journal of the European Union CE L 93, 31/03/06.

[2] Renou, J.P., Depongea, C., Gachonb, P., Bonnefoyc, J. C., Coulonc, J. B., Gareld, J. P., Veritee, R.,&Ritzf, P. (2004). Characterization of animal products according to its geographic origin and feeding diet using Nuclear Magnetic Resonance and Isotope Ratio Mass Spectrometry Part I: Cow milk. *Food Chemistry*, 85, 63-66.

[3] Pillonel, L., Badertscher, R., Froidevaux, P., Haberhauer, G., Horn, P., Jacob, A., Pfammatter, E., Piantini, U., Rossmann, A., Tabacchi, R.,&Bosset, J. O. (2003). Stable isotope ratios, major trace and radioactive elements in Emmental cheeses of different origins. *LebensmittelWissenschaft und Technologie*, 36, 615-623.

[4] Brescia, M. A., Caldarola, V., Buccolieri, G., Dell'Atti, A., &Sacco, A. (2003).Chemometric determination ofthe geographical origin of cow milk using ICP-OES data and isotopic ratios.*Italian Journal of Food Science*,3, 329-336.

[5] Brescia, M. Av Monfreda, M., Buccolieri, A.,&Carrino, C. (2005). Characterization of the geographical originof buffalo milk and mozzarella cheese by means of analytical and spectroscopic determinations.*Food Chemistry,*89, 139-147.

[6] Clark, I.,&Fritz, P. (1997)"Environmental isotopes in Hydrogeology". New York: Lewis Publishers; 1997.

[7] Kornexl, B. E., Werner, T., Rossmann, A.,&Schmidt, H. L. (1997). Measurement of stable isotope abundances in milk and milk ingredients - a possible tool for origin assignment and quality contro. *Food Research and Technology*,205, 19-24.

[8] Rossmann, Av Kornexel, B. E., Versini, G., Pichlmayer, F.,&Lamprecht, G. (1998). Origin assignment of milk from alpine regions by multielement stable isotope ratio analysis (Sira) ". La *Rivista di Scienza dell'Alimentazione*, 1, 9-21.

[9] Manca, G., Camin, F., Coloru, G., Del Caro, A., Depentori, D., Franco, M. A., & Versini, G. (2001). Characterization of the geographical origin of Pecorino Sardo cheese by casein stable isotope (13C/12C and 15N/14N) ratios and free amino-acid ratios. *Journal of Agricultural and Food Chemistry*, 2001, 49, 1404-1409.

[10] Chiacchierini, E., Bogoni, P., Franco, M. A.,Giaccio, M.,&Versini, G. (2002). Characterization of the regional origin of sheep and cow cheeses by casein stable isotope13C/12C and 15N/14N ratios. *Journal of Commodity Science,*41 (IV), 303-315.

[11] Kreitler, C.,& Jones, D. C. (1975). Natural soil nitrate: the cause of the nitrate contamination of groundwater Runnels Country, Texas. *Ground Water*, 13, 53-62.

[12] Heaton, H. T. E. (1987). The 15N/14N ratios of plants in South Africa and Namibia: relationship to climate and coastal/saline environments. *Oceologia* (Berlin),74, 236-246.

[13] Shintu, L.,&Caldarelli, S. (2005). High-Resolution MAS NMR and Chemometrics: Characterization of the Ripening of Parmigiano Reggiano Cheese". Journal*of Agricultural and Food Chemistry*, 53, 4026-4031.

[14] De Angelis Curtis, S; Curini, R; Delfini, M, Brosio, E, D'Ascenzo, G, Bocca, B.(2000) "Amino acid profile in the ripening of Grana Padano cheese: a NMR study". *Food Chemistry*, 71, 495-502.

[15] Modern Magnetic Resonance Part 3: Applications in Materials Science and Food Science, G.A. Webb (ed.), Berlin: Springer, 2006.

[16] Addeo, F., Trieu-Cout, P., Chianese, L.,&Ameno A. (1981). Qualitative and Quantitative Electrophoretic Methods for Evaluating Bovine and Buffalo Milks in Mixtures. *Scienza e Tecnica Lattiero-Casearia,* 32, 95-108.

[17] Addeo, F., Stingo, C., Chianese, L., Petrilli, P., Scudiero A.,&Anelli, G. (1983). Focalizzazione Isoelettrica delle Caseine per il Riconoscimento del Latte Bovino nella Mozzarella di Bufala. *Il latte*, 8, 795-803.

[18] Addeo F., Moio,L., Chianese,L.,&Nota,G. (1989). Evaluation of Bovine and Water-Buffalo Milk in Mixtures of Liquid and Mozzarella Cheese by Gel Isoelectric Focusing.*Italian Journal of Food Science*, 3, 71-80.

[19] Pellegrino, L., De Noni, I., Tirelli, A.,&Resmini, P. (1991). Determinazione del Latte di Vacca nei Formaggi di Specie Minori Mediante HPLC delle Sieroproteine. Nota 1.

Applicazione alla Mozzarella di Bufal. *Scienza e Tecnica Lattiero-Casearia*, 42, 87-101.

[20] European Communities. Regulation 213/2001, Off. J. Eur. Communities, L 03, 2001.

[21] Cartoni, G., Coccioli, F., Jasionowska R., Masci, M. (1998). Determination of Cow Milk in Buffalo Milk and Mozzarella Cheese by Capillary Electrophoresis of the Whey Protein Fractions.*Italian Journal of Food Science*, 10, 127-135.

[22] Tunick, MH; Malin, EL. "Differential Scanning Calorimetry of Water Buffalo and Cow Milk Fat in Mozzarella Cheese".JAOCS, 1997, 74,1565-1568.

[23] Klotz A.,&Einspanier, R. (2001). Development of a DNA-based Screening Method to Detect Cow Milk in Ewe, Goat and Buffalo Milk and Dairy Products Using PCR-LCR-EIA Technique *Milchwissenschaft*, 56, 67-70.

[24] Cozzolino R., Passalacqua, S., Salemi, S., Malvagna, P., Spina, E., &Garozzo, D. (2001). Identification of Adulteration in Milk by Matrix-Assisted Laser Desorption/Ionization Time-of-Flight Mass Spectrometry". *Journal of Mass Spectrometry*, 36, 1031-1037.

[25] Andreotti, G., Trivellone, E., Lamanna, R., Di Luccia A.,&Motta, A. (2000). Milk Identification of Different Species: 13C NMR Spectroscopy of Triacylglycerols from Cows and Buffaloes' Milks. *J. Dairy Sci.*, 83, 2432-2437.

[26] Brescia, M. A., Caldarola, V., De Giglio, A., Benedetti, D., Fanizzi, F. P.,&Sacco, A. (2002). Characterization of the geographical origin of Italian red wines based on traditional and nuclear magnetic resonance spectrometric determinatio. *Analytica Chimica Acta*, 458, 177-186.

[27] Folch, J., Lees, M.,&Sloane Stanley, G. H. (1957). A simple method for the isolation and purification of total lipids from animal tissues". *J. Biol. Chem*, 226, 497-509.

[28] Fan, TWM. "Metabolite profiling by one-and two-dimensional NMR analysis of complex mixture". Progress in Nuclear Magnetic Resonance Spectroscopy. 1996, 28, 161-219.

[29] Piasenter, E., Valusso, R., Camin, F.,&Versini, G. (2003). Stable isotope ratio analysis for authentication of lamb meat. *Meat Science*, 64, 239-247.

[30] Sacchi, R., Patumi, M., Fontanazza, G., Barone, P., Fiordiponti, P., Mannina, L., Rossi, E., & Segre, AL. (1996).A High Field 1H Nuclear Magnetic Resonance Study of the Minor Components in Virgin Olive Oil.*JAOCS*, 73, 747-758.

[31] Fauhl, C., Reniero, F., & Guillou, C. (2000). 1H-NMR as a Tool for the Analysis of Mixtures of Virgin Olive Oil with Oils of Different Botanical Origin. *Magnetic Resonance inChemistry*, 38, 436-443.

[32] Kalo, P., Kemppinen A.,& Kilpeläinen, I. (1996). Determination of Positional Distribution of Butyryl Groups in Milkfat Triacylglycerols, Triacylglycerol Mixtures, and Isolated Positional Isomers of Triacylglycerols by Gas Chromatography and 1H-Nuclear Magnetic Resonance Spectroscopy". *Lipids*, 1996, 31, 331-336.

In: Milk Production
Editor: Boulbaba Rekik, pp. 125-157
ISBN 978-1-62100-061-7

Chapter VII

Alternative Approaches for the Prevention of Bovine Mastitis. Probiotics, Bioactive Compounds and Vaccines

María Elena Fátima Nader-Macías[1*], Cristina Bogni[2], Fernando Juan Manuel Sesma[1], María Carolina Espeche[1], Matías Pellegrino[2], Lucila Saavedra[1] and Ignacio Frola[2]

[1] Cerela-Conicet. Chacabuco 145. 4000. Tucumán. Argentina.

[2]Department of Microbiology and Immunology, Universidad Nacional de Rio Cuarto. Ruta 36 Km 601, 5800 Río Cuarto, Córdoba. Argentina

Abstract

Bovine mastitis (BM), both clinical and subclinical, is the most important infectious disease in the dairy farm, and basically affects the quality and the quantity of milk production. They are responsible for the major economic losses in the dairy farm due to treatment, veterinary service, animal replacement, discarded milk and mainly reduced milk production. The main causative microorganisms are *Staphylococcus aureus*, *Streptococcus agalactiae*, *Streptococcus dysgalactiae*, *Streptococcus uberis* and *Escherichia coli.* Even though, current practices as proper milking hygiene and reduced exposure to pathogens contributes to decrease the occurrence of the disease, the treatment for BM relies heavily on the use of antibiotics, both for prophylaxis and therapy. These treatments have proved to be frequently ineffective to control cronic mastitis, causing a persistent bacterial reservoir within a herd with recurrent infections and the culling of infected animal from the productive circuit. During the last decades, world tendency to limit the use of antibiotics in dairy cattle, has lead researchers toward the study of cows natural defence mechanisms in order to ensure their absence in dairy products with the

[*] E-mail: fnader@cerela.org.ar

aim of satisfy consumers demand for "organic products". At the same time, the increase of functional products or "health-preventing foods" consumption supports the need to increase some other aspects of the animal welfare and breeding. Alternatives approaches are being applied for the prevention of this disease and highly promoted, that includes the application of vaccines, the use of probiotics or beneficial microorganisms, their metabolic products as organic acids and hydrogen peroxide, antimicrobial peptides, and/or other more physiological treatments.This chapter reviews the last advances in this area, the rational to support each one of the preventive applications, together with the products available in the international market. Also, the clinical trials supporting the use of the different approaches are included, to encourage the advances and studies in this area. In this way, is possible to help to the control of mastitis, the improvement of the health status of the animals, and in this way in the milk production.

Introduction

Bovine milk and dairy products have very long traditions in human nutrition. There is no doubt that all the components of milk provide a fundamental source of energy and nutrients to rapidly growing infants or adults. Milk components take part on human metabolism in several ways, providing essential aminoacids, vitamins, minerals and fatty acids, or by affecting absorption of nutrients. Consumption of 0.5 liter milk daily supplies a significant amount of many of the nutrients that are required daily (Haug et al., 2007).

The consumption of milk and milk products vary considerably among regions, and depend basically of the main animal species available in the specific geographical areas. But in the case of bovine milk, the drinking milk range from about 180 Kg yearly per capita in Island and Finland to less than 50 Kg in Japan and China. There are broad differences between the developing regions and developed countries, and also between urbanized areas. In south and south west Asia the milk consumptions has increases substantially to ten times higher in the last years (FAO, 2009). Brazil also has experiences a rapid expansion in the consumption of milk that has increased around 40%. According to the opinion of international and national market experts, the milk consumption will continue growing, and this will be higher than the offer. This higher demand is supported by the economic growth of western countries, as China and India, which even thought they have a higher production, they cannot self-sufficient, and by the higher urbanization of many Asiatic towns. The decrease of grants for the milk production in some countries as United States, Canada and UE, and the lower number of countries that produce milk in ecological conditions becomes Argentina in one of the countries with higher growth potential in the area (FAO, 2009).

Bovine milk is a complex food with a long list of components (lipids, proteins, aminoacids, vitamins, minerals, immunoglobulins, hormones, growth factors, cytokines, nucleotides, peptides, polyamines, enzymes and other bioactive peptides) which may have negative or positive health effect, and can be altered by many factors, including the animal-feeding regime (Lönnerdal, 2003). It is also important to consider that the association between food and health is well established, and recent studies have shown that modifiable risk factors seem to be of greater significance for health than previously anticipated (Haug et al., 2007). Many consumers are highly aware of the health-properties of food, and the market for healthy food and food with special health benefits is increasing.

The higher level of consumption of milk and milk based products with health properties, has been also supported by their lower cost originated by different sources of global changes, as the increase of production, the lower cost of the inputs, the technological changes and the higher scale efficiency. These technological changes are related with the advances and innovations in all the aspects of the animal production (breeding, food, health-related aspects) and also to the products release into the market (elaboration, transport and marketing). The application of modern technologies for growth and feed has originated an increase in productivity, supported mainly by the intensive animal breeding in feedlots where the farmers face a number of challenges to improve the sustainability of the system (FAO, 2009). In highly competitive times, farmers need to remain profitable whilst producing high quality, safe, nutritious and enjoyable foods at the same time as reducing the negative impacts of livestock farming on the environment (greenhouse gases, ammonia, nitrate pollution of water, odor nuisance) and also respond to societal concerns about the welfare of farmed animals. Improvements achieved through animal breeding are arguably the single best way to improve the sustainability of animal agriculture with the added benefit that the improvements made are cumulative and permanent. In the same way, the market is being more competitive and highly strict in terms of warranty of safety, environmental sanitation, and animal welfare, with a higher worth of some specific attributes of the products (natural, functional foods, with identity etc.).

Animal breeders have made considerable progress in recent decades in improving the economic efficiency of milk production, and probably this is one of the reasons of the fall of the real price of milk, but in recent years, animal breeding has become more complex with breeders needing to broaden their breeding objectives. Nowadays breeders want to improve a wide range of traits, such as product quality, welfare related fitness traits and disease resistance. Many of these traits are difficult or expensive to measure. And also the maintenance of animals in intensive-production feedlots and the current milking machines, together with a long list of factors, shows the importance of the emergency of many syndromes, as for example, cow mastitis.

Bovine Mastitis

Mastitis is an infectious bacterial disease which causes inflammation of the mammary gland, and is the most frequent disease in dairy cows, far ahead of feet and leg or metabolic disorders. This causes discomfort to the animal, makes milking difficult and reduces the yield and quality of milk and increases the rate of culling and veterinary costs (Miles et al., 1992; Calvinho, 1999; Bradley, 2002; Seegers et al., 2003; Halasa et al., 2007). Also reduces the profitability of farm milk production, but the calculation of the extent of this economic loss is complex because of the many factors involved and deficiencies in the evidence on the relationship between the disease and various production factors. The quantitative effects of mastitis on dairy cow performance are not easy to assess. These effects, however, have great economic consequences mainly because direct production losses must be considered along with milk withheld from the market following medicinal treatments, the cost of treatments, and various other costs, including additional labor and increased milling time (Miles et al., 1992; Seegers et al., 2003; Halasa et al., 2007).

Economic calculation of the mastitis costs or economic implications vary between countries and even between regions within the country. The results of these calculations change with time, based on the modifications of the milk quality regulations and market circumstances (Lescourret et al., 1994). There are not recent data available, but some reports have estimated that mastitis infections affected 30% of dairy cattle and cost the EU dairy industry about €1.55 billion in 2005. In 2005 there were nearly 23 million dairy cattle in the EU, producing over 500 million tones of milk. Over 600,000 people were involved in the dairy industry which has a turnover of €117 billion (SABRE, 2006). According to older data, the average cost of mastitis in the New York State is found to be $125 per cow from reduced milk production, treatment, and increased culling. At the 1988 cow inventory, this translates to approximately $100 million annually for the entire dairy farm sector. When quality and production losses for the processing sector are added, the cost to the New York industry alone is nearly $150 million annually (Miles et al., 1992; Halasa et al., 2007).

In Argentina, there are not recent data available on the economical implications of mastitis. According to the 1970´stimations, the annual losses produced only by the decrease in milk production were around U$S 115 millions, while in the 80´s were higher than U$S220 million (Calvinho and Tirante, 2005). Taking into account that the milk production has increases in the last decade, being Argentine the 16th producer country with 10325462 million tons, according to the FAOSTAT report the economical implications of mastitis is really surprising.

The cited data represent only some of the evaluations published. A revision performed by Halasa et al. (2009) explain the long list of economic factors associated with mastitis, and perform a comparative study of the papers published from 1990 on the economics of mastitis and management, showing a very large variation between them. It is clear from these data that the dairy sector is very important to the agricultural output of many countries and that any problem which harms milk production can negatively affect a lot of people and animals. In some way mastitis affects also the animal fertility, which is important for sustainable livestock production. Infertile dairy cattle, for example, are not cost-efficient as they need to regularly deliver a calf in order to maintain high milk yield. There has been a decline in fertility in recent years, probably in part due to side effects of selective breeding for increased milk production (Hansen et al., 2004; Vangroenweghe et al., 2005; Nava-Trujillo et al., 2010).

The data cited before evidence the importance of mastitis in the field of milk production worldwide. Mastitis remains a major challenge to the worldwide dairy industry despite the widespread implementation of mastitis control strategies. It is estimated that since 1970 the farms that have followed the recommended control procedures have reduced the average annual number of cases of clinical mastitis from 135 to 40 cases/100 cows each year, while the quarters remaining uninfected for a whole year has increased from 65 to 80% of the total quarters.

The costs of the main control procedures are broadly covered by the reduction in clinical mastitis, leaving the benefits of reduced subclinical infection. Also is interesting to consider that even though in the last forty years have seen a dramatic decrease in clinical mastitis incidence, this has been also accompanied by a change in the relative and absolute importance of different pathogens.

Mastitis Related Pathogens

The microorganisms that cause mastitis in dairy cows live in the udder and its surroundings. They are usually bacteria, but it is also possible to isolate mycoplasma, yeasts, algae and fungi, either. Over 140 infectious agents have been associated with clinically apparent episodes of mastitis, but the vast majority of cases are due to infection with a few types of bacteria (staphylococci, streptococci and several Gram-negative species (O´Grady and Doherty, 2009). These infectious agents fall into three categories (Table 1), and can be isolated by conventional cultivable techniques, or either by genetic identification of the microorganisms (Reinoso et al., 2004, 2008, Taponen et al., 2007).

a) *Contagious bacteria*: classified according to the pathogens inflammatory reaction in major and minor pathogens. The common contagious bacteria are *Staphylococcus aureus, Streptococcus agalactiae*, known as major pathogens *and Corynebacterium bovis* considered as minor pathogen. These pathogens are adapted to survive in the udder and are able to developed clinical, subclinical and long last period (chronic) infections. The reservoirs of these pathogens are the infected quarters and the transmission occurs mainly during milking, by their presence in the milking machines, washing water or hands of the operators.
b) *Environmental bacteria*: are commonly present in the cow's environment and may reach the teat end from that source. They cannot be removed from the environment, and although they can move from one quart to another, this route of transmission does not contribute to the development of new infections.The largest number of new infections with these organisms occurs in the dry period and around delivery. The main environmental bacteria associated with mastitis are streptococcus spp and coliform (Table 1). Approximately 40-50% of streptococcal infections and 80% of coliform infections can result in clinical mastitis. *S. uberis* is found in the udder, in the intestine, and on the cow's skin and teats. The particularity of the *S. uberis* is its extraordinary ability to develop in an external environment, i.e. in the bedding or anywhere on the animals. The contamination can take place during milking or in the environment (Odierno et al., 2006).
c) *Other pathogens:* Several types of bacteria commonly found in the udder are considered opportunist pathogens. These include a group of staphylococcal species, named coagulase negative *Staphylococcus* (CNS). These species have gained importance in recent years and are currently prevalent in infected glands and may cause an increase of two or three times the SCC. Mastitis caused by CNS usually remains sub clinical or mildly clinical, but may slightly decrease milk production.

The impact of pathogens on milk quality and productivity and cow health vary considerable among the different countries and through out a period of time. The incidence of the different microorganisms is constantly modifying. This variation is mainly related to the implementation of mastitis prevention programmes. As prevalence of major pathogens decrease, the relative importance of other organism increase. In some European countries (Belgium, Norway, and U.K), control programs for contagious mastitis have been developed for decades, resulting in a decrease in occurrence of *S. agalactiae*(sporadic occurrence) and *S. aureus* mastitis and an increase of *S. uberis* and *Escherichia coli* mastitis.

Table 1. Bovine mastitis infectious microorganisms

Contagious bacteria	Environmental bacteria	Others pathogens
Staphylococcus aureus	*Escherichia coli*	*Coagulase-negative staphylococci (CNS)*
Streptococcus agalactiae	*Klebsiella spp.*	
*Corynebacterium bovis**	*Citrobacter spp.*	
	Enterobacter spp.	
	Streptococcus uberis	
	Streptococcus dysgalactiae	
	Streptoccocus spp.	
	Enterococcus faecalis	
	Enterococcus faecium	

* Corynebacterium bovis, classified as minor pathogen, can disseminate cow to cow.

On the other hand, in America, i.e. in Brazil the prevalence of S. *agalactiae* is 60% of herds positive during 2004 and *S. aureus* remains being the main pathogen associated with clinical mastitis in several countries as The Netherlands, and Southern countries as Argentina, where the prevalence of IMI associated with *S. aureus* was 20%. During the last decade, the genomic information on pathogens have improved the knowledge and strain adaptation to human and bovine hosts has been recognised. Molecular epidemiological studies of bovine mastitis pathogens allowed a better knowledge of infective strains distributions in dairy herds. These developments contribute to novel approaches in formulating strategies to bovine mastitis control.

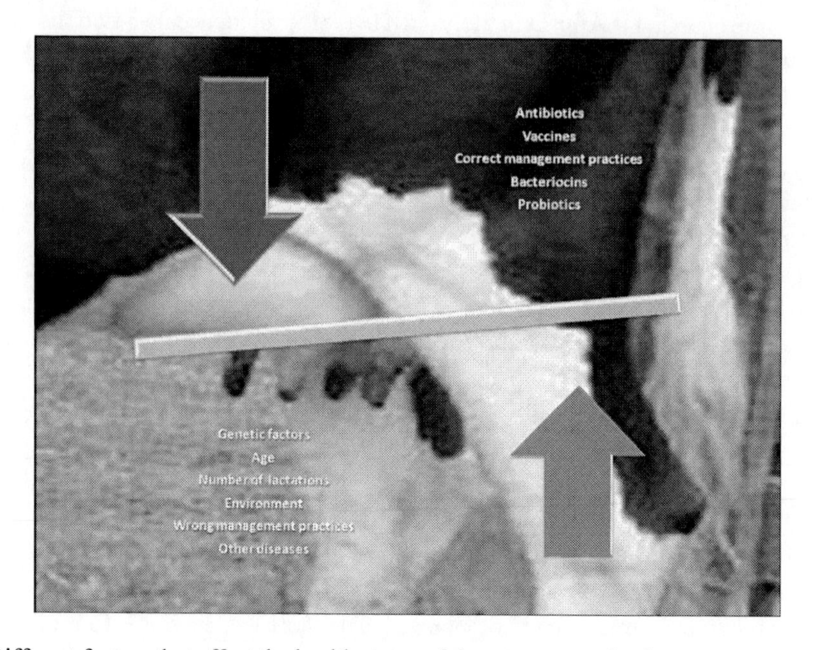

Figure 1. Different factors that affect the health status of the mammary gland.

The global dairy industry, the predominant pathogens causing mastitis, our understanding of mastitis pathogens and the host response to intramammary infection are changing rapidly. Globalization, energy demands, human population growth and climate change all affect the dairy industry. And in this way, mastitis can be considered a complex syndrome, where there are long lists of factors that have positive or negative incidence in the production or in the prevention, as summarized in Figure 1.

Preventive Control Strategies

Control methods to reduce mastitis in dairy herds must be directed to decrease the exposure of the teat ends to contagious pathogens (Smith and Hogan, 1997). This can be achieved by reducing the spread from cow to cow during the milking operation, eliminating the reservoir of the pathogens in the dairy herd and/or working on the genetic selection of dairy cattle resistant to mastitis (Gonzalez et al. 1993). The conventional methods of controlling mastitis are based upon adoption of preventive control strategies including diagnosis, segregation of the animals and the use of improved hygiene and therapeutic protocols. These strategies are summarised in the "Five Point Plan" proposed for the National Mastitis Council (2004). The main objective of this plan is to reduce the degree and the durability of the infection. The "Five Point Plan" includes:

1) The use of measures to improve the milking routine
    - Correct use of milking equipment and its periodic evaluation,
    - Clean and dry the teats before milking,
    - Teat disinfection pre and post milking (pre-dipping and teat-dipping), and
    - Washing and disinfection of the milking machine and the production line before and after milking
2) Quarter s*antibiotic treatment at the dry period:* Dry cow therapy continues to be effective in reducing the number of both contagious infections during the dry period and in preventing environmental streptococcal infections that can also occur in the early dry period.
3) *Clinical bovine mastitis treatment:* Before clinical bovine mastitis treatment, the selection of an initial treatment regimen should be based upon historical microbiological cultures and antibiotic-sensitivity results, severity of infection and documented success of previous treatments. Antibiotic treatment of subclinical mastitis during lactation is not cost effective and this practice increases the opportunities for drug residues in milk. According to the Food and Drug Administration/ Center for Veterinary Medicine (http://www.fda.gov/AnimalVeterinary) the approved antibiotics for the treatment of bovine mastitis are: Pirlimycin, Hetacillin, Cloxacillin, Amoxicillin, Novobiovin, Penicillin G, Dihydrostreptomycin, Cephapirin and Erythromycin. The choice of the antimicrobial agents and the route of administration will be directed by the characteristics of the drug and regulatory issues.
4) Replacement of cows with chronic mastitis

5) *Vaccination plan:* These control strategies have proved to decrease St. agalactiae and St. dysgalactiae bovine mastitis. Even though these control strategies contribute to a decrease in the occurrence of the disease, the treatment for bovine mastitis still relies heavily on the use of antibiotics, both for prophylaxis and therapy (Mc Dougall et al., 2009). The prevention of bovine mastitis by using antibiotics at the dry period have demonstrated that this practice eliminates up to 80% of new infections caused by different pathogens during the dry period, but is not completely effective in animals infected with S. aureus. Less than 15% of the dairy cows infected with S. aureus respond to the antibiotic therapy. This is due to the poor penetration of the drug in the mammary gland that originated that a significant survival percentage of bacteria in the host cells that increase the recurrent infections after the treatment (Yancey et al., 1991). For this reasons, the use of antibiotics is ineffective for the control of S. aureus´s chronic and recurrent infections within a herd. While antibiotics have had a major impact on health of dairy cows and consequently in the production of milk, its use is very questioned because of the possibility of producing anaphylactic reactions in consumers due to the presence of traces of antibiotics in milk for human consumption (Errecalde et al., 2004). In addition, during the last few decades, the global pressure to limit the use of antibiotics in dairy cattle has lead researchers to focus on cows' natural defence mechanisms in order to reduce potential for antibiotic residues and in this way to support their absence in dairy products with the aim to satisfy consumers demand for "organic products".

1. Vaccines

Several studies aimed at preventing bovine mastitis are primarily focused at the first step of the innate immune response, and to obtain its high stimulation. One of the component of this immune response is the migration of large numbers of white blood cells (in the udder called somatic cells) to the infected gland, but the presence of somatic cells in the milk is not considered a positive outcome, as somatic cells are evidence of mastitis and reduce milk quality. Effective immunization is difficult because of the nature of milk. For this reason, the immunization schedule employed for the prevention of this disease should promote the migration of neutrophils to the site of the infection through the production of inflammatory mediators and should stimulate in this step the production of specifics antibodies for the opsonization of bacteria.

This preventive method must also promote neutrophils phagocytosis, the neutralization of toxins and at the same time induce bacterial cell lysis (Tenhagen et al., 2001). Both, the level of antibodies and the number of neutrophils are low in healthy mammary glands. Because a large number of these cell types are required ($9x10^5$/ml) to prevent the intrammamary infections, and having in mind that this value exceeds the number present in a healthy gland, the emphasis of prevention methods is mainly directed to increase the opsonics antibodies in milk, and to reduce the number of neutrophils needed to protect the gland (Sear et al., 2004). Vaccination is one of the main strategy to control mastitis, as its principal objective is to increase the concentration of opsonics antibodies in blood and milk against a particular microorganism to inhibit their growth and toxin production.

Since the disease was first diagnosed, veterinarians and dairy herd's producers claim for the development of an efficient vaccine against bovine mastitis. The main obstacles to achieve this objective are the large number of microorganisms (bacteria, viruses, fungi) involved in the production of the disease, to an incomplete knowledge of the mammary gland's immunity and the virulence factors of the main pathogens, the failure to maintain a high level of immunoglobulins in milk and the lack of proper vaccination schedule.

In recent years, significant progress has been performed in the development of vaccines against some microorganisms that cause mastitis.Perhaps the greatest progress has been achieved by the use of vaccines against *E. coli* from which the best known and widely used are the mutant bacterin (J5 strain) against *E. coli* infection currently available in several countries including U.S., France and Denmark, among others (Table 2). This vaccine showed to reduce the duration and severity of the symptoms of clinical mastitis after challenge with a virulent strain of *E. coli*.

Mastitis caused by coliforms (*E. coli, Klebsiellaspp.* and others) is usually of short duration and <15% of affected animals usually develop chronic infections. Coliform mastitis is generally clinical and many affected animals exhibit systemic signs of disease. The clinical symptoms associated with coliform infections are the result of endotoxin released from the cell wall of dying gram-negative bacteria. There is rarely a long-term impact of coliform infections on SCC. Losses attributable to coliform mastitis are associated with the clinical episode and are the result of reduced milk yield, discarded milk, treatment costs, death and culling. The highest risk period for coliform mastitis is during the immediate periparturient period. Therefore, a vaccine may be judged effective if it successfully reduces symptoms of coliform mastitis during this limited "at-risk" period. For many years, the main cause for mastitis in dairy cows was identified as *S. aureus, St. agalactiae* and *St. dysgalactiae*. The presence of these species is generally due to deficiencies in the progress of hygiene in milking and dry cow routines. Although control recommendations have effectively reduced the incidence of mastitis due to these contagious pathogens, they have had little impact on the incidence of mastitis caused by environmental pathogens such as *St. uberis*. The first indication that supports vaccination against *St. uberis* was the demonstration that previous exposure could provide some resistance to infection. Sense that many research have been conducted to developed a *St. uberis* vaccine (Table 2). One of the most frustrating mastitis pathogens is *S. aureus*. This organism is a highly successful mastitis pathogen that it has evolved to produce infections of long duration with limited clinical signs. Most infections with this pathogen are subclinical and are detected by the production of poor quality milk. While clinical mastitis may occur sporadically, affected animals rarely become seriously ill and the major economic effect of this disease is reduced milk yield and quality premiums received by the producer. Animals are at risk for this organism throughout lactation and often becoming infected after prolonged periods of exposure.

S. aureus vaccines currently available or under study, point not only to reduce the frequency and severity of clinical mastitis, but also to eliminate chronic mastitis and prevent the establishment of new IMI. Several research works are conducted to avoid *S. aureus* bovine mastitis by using vaccines made on the base of wall cellular component, capsular exopolisacarides, bacterins and DNA technology. Many of them, mainly those that are the result of studies of the last two decades are summarized in table 2, where also is included some data of the specific clinical or experimental trials.

Table 2. Characteristics of the major vaccines released to the market or under study for the last 2 decades for the prevention of bovine mastitis

Vaccine type/ comercial name	Composition	Animal assays	Doses/ administration route	Vaccine Efficacy	Reference
Capsular exopolysaccharides/exoproteins					
Capsule-Inactivated cells/ REDUMAST	Crude exopolysaccharides extract and inactivated cells of S. aureus. Inactivated cells of S. agalactiae and S. uberis. Sodium azide preservative, thimerosal and formaldehyde. Aluminum Hydroxide	Field trial in hiefers N=30	Two doses: 30 and 10 days before calving. Subcutaneous	Reduces the incidence of clinical and subclinical mastitis against S. aureus. Does not decrease the frequency of mastitis caused by Streptococcus sp. Improves the quality of milk by lowering the bacterial count and somatic cell count and increases oil production.	Giraudo et al., 1997 Calzolari et al., 1997
Exoproteins	S. uberis PauA (total antigen) or PauA selectively removed with either Freund's incomplete adjuvant (FIA) or a commercially used adjuvant (SB62)	Experimental challenge trial in cows N=16	Five doses of the total antigen in FIA or five doses of the depleted antigen in FIA at 3, 5, 8, 11 and 13 weeks into lactation. Or two doses of total antigen combined with SB62 at 3-week intervals during early lactation.	PauA alone is not able to induce a protective immune response. However, some observations suggest that an inhibitory response to this protein was important to prevent colonisation of the bovine mammary gland and thus to protect against bovine mastitis due to S. uberis.	Leigh et al., 1999
Conjugate	S. aureus exopolysaccharide 5, 8 and 336 serotypes conjugated to a protein and incorporated in poly (DL-lactide-co-glycolide) microspheres.	Experimental trial in cows N=10	Two doses before calving. Supramammary lymph node and intramuscularly	Long-lasting antibodies titter to capsule that promoted neutrophil phagocytosis and prevented adherence to bovine epithelial cells. n=	O´Brien et al., 2000
Conjugate	Different combinations of S. aureus Reynolds and CP5-HSA conjugate	Experimental trial in heifers N=20	Three doses: one at 3-11 month and two doses at 23 days and 11 months after dose 1. Supramammary lymph node	CP5 vaccination gives a strong and long lasting immune response in cattle. Conjugation technology could not be necessary to achieve an immune response against S. aureus CP5 in cattle.	Tollersrud et al., 2001

Vaccine type/ comercial name	Composition	Animal assays	Doses/ administration route	Vaccine Efficacy	Reference
Exoproteins	(6 X His)GapC of S. uberis or S. dysgalactiae, or chimeric CAMP-factor antigen, CAMP-3.	Experimental challenge trial in cows N=32	Two doses at 36 (day 0) and 15 days prior to challenge. Subcutaneous	Vaccination with S. uberis (6 X His)GapC or CAMP-3 results in a significant reduction in inflammation on several days post-challenge, most significantly for the former antigen. Inflammation was not reduced in S. dysgalactiae (6×His)GapC vaccinates, suggesting that it does not confer cross-species protection.	Fontaine et al., 2002
Enterotoxin type C mutant/ MastaVac	Soybean oil, staphylococcal enterotoxin type C mutant (SEC-SER) and lecithin	Experimental challenge trail in cows N=11	Three doses: one at day 0 and later 2 and 6 weeks after initial vaccination. Intramuscularly	The vaccine was effective in preventing new infections with a virulent S. aureus strain that secreted A, B, C and D enterotoxin.	Chang et al., 2008
Whole cells based					
Bacterin/Lysigin Somato-Staph	Lysed culture of highly antigenic polyvalent somatic antigen containing types I, II, III, IV phages and miscellaneous groups of S. aureus.	Experimental trial in heifers N=12	Six months age followed by a booster dose two weeks later and subsequent vaccinations every 6 months until calving	Reduces 45% of both new intramammary S. aureus infections (IMI) during pregnancy and new IMI at calving. Reduces 30% of new CNS IMI which becames chronic and 31% reduction in new CNS IMI at calving	Nickerson et al., 1993
Bacterin	Inactivated S. aureus bacteria with pseudocapsule and α and β toxoids with a mixture of mineral oil and detergent as adjuvant.	Experimental trail in heifers N=108	Two doses: 78 and 18 days before calving. Supramammary lymph node	When all the parameters on udder health were considered together, the results indicate a potential protective effect during the entire lactation.	Nordhaug et al., 1994
Live strain	Live S. uberis 01405 strain and bacterial surface extract with Freund's incomplete adjuvant and Tween 20	Experimental challenge trial in cows N=12	Two doses: one 14 days before drying off and other within 48 h of calving. Subcutaneous	Vaccination protects against challenge with the homologous strain, but was less effective against an heterologous strain.	Finch et al., 1997

Table 2. (Continued)

Vaccine type/ comercial name	Composition	Animal assays	Doses/ administration route	Vaccine Efficacy	Reference
Bacterin	E. coli mutant (01 11:B4) J5 formalin killed in oil adjuvant	Experimental challenge trial in cows N=225	Three doses: one at dryingoff, one 30 days after drying off and other within 48 h after calving. Subcutaneous	Vaccination with the E. coli J5 bacterin does not prevent IMI, but reduces severity of clinical signs following intramammary experimental challenge with an heterologous E. coli strain.	Hogan et al., 1999
Bacterin	S. aureus Df lysate incorporated into microsphere	Experimental trial in cows N=8	Drying off immunization with two doses: one intramuscular and one in the area of the supramammary lymph node	Produces long-term opsonic antibodies for neutrophils and blocks adherence to mammary epithelium	O´Brien et al., 2001
Bacterin/MASTIVAC I	Exosecretion of S. aureus VLVL8407 strain and bacterial fragments of other two S. aureus strains (ZO3984 and BS449) (1:1:1 by protein concentration) and mixed 1:1 with incomplete Freund´s adjuvant.	Experimental challenge trial N=19	Non-pregnant cows in the first to second mild-lactation period inoculated with 3 doses: one with the adjuvant vaccine under the tail root and without adjuvant in the area of the supra-mammary lymph node. Similar doses were administered 36 and 56 days after the first dose. Intramuscular	The vaccine provides about 70% protection from S. aureus infection and complete protection from inflammatory reactions as expressed by the somatic cells count	Leitner et. al., 2003
Bacterin	Type 5, 8 and 336 S. aureus capsular polysaccharide	Experimental trials in heifers N=20	Three doses: the first 1 month before the expected calving, followed by 2 boots in a 2-week interval (days 14 and 28). Intramuscular and Supramammary lymph node	Increases the amount of antigen-specific antibodies in serum. The effect on lymphocyte subsets and neutrophil phagocytosis were minimal.	Lee et al., 2005

Vaccine type/ comercial name	Composition	Animal assays	Doses/ administration route	Vaccine Efficacy	Reference
Live mutant strain	S. aureus aroA mutant	Experimental challenge trials in mice N=10	Two doses: 7 days before parturition and 7 days after the first dose. Intramammary	Mice immunized with the vaccine are protected from intramammary heterologous challenge.	Buzzola et al., 2006
Bacterin/ STARVAC	Inactivated E. coli J5 strain, inactivated S. aureus (CP8) SP 140 strain, expressing Slime Antigenic Associated Complex (SAAC)	Experimental trial in multiparous cows N=386	Three doses: 45 and 10 days before calving and 52 days after calving.	Reduces the incidence of clinical or sub-clinical mastitis and the severity of the clinical signs of mastitis caused by S. aureus, coliforms or coagulase negative staphylococci. Vaccination also increases the spontaneous cure rate of the infected cows.	Prenafeta et al., 2010
Live mutant strain	S. aureus RC122 avirulent mutant strain	Experimental challenge trial in heifers N= 9	Three subcutaneous doses: one at 14-16 months age and two doses 30 days after pregnancy and 10 days before calving. One intramammary dose 20 days after calving.	Induces specific and significant opsonic antibody response in blood and milk and provides protection through a significant reduction of post challenge milk bacterial shedding.	Pellegrino et al., 2010
DNA based					
DNA	Fc binding region of Protein A in pcDNA3 vector	Experimental trial in cows N=15	Three doses: 6, 4 and 2 week's pre-partum. Intramuscular, Intradermal or intravulvarmucosal	Low antibody titters in response to the vaccine	Carter and Kerr, 2003
DNA	FnBP and ClfA (aa 221-550) of S. aureus in pCI-D_1D_3-RES-ClfA	Experimental trial in heifers N=8	Three doses: 30 and 10 days before calving and one dose of the recombinant protein within 6 days of calving	Causes lymphoproliferative and humoral immune responses and provides partial protection of mammary gland	Shkreta et al., 2004.
DNA	Clf A of S. aureus cloned into mammalian expression vector (pCI) and recombinant ClfA protein.	Experimental trial in cows N=35	Three doses of the pCI-ClfA 21 days apart. One boost with the recombinat ClfA protein three months after the third pCI-ClfA dose/ Intramusculary	The DNA-prime/protein boost strategy induces antibodies that improved phagocytosis and decreased adhesion of the bacteria.	Nour El-Din et al., 2006

Even though these vaccines increased the level of specific antibodies in blood, the levels reached in milk are very low and it is very difficult to prevent new infections and to induce both humoral and cellular immune response. The use of attenuated vaccines could enhance both cellular and humoral protective responses, simulating natural infection without causing the disease. Also, attenuated vaccines may contain native antigens (proteins, polysaccharides, lipids, nucleic acids, etc.) that are expressed for extended periods of time and simulates natural infection, as resumed in Table 2.

2. Novel Preventive Approaches for theTreatment of Mastitis

Probiotics, Bioactive Peptides and Bacteriocins

During the last years, there is an increasing tendency and recommendation of all the governmental and regulation organisms on the need to apply preventive strategies in all the human and animal areas. In this way, the use of different and novel preventive approaches are being suggested and assayed, which includes the use of probiotic microorganisms or some of the metabolic or bioactive compounds they are able to produce.

Probiotics are "*live microorganisms which when administered in adequate amounts confer a health benefit on the host*" (FAO-WHO, 2002). These products have been aimed at both the health of man and animals. Most studies published to date have focused mainly on studying probiotics for the prevention of infections in the gastrointestinal tract. However, any part of the body with a normal microbiota could be a potential target for specific probiotics (Ouwehand et al., 2002).The progressive decrease in the use of antibiotics as growth promoters increased the interest in the use of beneficial microorganisms in animal feed. These veterinary probiotics improve the health status, increase the weight gain and inhibit pathogens (Guillot, 2003; Ewaschuk et al., 2004; Draksler et al., 2004, Isolauri et al., 2004, Weese and Rousseau, 2005; Anadón et al., 2006, Nader-Macías et al., 2008, Azad and Al-Marzouk, 2008, Simpson et al., 2009; Herstad et al., 2010). Recently, the target of these products was extended to the bovine reproductive tract in order to prevent metritis and also to the mammary gland to prevent mastitis. Both of them are diseases that affect reproduction and productivity in the dairy herd and are responsible for important economic losses (Otero et al., 1999, 2006, Nader-Macías et al., 2008; Espeche et al., 2009). Lactic acid bacteria, bifidobacteria, spore-forming bacilli and yeast have been described as probiotic to date. However, lactic acid bacteria are the main group because they have been included in the GRAS (Generally Regarded as Safe) or Food-Grade Microorganisms (FGM) cathegories. Some specific species are being also included in the list of QPS microorganisms (qualified presumption of safety).

Mechanisms of Action of Probiotic Microorganisms

Probiotics may exert their beneficial effects on the host health by different and several mechanisms: adhesion to epithelial cell, colonization, biofilm formation, production of biosurfactants, aggregation and coaggregation, production of antagonistic metabolites (organic acids, hydrogen peroxide, bacteriocins), competition for nutrients, production of enzymes and/or immune system modulation. It is likely that microorganisms may exert their effects as the result of one or more of these mechanisms, as explained later.

a) *Adhesion and colonization:* adhesion to epithelial cells is a prerequisite for colonization and therefore to the probiotic action. Adherent bacteria may persist longer and thus are more likely to exert its effect compared with those non-adherent microorganisms. This property can also block the adhesion of pathogens to the epithelium, as a result of competition for the same receptor, or by steric blockage. Also, adhesion to the intestinal mucosa could modulate the immune system. In an initial step, the process is reversible and is based on non-specific interactions. Later it includes specific interactions between adhesins and surface receptors and the formation of a biofilm that protects against the entry of pathogens (van Loosdrecht et al., 1990, Saarela et al. 2000; Ouwehand et al., 2002; Kubota et al., 2008; Vesterlung et al., 2005 and 2006, Oelschlaeger, 2010, Eberl et al., 2010).

b) *Production of biosurfactants:* surfactants are amphiphilic substances capable of reducing surface tension. Their physiological function is to increase the solubility of hydrophobic compounds or metal binding that affect the bacterial aggregation and adhesion, and in this way the degree of pathogenesis. They also act as signaling molecules in quorum sensing and biofilm formation. In addition, these products have antimicrobial properties and inhibit the adhesion of some specific species such as *Enterococcus faecalis, Str mutans, E. coli, S. epidermidis* and *Candida albicans* (Ron and Rosenberg, 2001, Salminen et al., 2004; Golek et al., 2009; Gudiña et al., 2010).

c) *Aggregation and coaggregation*: both properties are required for the persistence of the microorganism in the epithelium. Aggregation involves the interaction between two organisms of the same species, whereas the coaggregation represents the interaction between microorganisms of different species. The aggregation of the beneficial bacteria can promote adhesion and biofilm formation and therefore protective colonization of mucosal surfaces. On the other hand, bacteria capable of coagreggate can produce a barrier that prevents colonization by pathogenic microorganisms (Boris et al., 1997, Cesena et al., 2001; Schachtsiek et al., 2004, Collado et al., 2007).

d) *Production of antagonistic metabolites:* over time, the ability of lactic acid bacteria to produce antimicrobial substances was used to preserve dairy products, meat and vegetables. Lactic acid bacteria produce a wide range of compounds active against other microorganisms. The production of these substances provides a competitive advantage. Lactic acid bacteria can produce organic acids, hydrogen peroxide, carbon dioxide, diacetyl, bacteriocins, and substances of low molecular weight of protective nature.

e) *Competition for nutrients:* microorganisms need certain essential nutrients for growth. Lactic acid bacteria can compete with pathogens for vitamins, amino acids or other essential nutrients and in this way limit the growth of other organisms (Salminen and Von Wright, 1998, Salminen et al., 2004; Gueimonde and Salminen, 2006; Oelschlaeger, 2010).

f) *Production of enzymes:*β-galactosidase and bile salt hydrolase are two examples of the enzymes that produce beneficial microorganisms. The first hydrolyzes lactose. Thus, consumption of fermented dairy products does not produce the characteristic symptoms of lactose intolerance in people who lack the intestinal enzyme. The second enzyme would be responsible for the cholesterol-lowering effects of some strains of probiotic lactic acid bacteria (Salminen and Von Wright, 1998; Taranto et al., 1998 and 2000, Salminen et al., 2004; Parvez et al., 2006).

g) *Modulation of the immune system*: the results of many animals and in human studies are being a very strong evidence of the importance of the indigenous microbiota, mainly composed by bacteria, in the maturation and modulation of the immune system. To produce some type of immunomodulation, the probiotic bacteria needs to interact with the epithelia, either by adhering or to remain for long periods of time to contact different cell types of the host. Thus, the interaction with the host triggers a cascade of events leading to the modulation of the immune response. The probiotic lactic acid bacteria modulate host response to innate and specific immunity, local and systemic level, and exert their effect on the humoral and cellular immune system. The mechanisms by which they act have not yet been fully elucidated (Schiffrin et al., 1997; Salminen et al., 2004; Gueimonde and Salminen, 2006; Oelschlaeger, 2010). The study of the immunological mechanisms involved in the protective effect of lactic bacteria in the bovine mammary gland is more recent. Crispie et al. (2008) showed that the intramammary administration of *L. lactis* DPC3147 isolated from kefir triggers an immunomodulatory effect in the healthy gland, with recruitment of neutrophils and lymphocytes to the mammary quarter. Beecher et al. (2009) found that the administration of *L. lactis* DPC3147 in healthy quarter stimulates the local immune response and particularly the expression of the *IL1β*, *IL-8* and *CXCR1* genes. The inflammatory response produced in this case was intense, with visible signs of inflammation in the gland, increase of rectal temperature and blood clots in the milk produced. The somatic cells count reached values up to $1x10^{12}$ cells per milliliter of milk. The strain was recovered in milk up to 48 hours after the challenge. The inflammation was completely self-limited seven days after inoculation. The type of response that occurred with the inoculation of lactic bacteria was different to that produced by *S. aureus*, *St. dysgalactiae* and *St. uberis*.

Probiotics for the Mammary Gland

In 1987, Woodward et al. observed that the indigenous microbiota of the teat skin is able of inhibiting mastitis pathogens. In a more recent work, Al-Qumber and Tagg (2006) found that some *Bacillus* strains isolated from the teat skin of healthy animals can inhibit a wide range of pathogenic strains of mastitis by the action of broad spectrum antimicrobials. Is worth to suppose that certain strains of the indigenous microbiota can maintain or restore the ecological balance in the niche and this is one of the traits of probiotic microorganisms. The autochthonous microbiota of each ecological niche is in equilibrium and this state can be altered by intrinsic or extrinsic factors.

In these circumstances, there is an overgrowth of some potentially pathogenic or pathogenic microorganisms, which produce some kind of disease. Probiotic bacteria have the ability to maintain or restore the balance through various mechanisms that have been explained previously (Salminen et al., 2004, Oelschlaeger, 2010). In 1991, Greene et al. performed a comparative study between probiotic and antibiotic intramammary treatment in dairy cows. Lactating cows with subclinical mastitis were divided in two groups. One of them was inoculated with a commercial antibiotic and the other with a combination of two strains: *Lactobacillus acidophilus* and *Lactobacillus casei*.

Both strains belonged to a probiotic product approved for oral use in calves, but with no-specification of their origin or citation of previous studies to support or justify their employment. The cure rate of the probiotic treatment was significantly lower than the antibiotic therapy and produced a significant increase in somatic cell counts. In 2008, Klotermann et al. compared an antibiotic therapy and a potentially probiotic containing *L. lactis* DPC3147 in the treatment of chronic subclinical mastitis and clinical mastitis.

After the intramammmary infusion, the application of the organism was as effective as the antibiotic in the treatment of clinical mastitis. The mechanism of the probiotic effect is the stimulation of the host intramammary immune system (Crispie et al., 2008; Beecher et al., 2009). *L. lactis* DPC3147 was isolated previously from a kefir grain and produces a bacteriocin active against many mastitis pathogens (Ryan et al., 1996 and 1998).

Our research group focused on the prevention of diseases that affect dairy cattle by restoring the balance in the indigenous microbiota in a certain niche (Nader et al., 2008). The study of the lactic acid microbiota of the mammary gland was performed with the aim of restoring the ecological homeostasis with a probiotic product. Based on the host specificity, one hundred and two lactic acid bacteria from foremilk, stripping milk and teat canal scrapped were isolated.

The production of antagonistic substances and bacterial-surface properties, such as hydrophobicity and autoaggregation were evaluated. Safety aspects as virulence traits and antibiotic resistance were also taken into account. Finally, four strains: *Lactococcus lactis* subsp. *lactis* CRL1655, *Lactobacillus perolens* CRL1724, *L. plantarum* CRL1716 and *Enteroccus mundtii* CRL1656 were selected. Both lactobacilli showed some surface properties and inhibited mastitis pathogens by organic acid production. The adhesion of *L. perolens* to teat canal cells was also studied. *L. perolens* showed an interesting pattern of adhesion to bovine teat canal epithelial cells. Both *L. lactis* and *E. mundtii* are bacteriocinogenic strains and produce nisin Z and mundticin CRL1656 respectively (Espeche et al., 2009; Frola et al., submitted).

The hypothesis of our research group is that by promoting the colonization of the vaginal mucosa of pregnant cows and the mammary gland of lactating cows with indigenous probiotic strains for the prevention of post-partum metritis and mastitis, the neonate will be protected. The vaginal probiotics would increase the first contact of these microorganisms with the sterile, unprotected neonate intestinal tract, during the passage through the vaginal tract at birth. Probiotics administered to the cow for mastitis prevention would be also in contact with the calf during the first days of its life and would stimulate the immature immune system and prevent calf diarrhea, as shown in Figure 2.

Bacteriocins in Mastitis

Traditionally, in veterinary areas, medicine treatment therapies and prophylaxis have been based primarily on the use of antibiotics. It is known that the indiscriminate use of Antibiotics (ATB) for the treatment of bovine mastitis is a main public concern, as cited before. Some sensitive organisms might become resistant and they can often spread the resistance genes to other related microorganism through different processes. In addition, the presence of residues of ATB in milk and other dairy products may not be detected in time and get to the market chain to the consumer producing various disorders. It is also important to note that the presence of residues of ATB in milk is also a serious problem for the dairy industry because it can lead to the failure of the fermentation processes.

Figure 2. Proposed colonization of lactic acid bacteria as a probiotic in cattle.

Inrecentyears, new antimicrobial compounds such as those called bacteriocins are being considered for the use in bovine mastitis. Bacteriocinsarepeptideswith antimicrobial activity,gene-encoded, whichareproduced byanumber ofbacterial genera, andhaveawide spectrumofinhibitory activity. In particular,bacteriocinsof lactic acid bacteriahave receivedmuch of attentiondue to the great potential for their useinfoodto controlfood-borne pathogensorsaprophyticorganisms and toincreasethe shelf lifeof food products, as well asfor thedevelopmentof products forhuman and veterinary medicine by pharmaceutical industries (Jack et al.1995, Cotter et al 2005).

As the interest in these antimicrobial peptides was increasing in recent decades, the discovery of new bacteriocins emerged in the literature, and then the need of new and improved classification of them was also evident. Since bacteriocins are heterogeneous group of peptides, many classification systems have been proposed. Originally, Klaenhammer (1993) proposed four major classes for bacteriocins. Later, the advance in the knowledge of this kind of compounds made necessary a revision of this classification, and more recently Cotter *et al* (Cotter *et al.*, 2005) andHeng *et al.*(2007) suggested some modifications.

Class I the post-translationally modified bacteriocins, i.e., the lantibiotics. Members of this group is divided according to their structure in *type A* (elongated, amphipathic structures), *type B* (globular and more compact structure) and *type C* (comprising more than one component). *Type A* is further subdivided into subtypes AI and AII according to their size, charge and leader peptide sequence. The archetype of AI subtype is nisin. This bacteriocin has been used as an additive in the food industry for over 40 years. The *type B* lantibiotics are more globular and compact than type A and its prototype is the mersacidin which is active against Methicillin Resistant *Staphylococcus aureus* (Heng et al., 2007).

The *type C* or multicomponents consist of two peptides. Each of them has little or no activity, but when combined together they have synergistic action. Lacticin 3147 is the most well studied example, produced by *Lactococcus lactis* DPC3147, active against mastitis pathogens (Ryan et al., 1999).

Class II are small (<10 kDa) heat-stable membrane-active bacteriocins. This group is further divided in subgroups. Subclass *IIa* are "pediocin like peptides" since pediocin PA-1 is the best characterized member. They encompass strong anti-listerial activity and this group is further distinguished by having a conserved amino-terminal hydrophilic region (YGNGV) and two cysteine residues forming a disulfide bridge.

The C-terminal region is hydrophobic and more variable (DRID et al., 2006, Heng et al., 2007). Subtype *IIb* (two-component peptides) and *IIc* (thiol-activated peptides requiring reduced cysteine residues for activity) need two or more peptides to exert its inhibitory effect. These individual peptides lack antimicrobial activity by themselves. Examples are lactacin F produced by *Lb. johnsoni* F and lactocin 705 synthesized by *Lb. curvatus* (Cuozzo et al., 2000, Heng et al., 2007). In the case of salivaricin ABP-118 produced by *Lb. salivarius* UCC118 and and salivaricin CRL 1328 synthesized by *Lb. salivarius* CRL1328 (Flynn et al., 2002, Heng et al., 2007, Vera Pingitore et al., 2009) each component exhibit his own antimicrobial activity and when combined, present synergistic effect.

Class III, are larger (>30 kDa) heat-labile bacteriocins, and *Class IV*, complex bacteriocins (lipid or carbohydrate moieties besides the protein).

Inthe classification proposed byCotteretal. (2005), they suggest a division in twomain classes: *ClassI*, peptideswith post-translational modifications, such asunusual amino acidslanthionine, beta-metil lanthionine, dehydroalanine. *ClassII:*includes allpeptidesthat do not undergoposttranslational modifications.

On the other hand, Heng et al (2007*)* retained *Class III* and divided it into *IIIa* (bacteriolysins) and *IIIb* (non-lytic proteins) and added *Class IV* (cyclic bacteriocins known before as Class IIc). Nissen-Meyer *et al.* (2009) suggested a division of class II bacteriocins into four subgroups according to similarities of their C-terminal regions. More recently, Zouhir *et al* (2010) define new structure-based sequence fingerprints that support a subdivision of the bacteriocins from Gram positive bacteria into 11 groups.

Application of LAB Bacteriocins

An interesting aspect of nisin is its spectrum of action, being able to inhibit the major pathogens responsible for bovine mastitis (Broadbent et al 1989). The use of bacteriocins for the treatment of bovine mastitis data from the 40's, with a report published by Taylor et al. (1949) who evaluated the use of the lantibiotic nisin in an intramammary infection replacing the ATBs.

Sears et al. (1992) reported the development of a preparation of a sanitizer germicidal agent suitable for use in the cow teats. The nisin-based formulation produced a significant reduction in the number of *S. aureus*, compared with the agents commonly used for cleaning the teat after milking. This peptide is the only GRAS (Generally Recognized as Safe) bacteriocin that is in the market and is present in products such as Concept (Applied

Microbiology, Inc., New York, NY), Mast Out ®, Wipe-Out ® (ImmuCell, Portland, OR) (Sears et al, 1992; Sears and McCarthy, 2003).

A variant of the lantibiotics nisin, named nisin Z has been used for the treatment of subclinical mastitis in lactating cows (Wu et al., 2007). In this study 46 lactating cows with subclinical mastitis were treated with nisin Z by intramammary infusion and compared with untreated control groups.

The results showed that treatment with nisin Z significantly decreased the number of several associated pathogens as well as a the levels of the enzyme N-acetyl-beta-D-glucosaminidase and somatic cell count, indicative of inflammatory state of the mammary gland (Cao et al. 2007, Wu et al., 2007). Some products are currently known such as "pre-and post-milking dip and cleaning wipes commercial nisin incorporated in its formulation as an active ingredient.

In some cases it is possible to combine the action of two antimicrobial substances. For example the use of lysostaphin considered also as bacteriocin (Heng et al., 2007) can complement the action of nisin and the inhibitory action of the mixture of these antibiotics is synergistic (Oldham and Daley, 1991).Studies using genetically engineered mice and transgenic cows secreting lysostaphin in their mammary glands demonstrated both animals were resistant to infections caused by staphylocci (Kerr et al 2001; Wall et al, 2005).

Other lantibiotic that has received much of attention is Lacticin 3147. This bacteriocin is a two-component lantibiotic with a wide spectrum of antimicrobial activity, particularly for some of mastitis pathogens such as *S aureus, Strep agalactiae, Strep uberis*. Twomey et al (2000) evaluated the effectiveness of a treatment of *S.aureus* infection (teat, teat duct and teat sinus) by using a teat-seal based on bismuth salts and lacticin 3147.

The results of this study showed the significant potential of this bacteriocin to prevent infection by staphylococci, especially when the bacteriocin concentration used was relatively high. An interesting fact is that pure cultures of lantibiotics producer strains were as effective as antibiotics for intramammary treatment of diseases (Crispie et al. 2008; Klostermann et al. 2008).

Recently, Klostermann et al. (2010) developed "a natural teat dip" from a fermented product with lacticin 3147-producing strain (*L. lactis* DPC 3251). *In vitro* experiments showed an inhibition of mastitis-producing bacteria, being more important in *S. aureus, Strept dysgalactiae* strains and to a lesser extent on *Strept. uberis.In vivo* studies used teats of lactating dairy cow inoculated with the "challenge organisms" and later treated and washed with fermented products containing bacteriocins-producer strains and isogenic non-bacteriocinogenic strains.

The results showed that after 10 min exposure there was a reduction of 80-97% pathogens (*S. aureus, S. dysgalactiae and S. uberis).* The clinical trials published to assay some of these bacteriocins or bioactive compounds, are showed in Table 3, where some complementary information is included, as the characteristics of the peptide, the description of the animal trials, the results obtained and the reference of the publication. As some of these bioactive peptides are in the market, they were included in Table 4, where additional information is shown.

Table 3. Microorganisms and bacteriocins evaluated for prevention or treatment of bovine mastitis

Micro organism	Origin	Effects	*In vivo* study	Results	Reference
Lactobacillus acidophilus and *Lactobacillus casei,*	Lacto-bac, a commercial probiotic approved for oral use in calves as a microbial supplement	Intramammary infusion of quarters with *Lactobacillus* increases quarter SCC with no effect on infection rate	Twenty-six cows with subclinical mastitis	Treatment of cows with *Lactobacillus* cured 21.7% of infected quarters, whereas 73.7% of infections treated with antibiotic were eliminated. Treatment of quarters with antibiotic did not reduce quarter SCC unless infected quarters were cured. Intramammary infusion of quarters with *Lactobacillus* increased quarter SCC.	Greene et al., 1991
Lactococcus lactis DPC3147	Irish kefir grain	Probiotic effective in the treatment of subclinical and clinical bovine mastitis	Twenty-two quarters with subclinical mastitis and fifty quarters from Holstein-Friesian cows, New Zealand Friesians, Norwegian Reds, Normandes and Montbelliards.	Intramammary infusion of a live culture of *L. lactis* DPC3147 is as efficient as common antibiotic treatments	Klostermann et al., 2008
Lactococcus lactis DPC3147	Irish kefir grain	Modulation of the immune response	Six healthy dairy cows. Holstein Friesian in her sixth lactation and a Norwegian Red in her second lactation	Administration of a live cultureof *L. lactis* DPC 3147 leads to a rapid and considerable increase of *IL-1b* and *IL-8*gene expression	Beecher et al., 2009
Nisin A	Ambicin N®	Germicidal activity against mastitis pathogens (*S. aureus, Strep. agalactiae, Strept uberis , Klebsiella pneumoniae and E. coli)*	Eighty three quarters	Reduction of *S. aureus* 3.90 log and *E. coli* 4.22 log after exposure for 1 min to the germicide. Activity was significantly greater than that exhibited by the 0.1 and 0.5% iodophors and by the 0.5% chlorhexidine digluconate teat dips.	Sears et al., 1992

Table 3. (Continued)

Bacteriocin	Origin	Effects	*In vivo* study	Results	Reference
Nisin A	Ambicin N®	Germicidal activity against mastitis pathogens (*S. aureus, Strep. agalactiae, Strept uberis , Klebsiella pneumoniae and E. coli)*	Eighty three quarters	Reduction of *S. aureus* 3.90 log and *E. coli* 4.22 log after exposure for 1 min to the germicide. Activity was significantly greater than that exhibited by the 0.1 and 0.5% iodophors and by the 0.5% chlorhexidine digluconate teat dips.	Sears et al., 1992
Nisin Z	Nisin Z Silver-Elephant®	Germicidal activity against mastitis pathogens in clinical and subclinical mastitis	Ninety Holstein cows with subclinical mastitis.	Nisin therapy had bacteriological cure rates of 90.1% for *Strept agalactiae*, 50% for *S. aureus*, 58.8% for CNS in the subclinical mastitis trial. Nisin presented similar results to those with antibiotic treatments.	Wu et al., 2007; Cao et al., 2007
			Ninety two cows and one hundred and seven quarters with clinically mastitis		
Lacticin 3147	Lacticin 3147 produced by *L. lactis* DPC3147 isolated from a Irish kefir grain	Germicidal activity of bismuth and Lacticin 3147-based teat seal in dry cows challenged with*S aureus.*	Ninty-nine quarters	Teat seal plus lacticin 3147 reduced the number of *S.aureus*	Twomey et al., 2000

Bacteriocin	Origin	Effects	*In vivo* study	Results	Reference
Lacticin 3147	Lacticin 3147 and *L. lactis* DPC3251 isolated from a Irish kefir grain	Natural teat dip using a fermentate containing the live bacterium*L. lactis* DPC 3251	Eighteen teats of eight lactating Holstein-Friesian cows	After 10 min exposure with teat dip, staphylococci were reduced by 80% when compared with the un-dipped control teat. Streptococci were reduced between 90-97%.	Klostermann et al., 2010
Lacticin NK34	Partially purified form of lacticin NK34	*In vivo* preventive and therapeutic effects on mouse infection model using mastitis pathogens	A hundred ICR mice were used to evaluate the antimicrobial effects of lacticin NK34	Mice infected with *S.aureus* and CNS isolated from bovine mastitis and treated with lacticin NK34 demonstrate an 80% survival rate compared with a 7.5% in control mice.	Kim et al., 2010
Lysostaphin	Recombinant lysostaphin	Therapeutic effect against *S. aureus* by intramammary application	Thirty Holstein-Freisian dairy cattle in their first lactation were infected with *S. aureus.*	The cure rate of quarters receiving recombinant lysostaphin was 20% compared with 29% for sodium cephapirin and 57% for a commercial antibiotic formulation	Oldham and Daley, 1991
Consept®	Nisin A	Germicidal activity against mastitis pathogens (*S. aureus, Strep. agalactiae, Strept uberis , Klebsiella pneumoniae and E. coli)*	Eighty three quarters	Reduction of *S. aureus* 3.90 log and *E. coli* 4.22 log after exposure for 1 min to the germicide. Activity was significantly greater than that exhibited by the 0.1 and 0.5% iodophors and by the 0.5% chlorhexidine digluconate teat dips.	Sears et al., 1992

Table 3.(Continued)

Product	Bacteriocin	Application	Clinical study	Results	Reference
Mast out (Immucell, EEUU)	Nisin A	Intramammary infusion product containing Nisin for the treatment of subclinical mastitis in lactating dairy cows	Three hundred cows with subclinical mastitis	Mast Out® treatment group showed a statistically highly significant ($p < 0.0001$) overall cure rate in comparison to the placebo group	http://www.immucell.com/pdf/mastout.pdf
Wipe Out (Immucell, EEUU)	Nisin A	Dairy Wipes made from a non-woven material that has been soaked through with a quick-acting and effective antimicrobial solution containing Nisin	Eighty three quarters	Kills 99% of the most common mastitis-causing organisms	Sears et al., 1992 http://www.immucell.com/pdf/WIPEOUTResults.E%20Chartv6.pdf

Table 4. Examples of commercially available products containing nisin for bovine mastitis

Product	Bacteriocin	Application	Clinical study	Results	Reference
Consept®	Nisin A	Germicidal activity against mastitis pathogens (*S. aureus, Strep. agalactiae, Strept uberis , Klebsiella pneumoniae and E. coli)*	Eighty three quarters	Reduction of *S. aureus* 3.90 log and *E. coli* 4.22 log after exposure for 1 min to the germicide. Activity was significantly greater than that exhibited by the 0.1 and 0.5% iodophors and by the 0.5% chlorhexidine digluconate teat dips.	Sears et al., 1992
Mast out (Immucell, EEUU)	Nisin A	Intramammary infusion product containing Nisin for the treatment of subclinical mastitis in lactating dairy cows	Three hundred cows with subclinical mastitis	Mast Out® treatment group showed a statistically highly significant ($p < 0.0001$) overall cure rate in comparison to the placebo group	http://www.immucell.com/pdf/mastout.pdf
Wipe Out (Immucell, EEUU)	Nisin A	Dairy Wipes made from a non-woven material that has been soaked through with a quick-acting and effective antimicrobial solution containing Nisin	Eighty three quarters	Kills 99% of the most common mastitis-causing organisms	Sears et al., 1992 http://www.immucell.com/pdf/WIPEOUTResults.E%20Chartv6.pdf

Conclusion

Global milk production is increasing based in the higher level of consumption of milk and milk-derived products in many countries that supports more complex milk-production systems. At the same time there is a high requirement for functional and healthy foods, which must be obtained mainly from healthy animals. Bovine mastitis continues to be a cause of significant economic loss to the dairy industry internationally, and also, fundamental for the health related items, and is generally treated with antibiotics. Prevention strategies have decrease the incidence of some pathogens, modifying the pattern and appearance of some others, which encourage looking for different practices to decrease this syndrome. Some modified microorganisms and biological substances are being studied to help in the eradication of mastitis that will be difficult to get by the cultural, economic and geographical differences of the producer's countries. More governmental requirements should be applied to increase the safety of the food-products and animal welfare. Further studies should be performed to understand the mechanisms of action of each one of the preventive strategies, trying to get the synergistic or combined effect of them, to advance in the eradication, or at least in the decreasement of this syndrome.

Acknowledgements

The work was supported by grants from CONICET (PIP 632), ANPCyT (PICT 543).

References

Al-Qumber, M., & Tagg, J. R. (2006). Commensal bacilli inhibitory to mastitis pathogens isolated from the udder microbiota of healthy cows. *Journal of Applied Microbiology, 101*(5), 1152-1160.

Anadón, A., Rosa Martínez-Larrañaga, M., & Aranzazu Martínez, M. (2006). Probiotics for animal nutrition in the European Union. Regulation and safety assessment. *Regulatory Toxicology and Pharmacology, 45*(1), 91-95.

Azad, I. S., & Al-Marzouk, A. (2008). Autochthonous aquaculture probiotics - A critical analysis. *Research Journal of Biotechnology, 3*(SPEC. ISS.), 171-177.

Beecher, C., Daly, M., Berry, D. P., Klostermann, K., Flynn, J., Meaney, W., et al. (2009). Administration of a live culture of *Lactococcus lactis* DPC 3147 into the bovine mammary gland stimulates the local host immune response, particularly *IL-1* and *IL-8* gene expression. *Journal of Dairy Research, 76*(3), 340-348.

Boris, S., Suarez, J. E., & Barbes, C. (1997). Characterization of the aggregation promoting factor from *Lactobacillus gasseri*, a vaginal isolate. *Journal of Applied Microbiology, 83*(4), 413-420.

Bradley, A. J. (2002). Bovine Mastitis: an evolving disease. *The Veterinary Journal, 164*, 116-128.

Broadbent, J. R., Chou, Y. C., Gillies, K., & Kondo, J. K. (1989). Nisin inhibits several gram-positive, mastitis-causing pathogens. *J Dairy Sci, 72*(12), 3342-3345.

Buzzola, F. R., Barbagelata, M. S., Caccuri, R. L., & Sordelli, D. O. (2006). Attenuation and persistence of and ability to induce protective immunity to a *Staphylococcus aureus* aroA mutant in mice. *Infection and Immunity, 74*(6), 3498-3506.

Calvinho, L. F. (1999). El control de mastitis causado por *Staphylococcus aureus* a través de segregación. *Chacra & Campo Moderno, 828*, 10-11.

Calvinho, L. F., & Tirante, L. (2005). Prevalencia de microorganismos patógenos de mastitis bovina y evolución del estado de salud de la glándula mamaria en Argentina en los últimos 25 años. *FAVE - Ciencias Veterinarias, 4*(1-2), 29-40.

Calzolari, A., Giraudo, J. A., Rampone, H., Odierno, L., Giraudo, A. T., Frigerio, C., et al. (1997). Field trials of a vaccine against bovine mastitis. 2. Evaluation in two commercial dairy herds. *Journal of Dairy Science, 80*(5), 854-858.

Cao, L. T., Wu, J. Q., Xie, F., Hu, S. H., & Mo, Y. (2007). Efficacy of nisin in treatment of clinical mastitis in lactating dairy cows. *Journal of Dairy Science, 90*(8), 3980-3985.

Carter, E. W., & Kerr, D. E. (2003). Optimization of DNA-based vaccination in cows using green fluorescent protein and protein A as a prelude to immunization against staphylococcal mastitis. *Journal of Dairy Science, 86*(4), 1177-1186.

Cesena, C., Morelli, L., Alander, M., Siljander, T., Tuomola, E., Salminen, S., et al. (2001). *Lactobacillus crispatus* and its nonaggregating mutant in human colonization trials. *Journal of Dairy Science, 84*(5), 1001-1010.

Chang, B. S., Moon, J. S., Kang, H. M., Kim, Y. I., Lee, H. K., Kim, J. D., et al. (2008). Protective effects of recombinant staphylococcal enterotoxin type C mutant vaccine against experimental bovine infection by a strain of *Staphylococcus aureus* isolated from subclinical mastitis in dairy cattle. *Vaccine, 26*(17), 2081-2091.

Collado, M. C., Meriluoto, J., & Salminen, S. (2007). Development of new probiotics by strain combinations: is it possible to improve the adhesion to Intestinal mucus? *Journal of Dairy Science, 90*(6), 2710-2716.

Cotter, P. D., Hill, C., & Ross, R. P. (2005). Bacteriocins: developing innate immunity for food. *Nat Rev Microbiol, 3*(10), 777-788.

Craven, N., & Anderson, J. C. (1984). Phagocytosis of *Staphylococcus aureus* by bovine mammary gland macrophages and intracellular protection from antibiotic action in vitro and in vivo. *Journal of Dairy Research, 51*(4), 513-523.

Crispie, F., Alonso-Gómez, M., O'Loughlin, C., Klostermann, K., Flynn, J., Arkins, S., et al. (2008). Intramammary infusion of a live culture for treatment of bovine mastitis: Effect of live lactococci on the mammary immune response. *Journal of Dairy Research, 75*(3), 374-384.

Draksler, D., Gonzáles, S., & Oliver, G. (2004). Preliminary assays for the development of a probiotic for goats. *Reproduction Nutrition Development, 44*(5), 397-405.

Eberl, H. J., Khassehkhan, H., & Demaret, L. (2010). A mixed-culture model of a probiotic biofilm control system. *Computational and Mathematical Methods in Medicine, 11*(2), 99-118.

Errecalde, J. O. (2004). *Uso de antimicrobianos en animales de consumo. Incidencia del desarrollo de resistencias en la salud pública.* (Vol. 162). Roma.

Espeche, M. C., Otero, M. C., Sesma, F., & Nader-Macías, M. E. F. (2009). Screening of surface properties and antagonistic substances production by lactic acid bacteria isolated from the mammary gland of healthy and mastitic cows. *Veterinary Microbiology, 135*(3-4), 346-357.

Ewaschuk, J. B., Naylor, J. M., Chirino-Trejo, M., & Zello, G. A. (2004). *Lactobacillus rhamnosus* strain GG is a potential probiotic for calves. *Canadian Journal of Veterinary Research, 68*(4), 249-253.

FAO. (2009). *The state of food and agriculture*. Roma, Italia: Communication Division FAO.

FAOSTAT. (2008, 1/12/2010). Top Production - Cow milk, whole, fresh - 2008. from http://faostat.fao.org/DesktopDefault.aspx?PageID=339&lang=en

FAO-WHO. (2002). Guidelines for the Evaluation of Probiotics in Food. . *Report of a Joint FAO/WHO Working Group on Drafting Guidelines for the Evaluation of Probiotics in Food.*, from http://www.who.int/foodsafety/fs_management/en/probiotic_guidelines.pdf

Finch, J. M., Winter, A., Walton, A. W., & Leigh, J. A. (1997). Further studies on the efficacy of a live vaccine against mastitis caused by *Streptococcus uberis*. *Vaccine, 15*(10), 1138-1143.

Fontaine, M. C., Perez-Casal, J., Song, X. M., Shelford, J., Willson, P. J., & Potter, A. A. (2002). Immunisation of dairy cattle with recombinant *Streptococcus uberis* GapC or a chimeric CAMP antigen confers protection against heterologous bacterial challenge. *Vaccine, 20*(17-18), 2278-2286.

Giraudo, J. A., Calzolari, A., Rampone, H., Rampone, A., Giraudo, A. T., Bogni, C., et al. (1997). Field trials of a vaccine against bovine mastitis. 1. Evaluation in heifers. *Journal of Dairy Science, 80*(5), 845-853.

Golek, P., Bednarski, W., Brzozowski, B., & Dziuba, B. (2009). The obtaining and properties of biosurfactants synthesized by bacteria of the genus*Lactobacillus*. *Annals of Microbiology, 59*(1), 119-126.

González, R. (1993). Evaluación de técnicas y procedimientos utilizados en el diagnóstico, prevención y control de la mastitis bovina. In E. G. Pergamino (Ed.), *II Congreso Nacional de Lechería* (pp. 66). Venado Tuerto, Argentina: Estudio Ganadero Pergamino.

Greene, W. A., Gano, A. M., Smith, K. L., Hogan, J. S., & Todhunter, D. A. (1991). Comparison of probiotic and antibiotic intramammary therapy of cattle with elevated somatic cell counts. *Journal of Dairy Science, 74*(9), 2976-2981.

Gudiña, E. J., Rocha, V., Teixeira, J. A., & Rodrigues, L. R. (2010). Antimicrobial and antiadhesive properties of a biosurfactant isolated from *Lactobacillus paracasei* ssp. *paracasei* A20. *Letters in Applied Microbiology, 50*(4), 419-424.

Gueimonde, M., & Salminen, S. (2006). New methods for selecting and evaluating probiotics. *Digestive and Liver Disease, 38*(SUPPL. 2).

Guillot, J. F. (2003). Probiotic feed additives. *Journal of Veterinary Pharmacology and Therapeutics, 26*(1), 52-55.

Halasa, T., Huijps, K., Osterás, O., & Hogeveen, H. (2007). Economic effects of bovine mastitis and mastitis management: A review. *Veterinary Quarterly, 29*(1), 18-31.

Halasa, T., Nielen, M., Huirne, R. B. M., & Hogeveen, H. (2009). Stochastic bio-economic model of bovine intramammary infection. *Livestock Science, 124*(1-3), 295-305.

Hansen, P. J., Soto, P., & Natzke, R. P. (2004). Mastitis and fertility in cattle - Possible involvement of inflammation or immune activation in embryonic mortality. *American Journal of Reproductive Immunology, 51*(4), 294-301.

Haug, A., Høstmark, A. T., & Harstad, O. M. (2007). Bovine milk in human nutrition - A review. *Lipids in Health and Disease, 6.*

Heng, N. C., Wescombe, P. A., Burton, J. P., Jack, R. W., & Tagg, J. R. (2007). The Diversity of Bacteriocins in Gram-Positive Bacteria. In C. Eckey (Ed.), *Bacteriocins. Ecology and Evolution*. Berlin: Springer.

Herstad, H. K., Nesheim, B. B., L'Abée-Lund, T., Larsen, S., & Skancke, E. (2010). Effects of a probiotic intervention in acute canine gastroenteritis - A controlled clinical trial. *Journal of Small Animal Practice, 51*(1), 34-38.

Hogan, J. S., Bogacz, V. L., Aslam, M., & Smith, K. L. (1999). Efficacy of an *Escherichia coli* J5 bacterin administered to primigravid heifers. *Journal of Dairy Science, 82*(5), 939-943.

Isolauri, E., Salminen, S., & Ouwehand, A. (2004). Probiotics. *Best Practice & Research Clinical Gastroenterology, 18*(2), 299-313.

Jack, R. W., Tagg, J. R., & Ray, B. (1995). Bacteriocins of gram-positive bacteria. *Microbiological Reviews, 59*(2), 171-200.

Kerr, D. E., Plaut, K., Bramley, A. J., Williamson, C. M., Lax, A. J., Moore, K., et al. (2001). Lysostaphin expression in mammary glands confers protection against staphylococcal infection in transgenic mice. *Nat Biotechnol, 19*(1), 66-70.

Kim, S. Y., Shin, S., Koo, H. C., Youn, J. H., Paik, H. D., & Park, Y. H. (2010). In vitro antimicrobial effect and in vivo preventive and therapeutic effects of partially purified lantibiotic lacticin NK34 against infection by Staphylococcus species isolated from bovine mastitis. *J Dairy Sci, 93*(8), 3610-3615.

Klaenhammer, T. R. (1993). Genetics of bacteriocins produced by lactic acid bacteria. *FEMS Microbiol Rev, 12*(1-3), 39-85.

Klostermann, K., Crispie, F., Flynn, J., Meaney, W. J., Paul Ross, R., & Hill, C. (2010). Efficacy of a teat dip containing the bacteriocin lacticin 3147 to eliminate Gram-positive pathogens associated with bovine mastitis. *Journal of Dairy Research, 77*(2), 231-238.

Klostermann, K., Crispie, F., Flynn, J., Ross, R. P., Hill, C., & Meaney, W. (2008). Intramammary infusion of a live culture of *Lactococcus lactis* for treatment of bovine mastitis: Comparison with antibiotic treatment in field trials. *Journal of Dairy Research, 75*(3), 365-373.

Kubota, H., Senda, S., Nomura, N., Tokuda, H., & Uchiyama, H. (2008). Biofilm formation by lactic acid bacteria and resistance to environmental stress. *Journal of Bioscience and Bioengineering, 106*(4), 381-386.

Lee, J. W., O'Brien, C. N., Guidry, A. J., Paape, M. J., Shafer-Weaver, K. A., & Zhao, X. (2005). Effect of a trivalent vaccine against Staphylococcus aureus mastitis lymphocyte subpopulations, antibody production, and neutrophil phagocytosis. *Canadian Journal of Veterinary Research, 69*(1), 11-18.

Leigh, J. A., Finch, J. M., Field, T. R., Real, N. C., Winter, A., Walton, A. W., et al. (1999). Vaccination with the plasminogen activator from *Streptococcus uberis* induces an inhibitory response and protects against experimental infection in the dairy cow. *Vaccine, 17*(7-8), 851-857.

Leitner, G., Lubashevsky, E., Glickman, A., Winkler, M., Saran, A., & Trainin, Z. (2003). Development of a Staphylococcus aureus vaccine against mastitis in dairy cows: I. Challenge trials. *Veterinary Immunology and Immunopathology, 93*(1-2), 31-38.

Lescourret, F., & Coulon, J. B. (1994). Modeling the impact of mastitis on milk production by dairy cows. *Journal of Dairy Science, 77*(8), 2289-2301.

Lönnerdal, B. (2003). Nutritional and physiologic significance of human milk proteins. *The American journal of clinical nutrition, 77*(6), 1537S-1543S.

McDougall, S., Parker, K. I., Heuer, C., & Compton, C. W. R. (2009). A review of prevention and control of heifer mastitis via non-antibiotic strategies. *Veterinary Microbiology, 134*(1-2), 177-185.

Miles, H., Lesser, W., & Sears, P. (1992). The economic implications of bioengineered mastitis control. *Journal of Dairy Science, 75*(2), 596-605.

Nader-Macías, M. E. F., Otero, M. C., Espeche, M. C., & Maldonado, N. C. (2008). Advances in the design of probiotic products for the prevention of major diseases in dairy cattle. *Journal of Industrial Microbiology and Biotechnology, 35*(11), 1387-1395.

Nava-Trujillo, H., Soto-Belloso, E., & Hoet, A. E. (2010). Effects of clinical mastitis from calving to first service on reproductive performance in dual-purpose cows. *Animal Reproduction Science, 121*(1-2), 12-16.

Nickerson, S. C. (1993). *Vaccination programs for preventing and controlling mastitis.* Paper presented at the National Mastitis Council Regional Meeting Proceeding Syracuse, New York.

Nissen-Meyer, J., Rogne, P., Oppegard, C., Haugen, H. S., & Kristiansen, P. E. (2009). Structure-function relationships of the non-lanthionine-containing peptide (class II) bacteriocins produced by gram-positive bacteria. *Curr Pharm Biotechnol, 10*(1), 19-37.

NMC. (2004). *Current Concepts of Bovine Mastitis* (4th ed.). Arlington, VA: National Mastitis Council.

Nordhaug, M. L., Nesse, L. L., Norcross, N. L., & Gudding, R. (1994). A field trial with an experimental vaccine against *Staphylococcus aureus* mastitis in cattle. 2. Antibody response. *Journal of Dairy Science, 77*(5), 1276-1284.

Nour El-Din, A. N. M., Shkreta, L., Talbot, B. G., Diarra, M. S., & Lacasse, P. (2006). DNA immunization of dairy cows with the clumping factor a of *Staphylococcus aureus*. *Vaccine, 24*(12), 1997-2006.

O'Brien, C. N., Guidry, A. J., Douglass, L. W., & Westhoff, D. C. (2001). Immunization with *Staphylococcus aureus* lysate incorporated into microspheres. *Journal of Dairy Science, 84*(8), 1791-1799.

O'Brien, C. N., Guidry, A. J., Fattom, A., Shepherd, S., Douglass, L. W., & Westhoff, D. C. (2000). Production of antibodies to *Staphylococcus aureus* serotypes 5, 8, and 336 using poly(DL-lactide-co-glycolide) microspheres. *Journal of Dairy Science, 83*(8), 1758-1766.

Odierno, L., Calvinho, L., Traverssa, P., Lasagno, M., Bogni, C., & Reinoso, E. (2006). Conventional identification of *Streptococcus uberis* isolated from bovine mastitis in Argentinean dairy herds. *Journal of Dairy Science, 89*(10), 3886-3890.

Oelschlaeger, T. A. (2010). Mechanisms of probiotic actions - A review. *International Journal of Medical Microbiology, 300*(1), 57-62.

O'Grady, L., & Doherty, M. (2009). Focus on bovine mastitis: knowledge into practice. *Irish Veterinary Journal, 62*, 4-5.

Oldham, E. R., & Daley, M. J. (1991). Lysostaphin: Use of a recombinant bactericidal enzyme as a mastitis therapeutic. *Journal of Dairy Science, 74*(12), 4175-4182.

Otero, C., Silva De Ruiz, C., Ibañez, R., Wilde, O. R., De Ruiz Holgado, A. A. P., & Nader-Macías, M. E. (1999). Lactobacilli and enterococci isolated from the bovine vagina during the estrous cycle. *Anaerobe, 5*(3-4), 305-307.

Otero, M. C., Morelli, L., & Nader-Macías, M. E. (2006). Probiotic properties of vaginal lactic acid bacteria to prevent metritis in cattle. *Letters in Applied Microbiology, 43*(1), 91-97.

Ouwehand, A., Salminen, S., & Isolauri, E. (2002). Probiotics: an overview of beneficial effects. *Antonie van Leeuwenhoek, 82*, 279-289.

Parvez, S., Malik, K. A., Ah Kang, S., & Kim, H. Y. (2006). Probiotics and their fermented food products are beneficial for health. *Journal of Applied Microbiology, 100*(6), 1171-1185.

Pellegrino, M., Giraudo, J., Raspanti, C., Odierno, L., & Bogni, C. (2010). Efficacy of immunization against bovine mastitis using a *Staphylococcus aureus* avirulent mutant vaccine. *Vaccine, 28*(28), 4523-4528.

Prenafeta, A., March, R., Foix, A., Casals, I., & Costa, L. (2010). Study of the humoral immunological response after vaccination with a *Staphylococcus aureus* biofilm-embedded bacterin in dairy cows: possible role of the exopolysaccharide specific antibody production in the protection from *Staphylococcus aureus* induced mastitis. *Veterinary Immunology and Immunopathology, 134*(3-4), 208-217.

Pyörälä, S. (2009). Treatment of mastitis during lactation. *Irish Veterinary Journal, 62*(40), 40-44.

Reinoso, E., Bettera, S., Frigerio, C., DiRenzo, M., Calzolari, A., & Bogni, C. (2004). RAPD-PCR analysis of *Staphylococcus aureus* strains isolated from bovine and human hosts. *Microbiological Research, 159*(3), 245-255.

Reinoso, E. B., El-Sayed, A., Lämmler, C., Bogni, C., & Zschöck, M. (2008). Genotyping of *Staphylococcus aureus* isolated from humans, bovine subclinical mastitis and food samples in Argentina. *Microbiological Research, 163*(3), 314-322.

Ron, E. Z., & Rosenberg, E. (2001). Natural roles of biosurfactants. *Environmental Microbiology, 3*(4), 229-236.

Ryan, M. P., Meaney, W. J., Ross, R. P., & Hill, C. (1998). Evaluation of lacticin 3147 and a teat seal containing this bacteriocin for inhibition of mastitis pathogens. *Applied and Environmental Microbiology, 64*(6), 2287-2290.

Ryan, M. P., Rea, M. C., Hill, C., & Ross, R. P. (1996). An application in cheddar cheese manufacture for a strain of *Lactococcus lactis* producing a novel broad-spectrum bacteriocin, lacticin 3147. *Applied and Environmental Microbiology, 62*(2), 612-619.

Saarela, M., Mogensen, G., Fondén, R., Mättö, J., & Mattila-Sandholm, T. (2000). Probiotic bacteria: Safety, functional and technological properties. *Journal of Biotechnology, 84*(3), 197-215.

SABRE. (2006). A European Integrated Research Project on Cutting Edge Genomics for Sustainable Animal Breeding. http://www.sabre-eu.eu/Portals/0/SABRE%20Publications/SABRE_Brochure_11p.pdf.

Salminen, S., & Von Wright, A. (1998). *Lactic acid bacteria. Microbiology and functional aspects, second edition. Revised and expanded.* New York: Marcel Dekker.

Salminen, S., von Wright, A., & Ouwehand, A. (2004). *Lactic acid bacteria. Microbiological and functional aspects. Third edition, revised and expanded.* New York: Marcel Dekker, Inc.

Schachtsiek, M., Hammes, W. P., & Hertel, C. (2004). Characterization of *Lactobacillus coryniformis* DSM 20001T surface protein Cpf mediating coaggregation with and aggregation among pathogens. *Applied and Environmental Microbiology, 70*(12), 7078-7085.

Schiffrin, E. J., Brassart, D., Servin, A. L., Rochat, F., & Donnet-Hughes, A. (1997). Immune modulation of blood leukocytes in humans by lactic acid bacteria: Criteria for strain selection. *American Journal of Clinical Nutrition, 66*(2).

Sears, P. M., & McCarthy, K. K. (2003). Management and treatment of staphylococcal mastitis. *Veterinary Clinics of North America - Food Animal Practice, 19*(1), 171-185.

Sears, P. M., Norcross, N. L., Kenny, K., Smith, B., González, R. N., & Romano, M. N. (1990). *Resistance to Staphylococcus aureus infections in staphylococcal vaccinated heifers.* Paper presented at the Proceedings of the International Symposium on Bovine Mastitis, Indianapolis, IN.

Sears, P. M., Smith, B. S., Stewart, W. K., Gonzalez, R. N., Rubino, S. D., Gusik, S. A., et al. (1992). Evaluation of a nisin-based germicidal formulation on teat skin of live cows. *Journal of Dairy Science, 75*(11), 3185-3190.

Seegers, H., Fourichon, C., & Beaudeau, F. (2003). Production effects related to mastitis and mastitis economics in dairy cattle herds. *Veterinary Research, 34*(5), 475-491.

Shkreta, L., Talbot, B. G., Diarra, M. S., & Lacasse, P. (2004). Immune responses to a DNA/protein vaccination strategy against *Staphylococcus aureus* induced mastitis in dairy cows. *Vaccine, 23*(1), 114-126.

Simpson, K. W., Rishniw, M., Bellosa, M., Liotta, J., Lucio, A., Baumgart, M., et al. (2009). Influence of *Enterococcus faecium* SF68 probiotic on giardiasis in dogs. *Journal of Veterinary Internal Medicine, 23*(3), 476-481.

Smith, K. L., & Hogan, J. S. (1995). *Epidemiology of mastitis.* Paper presented at the Proceeding of the 3rd International Mastitis Seminar Tel Aviv, Israel.

Taponen, S., Koort, J., Björkroth, J., Saloniemi, H., & Pyörälä, S. (2007). Bovine intramammary infections caused by coagulase-negative staphylococci may persist throughout lactation according to amplified fragment length polymorphism-based analysis. *Journal of Dairy Science, 90*(7), 3301-3307.

Taranto, M. P., Medici, M., Perdigon, G., Ruiz Holgado, A. P., & Font de Valdez, G. (2000). Effect of *Lactobacillus reuteri* on the prevention of hypercholesterolemia in mice. *Journal of Dairy Science, 83*(3), 401-403.

Taranto, M. P., Medici, M., Perdigon, G., Ruiz Holgado, A. P., & Valdez, G. F. (1998). Evidence for hypocholesterolemic effect of *Lactobacillus reuteri* in hypercholesterolemic mice. *Journal of Dairy Science, 81*(9), 2336-2340.

Taylor, J. I., Hirsch, A., & Mattick, T. R. (1949). The treatment of bovine streptococcal and staphylococcal mastitis with nisin. *The Veterinary Records, 61*, 197-198.

Tenhagen, B. A., Edinger, D., Baumgärtner, B., Kalbe, P., Klünder, G., & Heuwieser, W. (2001). Efficacy of a herd-specific vaccine against *Staphylococcus aureus* to prevent post-partum mastitis in dairy heifers. *Journal of Veterinary Medicine Series A: Physiology Pathology Clinical Medicine, 48*(10), 601-607.

Tollersrud, T., Zernichow, L., Andersen, S. R., Kenny, K., & Lund, A. (2001). Staphylococcus aureus capsular polysaccharide type 5 conjugate and whole cell vaccines stimulate antibody responses in cattle. *Vaccine, 19*(28-29), 3896-3903.

Twomey, D. P., Wheelock, A. I., Flynn, J., Meaney, W. J., Hill, C., & Ross, R. P. (2000). Protection against *Staphylococcus aureus* mastitis in dairy cows using a bismuth-based teat seal containing the bacteriocin, Lacticin 3147. *Journal of Dairy Science, 83*(9), 1981-1988.

Van Loosdrecht, M. C. M., Lyklema, J., Norde, W., & Zehnder, A. J. B. (1990). Influence of interfaces on microbial activity. *Microbiological Reviews, 54*(1), 75-87.

Vangroenweghe, F., Lamote, I., & Burvenich, C. (2005). Physiology of the periparturient period and its relation to severity of clinical mastitis. *Domestic Animal Endocrinology, 29*(2), 283-293.

Vesterlund, S., Karp, M., Salminen, S., & Ouwehand, A. C. (2006). *Staphylococcus aureus* adheres to human intestinal mucus but can be displaced by certain lactic acid bacteria. *Microbiology, 152*(6), 1819-1826.

Vesterlund, S., Paltta, J., Karp, M., & Ouwehand, A. C. (2005). Measurement of bacterial adhesion-in vitro evaluation of different methods. *Journal of Microbiological Methods, 60*(2), 225-233.

Wall, R. J., Powell, A. M., Paape, M. J., Kerr, D. E., Bannerman, D. D., Pursel, V. G., et al. (2005). Genetically enhanced cows resist intramammary Staphylococcus aureus infection. *Nature Biotechnology, 23*(4), 445-451.

Weese, J. S., & Rousseau, J. (2005). Evaluation of *Lactobacillus pentosus* WE7 for prevention of diarrhea in neonatal foals. *Journal of the American Veterinary Medical Association, 226*(12), 2031-2034.

Woodward, W. D., Besser, T. E., Ward, A. C., & Corbeil, L. B. (1987). In vitro growth inhibition of mastitis pathogens by bovine teat skin normal flora. *Canadian journal of veterinary research, 51*(1), 27-31.

Wu, J., Hu, S., & Cao, L. (2007). Therapeutic effect of nisin Z on subclinical mastitis in lactating cows. *Antimicrobial Agents and Chemotherapy, 51*(9), 3131-3135.

Yancey, R. J., Sanchez, M. S., & Ford, C. W. (1991). Activity of antibiotics against *Staphylococcus aureus* within polymorphonuclear neutrophils. *European Journal of Clinical Microbiology and Infectious Diseases, 10*(2), 107-113.

Zouhir, A., Hammami, R., Fliss, I., & Hamida, J. B. (2010). A new structure-based classification of gram-positive bacteriocins. *Protein J, 29*(6), 432-439.

In: Milk Production
Editor: Boulbaba Rekik, pp. 159-170
ISBN 978-1-62100-061-7

Chapter VIII

Milk Production of Holsteins under Mediterranean Conditions: Case of the Tunisian Population

A. Ben Gara[1*], B. Jemmali[1], H. Hammami[2], H. Rouissi[1], M. Bouallegue[3] and B. Rekik[1]

[1]Ecole Supérieure d'Agriculture de Mateur, 7030, Mateur, Université Carthage, Tunisia.
[2]Animal Science Unit, Gembloux Agricultural University, B-5030 Gembloux, Belgium.
[3]Faculté de Sciences de Tunis, Tunisia

Abstract

Continued selection for improved milk production has resulted in sizeable genetic gains in Holsteins within countries with leading dairy industries, e.g., North America and some European countries. Recognition of the Holstein as the top yielding breed led to the spread of its genes worldwide. The main breeding objective in Tunisia has long been milk yield and breeding strategies to improve milk production were based on semen and heifer imports. Actually, Holsteins account around 223000 cows of which only 10% are officially enrolled in the national milk recording system that dates back to the early 1970's. Studies using data recorded on the Tunisian Holstein population focused on milk production, reproductive performances, and genetic parameters. A cow spends roughly 3.22 years in the herd. Average national production level from all completed records was around 6220 kg of milk over a 305-d lactation period. Maximum yields were reached during third lactation where mean test- day yields were 20.7, 0.70, and 0.65 kg for milk, fat, and protein yields, respectively. Mean test-day somatic cell score (mean = 3.8 over the first three lactations) was relatively high and fertility measures were below satisfactory levels. In addition to annual and seasonal variations originating mainly from climatic changes, Holstein performances varied with herd ownership (state, cooperative or farmers' herds) and the region (from the North to the Centre of the country) that are determinant of the management level and production system (range of 305-d mean milk

* Email : rekik.boulbaba@iresa.agrinet.tn

yield by management level: 4,623 to 7,375 kg). Genetic parameter estimates mirrored management levels and were overall moderate in the range of those reported in the literature for populations in countries with emerging dairy industries. The lack of a stable genetic improvement policy coupled with a national routine genetic evaluation remains the main weakness limiting Holstein performances in the country.

Keywords: Holstein, milk production, management system, Tunisian population.

Introduction

Important developments in the history of agriculture are specialized dairy breeds and technological advances in the fields of milk processing, ration formulation, and feed conditioning. Milk is a highly nutritious food, a perfect complement to cereal based diets. The world dairy market is constantly growing and evolving. Every year, global production consistently increases on average to fulfill new needs and requirements. The purpose of dairy breeding programs is to produce a cow that, through her genetic makeup and managed under optimal conditions would maximize net income for the farmer. It was consistently shown that improving the cow's genetic merit, in addition to nutrition, health and proper management, are vital for improved productivity. European dairy cattle breeds and those from North American origins are of great merit under suitable environments where feed are readily available in quality and quantity and where diseases are efficiently challenged. In developing countries, the performances of high-yielding breeds imported from regions with developed dairy industries are unfavorably affected by genotype x environment interactions (Bondoc et al., 1989; Ojango et al., 2005; Hammami et al., 2008). The Holstein-Friesian is the most widely used exotic dairy breed in all farming in the tropics. The breed is popular for its potentially high milk producing ability and has attractive capabilities for countries where milk supply is not yet able to meet demands of growing populations (Hammami et al., 2007).

Tunisia is characterized by a Mediterranean climate showing irregularities within and between years. The summer season is hot and dry while the winter season is temperate and rainy in the North of the country but arid elsewhere. Actually, hot temperatures and dry periods are dominating during 2 thirds of the whole year. Rain fall varies with the latitude to divide the country in various bioclimatic stages, i.e., ecosystems. Forages are cultivated in the north and in irrigated few areas in the center and south of the country. Dairy farms, relatively in large herds, are essentially integrated in large agriculture enterprises located in the North. Although limited, roughages are distributed in green, silage and hay while in the center and the south of the country, off land management of small herds of dairy cows is ruling. Rations of farm animals are made of by products and, hey from the north and concentrates.

The Tunisian cow population size reached 454000 cows in recent years. Around 223000 of these cows are Holsteins. Tunisia imported purebred pregnant Friesian heifers from the Netherlands in 1970 (Djemali and Berger, 1992). Holsteins were then imported from Canada, the United States, and some European countries (Hammami et al., 2007; Rekik et al., 2009). Breeding strategies to improve milk production, based on the import of purebred Holstein heifers and semen, have been implemented as it is the case in many developed and developing countries over the last 40 years (Hammami et al., 2007; Rekik et al.; 2009).In Tunisia, the focus has long been on increased yield by means of an intra herd index used to select cows in addition to semen imports. Gene origins of the cattle population and its genetic links with other populations were investigated by Hammami et al. (2007).

Various factors affect milk production, in addition to the breed, fixed environmental factors and management practices such as the calving season, farm operation and parity (Rekik et al., 2003; Rekik and Ben Gara, 2004; Ben Gara et al., 2006; Hammami et al., 2008; Hammami et al., 2008b; Hammami et al., 2009; Hammami et al., 2009b). There are 3 distinct production sectors in Tunisia (Rekik et al., 2003), the state herds, the cooperative herds, and farmers' herds. These herds differ with respect to management and milk production levels. Artificial insemination is the usual type of breeding. Rations for cows are made of oat hey and silage and occasionally green forage feedings (Egyptian Glover, alfalfa, and sorghum) in some farms. The objective of this chapter was to assess performances of the Tunisian Holstein population by summarizing results obtained with different approaches and analysis methods of data recorded on cows enrolled in the Tunisian recording system taking into account specificities of various production environments within the country.

Overview of Worldwide Milk Production

Milk is as ancient as mankind itself. All species of mammals, from man to whales, produce milk for this purpose. Many centuries ago, perhaps as early as 6000-8000 BC, ancient man learned to domesticate species of animals for the provision of milk to be consumed by them. These included cows (genus*Bos*), buffaloes, sheep, goats, and camels, all of which are still used in various parts of the world for the production of milk for human consumption. The role of milk in the traditional diet varies with regions of the world. The tropical countries have not been traditional milk consumers, whereas the more northern regions of the world, Europe (especially Scandinavia) and North America, have traditionally consumed far more milk and milk products in their diet. In tropical countries where temperatures are dominating and refrigeration capabilities are limited, there are difficulties to store fresh milk, and milk has traditionally been preserved through means other than refrigeration, including immediate consumption of warm milk after milking, by boiling milk, or by processing milk into more stable products such as fermented milks. Fermented products such as cheeses were discovered by accident, but their history has also been documented for many centuries, as has the production of concentrated milks, butter, and ice cream. Technological advances have only come about very recently in the history of milk consumption, and our generations will be the ones credited for having turned milk processing from an art to a science. The availability and distribution of milk and milk products today in the modern world is a blend of the centuries old knowledge.

Total milk consumption (as fluid milk and processed products) per person varies widely from highs in Europe and North America to lows in Asia and Africa. However, as the various regions of the world become more integrated through travel and migration, these trends are changing, a factor which needs to be considered by product developers and marketers of milk and milk products. Even within regions such as Europe, the custom of milk consumption has varied greatly. Consider for example the high consumption of fluid milk in countries like Finland, Norway and Sweden compared to France and Italy where cheeses have tended to dominate milk consumption. When we consider the climates of temperate regions, it would appear that the culture of producing more stable products (cheese) in hotter climates as a means of preservation is evident.

Continued selection for improved milk production has resulted in sizeable genetic gains in countries with leading dairy industries, e.g., North America and some European countries. Table 1 reports changing yield and composition of milk over the last two decades in Canada, by breed. The 2009 average actual production for the U.S. Holstein herds enrolled in dairy herd improvement program were 10500 kg of milk, 382 pounds of fat and 323 kg of protein per lactation. Top producing Holsteins milked three times a day produced up to 33000 kg of milk in 365 days. In Luxembourg, production level reached 7946 kg for recorded cows over completed lactations (Hammami et al., 2008) and was around 6220 kg in Tunisia in the same study. In table 2 are given total milks produced in selected countries from over the world.

Milk Production in Tunisia

Production Environments

Decision makers in dairy herds vary in Tunisia. Most of herds are detained by 4 sectors, the state, groups of investors, cooperative, and farmers' herds (Rekik et al., 2003). Most of herds are managed in small sized farms. Surfaces of cultivable lands in these farms are partitioned as follows: 51% are less than 5 ha, 21% less between 5 and 10 ha, 23% between 10 and 50 ha, 2.5% between 50 and 100 ha, and 1.5% greater than 100 ha. State and groups of investors' herds are usually large (>= 200 cows per herd) while farmers' herds are small sized. Fragmented lands with climate conditions that are varied from the north to the south of the country and marked by irregularities within and between years impact feed resources, essentially roughages. We then may distinguish different production systems and management levels ranging from off land low input, in the center and the south of the country, to few high input (found in the north of the country and are comparable to systems found in temperate regions) (Rekik et al., 2003; Hammami et al., 2009). There is a total absence of grazing for dairy cattle in the country because of essentially limited rain falls, average annual precipitations is around 422 mm and the maximum is recorded in the extreme North (Ain Drahem <= 800 mm).

Milk Production Performances

Total milk production was estimated at 1006 million liters in 2004 and was projected to reach 1173 million liters in the 2007-2011 period according to the 11th development plan (Ministry of Agriculture and environment, 2011). A considerable quantity is collected by 269 collection centers all over the country from small herd holders. There are also 10 dairy stations to where large milk quantities from large herds are delivered. In addition to UHT milk, milk is transformed by 25 cheese making factories and 7 yogurt fabrication units. Traditionally fermented milk, either "Raeib" or "Leben" in addition to a smooth cooked cheese "Rigota" are heavily consumed by Tunisians in the whole country, essentially in Ramadan.

Table 1. Changes in milk yield and contents over the last two decades in Canada by breed

Year	All Breeds			Holstein			Jersey			Guernsey			Brown Swiss		
	Yield (kg)[1]	Fat (%)	Protein (%)	Yield (kg)	Fat (%)	Protein (%)	Yield (kg)	Fat (%)	Protein (%)	Yield (kg)	Fat (%)	Protein (%)	Yield (kg)	Fat (%)	Protein (%)
2009	9,592	3.77	3.22	9,793	3.76	3.19	6,371	4.87	3.81	6,812	4.56	3.43	8,128	4.05	3.48
2008	9,642	3.78	3.23	9,836	3.74	3.20	6,435	4.84	3.81	6,820	4.51	3.45	8,366	4.04	3.48
2007	9,538	3.77	3.22	9,733	3.72	3.19	6,412	4.82	3.78	6,673	4.51	3.46	8,159	4.06	3.48
2006	9,481	3.75	3.21	9,677	3.71	3.18	6,331	4.83	3.77	6,540	4.55	3.43	8,064	4.06	3.46
2005	9,422	3.76	3.21	9,624	3.71	3.19	6,279	4.85	3.77	6,398	4.50	3.43	7,792	4.12	3.48
2004	9,458	3.72	3.22	9,658	3.67	3.19	6,291	4.85	3.77	6,435	4.54	3.45	8,048	4.07	3.47
2003	9,519	3.73	3.23	9,721	3.68	3.21	6,344	4.87	3.81	6,570	4.49	3.49	8,038	4.04	3.49
2002	9,511	3.72	3.25	9,717	3.67	3.22	6,407	4.86	3.84	6,347	4.44	3.51	8,215	4.03	3.50
2001	9,242	3.72	3.24	9,440	3.68	3.22	6,186	4.87	3.83	6,015	4.45	3.51	8,020	4.02	3.50
2000	9,152	3.70	3.23	9,350	3.67	3.21	6,203	4.90	3.83	5,949	4.45	3.48	7,920	3.96	3.48
1999	8,960	3.69	3.24	9,162	3.66	3.22	6,072	4.89	3.85	5,939	4.43	3.54	7,585	4.03	3.54
1998	8,738	3.70	3.24	8,946	3.65	3.22	6,002	4.88	3.84	5,991	4.48	3.55	7,105	4.04	3.50
1997	8,427	3.72	3.24	8,697	3.68	3.22	5,753	4.90	3.86	5,919	4.53	3.55	6,818	4.05	3.52
1996	8,424	3.76	3.25	8,633	3.72	3.23	5,720	4.93	3.86	5,984	4.56	3.54	6,910	4.08	3.53
1995	8,251	3.74	3.24	8,461	3.70	3.21	5,620	4.89	3.86	5,936	4.57	3.54	6,709	4.05	3.49
1994	8,103	3.74	3.26	8,309	3.69	3.21	5,501	4.94	3.94	5,867	4.76	3.56	6,718	4.06	3.51
1993	7,988	3.75	3.25	8,193	3.71	3.21	5,408	4.94	3.88	5,826	4.58	3.57	6,639	4.07	3.52
1992	7,807	3.73	3.24	8,028	3.67	3.21	5,244	4.92	3.91	5,755	4.55	3.58	6,483	4.03	3.48
1991	7,523	3.70	3.21	7,717	3.66	3.22	4,992	4.90	3.92	5,554	4.52	3.60	6,176	4.01	3.54

Source: Canadian Dairy Information Centre, www.dairyinfo.gc.ca, 2010.

Table 2. Cow milk production (10^3 tonnes) in selected countries in the world, 2006

Country	Milk production
United States	82,462
India	39,759
China	31,934
Russia	31,100
Germany	27,955
Brazil	25,750
France	24,195
New Zealand	15,000
United Kingdom	14,359
Ukraine	13,287
Poland	11,970
Italy	11,186
Netherlands	10,995
Mexico	10,352
Argentina	10,250
Turkey	10,000
Australia	9550
Canada	7854

Source: International Dairy Federation, Bulletin 423/2007.

Milk, Fat, and Protein Yields

The Tunisian Holstein population has strong genetic ties with North American and European Holstein populations (Hammami et al., 2007). For example, the Average additive relationships between the Tunisian and Luxembourg populations were as high as 2.2% in 2000. It was also found in the same study (Hammami et al., 2007) that the rate of inbreeding in the Tunisian population was up to 3.10 in 2000.

Total milk per recorded cow for all completed lactations was 6220 kg while average daily milk yield for these same cows was 18 kg. Peak yield was 23.9 kg that occurs 65 days following parturition (Hammami et al., 2008). Maximum productions of milk, fat, and protein yields were obtained at the third parity in Tunisia (Rekik et al., 2003, Ben Gara et al., 2006, Rekik et al., 2008; Hammami et al., 2008, Rekik et al., 2009). Mean daily performances from completed and in progress records (average daily records per cow = 7.4) were 18.2 kg, 0.59 kg, and 0.56 kg and 20.7 kg, 0.70 kg, and 0.65 kg for milk, fat, and protein yields in the second and third lactations, respectively (Hammami et al., 2008). Overall lactations daily performances were 18.8 kg, 0.61 kg and 0.58 kg for the same traits in the same order (Rekik et al., 2008; Rekik et al., 2009).

Milk production performances of Holsteins in Tunisia are lower than the breed performances in the countries of origin. Limited production levels of Tunisian Holsteins have always been explained by unsatisfactory overall management with no or little emphasis on health (Abdouli et al., 2008; Bouraoui et al., 2008, Rekik et al., 2008). The herd manager was an important source of variation of milk performances. The lowest levels were recorded in the cooperative herds while the highest levels were observed in investors and farmers herds (Rekik et al., 2003). In the private sector, compared to state herds, there is more flexibility in making breeding and management decisions (Harbaoui, 1999). They also have better feed resources and health coverage. Furthermore, environmental factors affected the shape of lactation curves (Rekik et al., 2003; Rekik and Ben Gara, 2004). Approximately 25% of the lactation curves were defined as atypical. Highest percentages of atypical curves were observed in the cooperative and state herds and in the spring and summer seasons. Changes in rations are frequently caused by lack of roughage in the second half of the spring season. In the summer season, herd conditions are aggravated by heat stress and resulting health disorders (Bouraoui, 2002; Bouraoui et al., 2008).

A problem that Tunisian breeders have to face when choosing semen is that bulls may rank differently for milk yield in country specific environments. Genotype by environment interaction may cause re-ranking of sires in the Tunisian production systems (Powell and VanRaden, 2002). Several studies have supported the fact that genotype by environment interaction is present when differences among environmental conditions exist (Costa et al., 2000; Ojango and Pollot 2002; Zwald et al., 2003; Bytyqi et al., 2007). Since Tunisia imports semen and heifers for genetic improvement and increased yield, Hammami et al. (2008 and 2009) investigated the reaction of similar genetic materials to varying genetic environments in Tunisia. These authors found evidence for genotype by environment interaction using herd management level as a criterion. In high management levels, yield performances and ranking of animals were comparable to those reported from production systems in temperate regions. Rekik et al. (2008) working on samples of imported heifers and Tunisian born cows reported that the origin (Tunisia, Europe, or the USA) of the cow was an important source of variation of milk production. Tunisian cows had the highest production levels in the first but imported heifers (mainly from the USA) produced more milk in the subsequent lactations. Imported cows showed difficulties to express their potential for milk production in the first lactation but were able to adjust to new management conditions with the rank of lactation and to maintain relatively low mastitis infection rates compared to Tunisian cows.

Genetic Parameters of Milk, Fat, and Protein Yields

Ben Gara et al. (2006) used a Markov Chain Monte Carlo Bayesian method via an animal model on 305-d milk yield to estimate genetic parameters from Tunisian dairy cattle data. Posterior means of heritability and repeatability were 0.17 and 0.39, respectively. Hammami et al., using random regression and an animal model (2008), reported similar heritability estimates for 305-d milk yield. Heritability estimates for 305-d fat and protein yields were moderate (0.12 to 0.18). However, the latter authors found that genetic parameters mirrored management levels and the stage of lactation. Highest heritability estimates of 305-d milk yield (0.21) were found in high HM levels, whereas lowest estimates (0.12) were associated with low HM levels. Furthermore, Heritabilities of test-day milk and protein yields for selected days in milk were higher in the middle than at the beginning or the end of lactation and were in the same range of parameters estimated in management systems with low to medium production levels. Inversely, heritabilities of fat yield were high at the peripheries of lactation. Genetic correlations among 305- d yield traits ranged from 0.50 to 0.86. The largest genetic correlation was observed between the first and second lactation. Hammami et al. (2008) questioned data quality especially for fat yield, a trait that seems not adequately recorded.

Age at First Calving, Fertility,and Longevity Measures in Tunisian Holsteins

The age at first calving affects the career of dairy cows (Abeni et al., 2000). The beginning of a cow's productive life depends on its pre- and post- pubertal management conditions and growth rate (Hoffman et al., 1996; Abeni et al., 2000; Silva et al., 2002). The age at first calving was found to affect milk performances (Abeni et al., 2000; Silva et al., 2002, reproductive traits (Hare et al., 2006; Evans et al., 2006), and longevity traits (Abeni et al., 2000; Silva et al., 2002). Mean age at first calving was 28.7 months in Tunisian Holsteins (Ben Gara et al., 2009). Heritability of this age was, as other reproductive traits, low (0.08). This trait was found to affect 305-d milk yield, true herd life which was up to 38.6 months, and life time milk production of a cow (Ben Gara et al., 2009). A 24 month age at first calving seemed optimal for improved yield and long herd life. In an earlier study, Ajili et al. (2007) reported that the optimal age at first calving ranged between 23 and 27 months. Cows with intermediate yields tended to stay longer in the herd than low or high producing cows. The mean lactation number was 2.6 with more than 57% of cows culled after the first two lactations and only 7.14% of them reached their fifth lactation. Phenotypic correlations of true herd life with milk, fat, and protein yields were 0.07, 0.11, and 0.09, respectively. Those with fertility parameters ranged from -0.04 to 0.06. Furthermore, these authors found that intervals between successive calving and from calving to first service, and days open were 427, 90, and 163 days, respectively. True herd life was 41.99 months.

Health and Animal Well Being in Tunisian Holsteins

Data on health are lacking in Tunisia. Because of insurance policies and subventions, reasons for culling and deaths are not routinely recorded. In recent years, the comfort of animals is being scrutinized not only because of respect for animals but also for enhancing productivity to ensure durability of production systems. Attempts to investigate animal welfare were limited in Tunisia. Bouraoui et al. (2008) evaluated animal welfare in a large dairy herd in the north of Tunisia. Housing conditions, reproductive and productive performances and health care were studied to assess comfort of cows. Barns were found not to meet the standards for cows comfort. There was a degradation of housing conditions because sheds were implemented in low ground, barns concrete coating was deteriorating, litter was in bad shape and there was a persistent draught. Reproductive and productive performances and health indicators reflected also discomfort of cows. In fact, milk production level was low. Mean somatic cell count was 634720 cells/ml. The SCC mean level for Holstein cows without clinical mastitis was around 200000 cells/ml (Coulon et al., 1996; Rupp et al., 2000). The odds of culling a cow with an average SCC of 700000 cells/ml, were 3.4 times greater than those of a cow with an average SCC between 200,000 and 250,000 cells/ml in herds with a low SCC average (Caraviello et al., 2005). Miller et al. (2004) reported a decrease in 305-day milk yield of 54.6 kg and 61.4 kg per somatic cell score unit increase on the first test-day for the first and second parities, respectively. Results on reproduction showed limited fecundity and fertility of cows. Mean calving interval and insemination per conception were 445 days and 2.43 (Bouraoui et al., 2008). Infertility, dystocia, post-partum calving, leg and metabolic disorders and lung diseases were the main causes of culling and death of cows. Rekik et al., (2008) determined somatic cell score levels in Holsteins under Tunisian management circumstances and examined the effect of these scores on milk and protein yields and on intervals from calving to first service or conception. The most important source of variation of somatic cells in milk was the herd in interaction with calving year or test-day date. That is, management practices vary with the herd between and within calving years. Contrary to other reports (Reents et al., 1995); the calving season had no effects on test day somatic cell scores levels even though the summer season in Tunisia is characterized by its high temperature–humidity index (Bouraoui et al., 2002). Direct losses in milk and protein yields from increased somatic cell scores are important (Rekik et al., 2008).

Test-day milk and protein yields were unfavorably affected by high somatic cell scores. Reduction in milk and protein productions from increased somatic cells varied from 0.23 to 1.76 kg and from 6 to 75 g, respectively. Likewise, increased test-day somatic cell scores lengthened both calving to first service and calving to conception intervals by 1.3 to 2.0 days for each unit increase in somatic cell scores. Cows born in Tunisia seemed to perform better than imported cows in the first lactation while imported cows showed clearly better performances in later lactations. North American cows produced the highest yields and had the lowest SCS among all cows in the second and third lactations. Imported high producing cows seemed able to adjust to Tunisian management conditions following their first lactation.

Conclusion

Climate Mediterranean conditions in Tunisia and holders of management making decisions resulted in varied production environments and production systems of the Holstein breed in Tunisia. Milk production performances mirrored management levels and were below the breed potential except in some private herds in the north of the country. Limits are varied and there are considerable challenges that ought to be overcome to improve productivity. Data recording, rations (mainly vetch-oat silage, vetch oat hay, occasionally green roughages, and concentrate), heat, age at first calving, and genetic resources management (culling) and health are areas where improvement might be made. All the latter aspects coupled with a routine genetic evaluation may provide solutions for a durable competitive dairy industry. In particular, the production of replacements of high genetic merit proved under the country production conditions.

References

Abdouli, H., Rekik, B. & Haddad Boubaker, A. (2008). Non-nutritional factors associated with milk urea concentrations in a commercial dairy herd in Tunisia. *World Journal of Agricultural Sciences*. 4 (2), 183-188.

Abeni, F., Calamari, L., Stefanini, L., & Pirlo, G. (2000). Effects of daily gain in pre-and postpubertal replacement dairy heifers on body condition score, body size, metabolic profile, and future milk production. J. Dairy Sci., 83: 1468-1478.

Ajili, N., Rekik B., Ben Gara A., & Bouraoui R., (2007). Relationships among milk production, reproductive traits, and herd life for Tunisian Holstein-Friesian cows. *African J. Agric. Research*, 2(2), 47-51.

Ben Gara, A., Rekik, B. & Bouallègue, M. (2006). Genetic parameters and evaluation of the Tunisian dairy cattle population for milk yield by Bayesian and BLUP analyses. Livest. *Prod. Sci.* 100:142–149.

Ben Gara, A., Bouraoui, R., Rekik, B., Hammami H., & Rouissi, H. (2009). Optimal age at first calving for improved milk yield and length of productive life in Tunisian Holstein cows. *American-Eurasian Journal of Agronomy,* 2 (3),162-167.

Bondoc, O.L., Smith, C. & Gibson, J.P. (1989). A review of breeding strategies for genetic improvement of dairy cattle in developing countries. *Anim. Breed.*, 57, 819-829.

Bouraoui, R. (2002). Impact of heat stress on physiological responses, milk yield, and reproduction of Friesian–Holstein cows in Tunisia, Ph.D. thesis, Institut National Agronomique de Tunis. Ariana, Tunisie.

Bouraoui, R., Lahmar, M., Majdoub, A., Djemali M., & Beleyea, R. (2002). The relationship of temperature-humidity index with milk production of dairy cows in a Mediterranean climate. *Anim. Res.* 51(6), 479-491.

Bouraoui, R., Rekik, B., Yozmane, R., & Haddad, B. (2008). Assessing animal welfare through management, productive and reproductive performances and health care in a large dairy herd in the North of Tunisia. *Research J. Dairy Sci.*, 2 (4), 68-73.

Bytyqi, H., Ødegörd, J., Mehmeti, H., Vegara, M., & Klemetsdal, G. (2007) Environmental sensitivity of milk production in extensive environments: A comparison of Simmental, Brown Swiss, and Tyrol Grey using random regression models. *J. DairySci.*, 90, 3883–3888.

Caraviello, D.Z., Weigel, K.A., Shook, G.E., & Ruegg, P.L. (2005). Assessment of the impact of somatic cell count on functional longevity in Holstein and Jersey cattle using survival analysis methodology. *J. Dairy Sci.* 88, 804–811.

Costa, C.N., Blake, R.W., Pollak, E.J., Oltenacu, P.A., Quaas, R.L. & Searle, S.R. (2000). Genetic analysis of Holstein cattle populations in Brazil and the United States. *J. Dairy Sci.* 83, 2963-2974.

Coulon, J.B., F. Dauver and G.P. Garel, (1996). Facteurs de variation de la numération cellulaire du lait chez des vaches laitières indemnes de mammites cliniques. *INRAProd. Anim.,* 9(2): 133-139.

Djemali, M., and Berger, P. J. (1992). Yield and reproduction characteristics of Friesian cattle under North African conditions. *J. Dairy Sci.* 75:3568–3575.Fikse et al., 2003.

Evans, R.D., Wallace, M., Garrick, D.J., Dillon, P., Berry, D.P., & Olori, V. (2006). Effects of calving age, breed fraction and month of calving on calving interval and survival across parities in Irish spring-calving dairy cows. *Livestock Sci.*, 100(2-3), 216-230.

Hammami, H., Croquet, C., Stoll, J., Rekik, B., & Gengler, N. (2007). Genetic diversity and joint-pedigree analysis of two importing Holstein populations. *J. Dairy Science.* 90:3530-3541.

Hammami, H., Rekik, B., Soyeurt, H., Bastin, C., Stoll, J., & Gengler, N. (2008). Genotype x Environment Interaction for Milk yield in Holsteins Using Luxembourg and Tunisian Populations. *J. Dairy Science*. 91:3661-3671.

Hammami, H., Rekik, B., Soyeurt, H., Ben Gara, A., & Gengler, N. (2008b). Genetic parameters for Tunisian Holsteins using a Test-Day Random Regression Model. *J DairyScience*. 91, 2118-2126.

Hammami, H., Rekik, B., Soyeurt, H., Bastin, C., Bay, E., Stoll, J. & Gengler, N. (2009). Accessing Genotype by Environment Interaction Using Within- and Across-Country Test-Day Random Regression Sire Models. *Journal of Animal Breeding and Genetics.* 126: 366-377.

Hammami, H., Rekik, B., Bastin, C., Soyeurt, H., Bormann, J., Stoll, J. & Gengler, N. (2009b). Environmental Sensitivity for Milk Yield in Luxembourg and Tunisian Holsteins by Herd Management Level. *J. Dairy Science*. 92:4604-4612.

Harbaoui, H. (1999). Management and health factors of culling cows in large dairy cattle herds in Tunisia, Thèse de doctorat en Me´decine ve´te´rinaire. ENMV, Sidi Thabet.

Hare, E., Norman, H. D. & Wright, J. R. (2006). Survival rates and productive herd life of dairy cattle in the United States. *J. Dairy Sci*. 89:3713.

Hill, W. G., Edwards, M. R., Ahmed, M. K. A., & Thompson, R. (1983). Heritability of milk yield and composition at different levels and variability of production. *Anim. Prod.* 36:59–68.

Hoffman, G. L., Salpeter, E. E., Farhat, B., Roos, T., Williams, H., & Helou, G. (1996). Arecibo H i Mapping of a Large Sample of Dwarf Irregular Galaxies. *ApJS*, 105, 269.

Miller, R.H., Norman, H.D., Wiggans, G.R., & Wright, J.R. (2004). Relationship of test day Somatic cell score with test day and lactation milk yields. *J. Dairy sci*. 87, 2299–2306.

Ojango, J. M. K., & Pollot, G. E. (2002). The relationship between Holstein bull breeding values for milk yield derived in both the UK and Kenya. *Livest. Prod. Sci*. 74:1–12.

Reents, R., Jamrozik, J., Schaeffer, L.R. and Dekkers, J.C.M. (1995). Estimation of genetic parameters for test day records of somatic cell score. J. Dairy Sci. 78, 2847.

Rekik, B., Ben Gara, A., Ben Hammouda M., & Hammami, H. (2003). Fitting lactation curves of dairy cattle in different types of herds in Tunisia. *Livest. Prod. Sci.*, 83 (2-3): 309-315.

Rekik, B. & Ben Gara, A. (2004). Factors affecting the occurrence of atypical lactations for the Holstein-Friesian cows. *Livest. Prod. Sci.,* 87(2-3): 245-250.

Rekik, B., Ajili, N., Bel Hani, H., Ben Gara A., and Rouissi, H. (2008). Effect of somatic cell count on milk and protein yields and female fertility in Tunisian Holstein cows. *Livestock Sci.,* 16 (1-3): 309-317.

Rekik, B., Bouraoui, R., Ben Gara, A., Hammami, H., & Hmissi, M. (2009). Milk production of imported heifers and Tunisian- born Holstein cows. *American-Eurasianjournal of Agronomy*. 2(1),36-42.

Rupp, R., Boichard, D., Bertrand, C., & Bazin, S. (2000). Bilan national des numérations cellulaires dans le lait des différentes races bovines laitières françaises. *INRA Prod.Anim.* 13 (4), 257–267.

Silva, L.F.P., VandeHaar, M.J., Weber Nielsen, M.S., & Smith, G.W. (2002). Evidence for a local effect of leptin in bovine mammary gland. *J. Dairy Sci.*,85:3277–3266.

Zwald, N. R., Weigel, K. A., Fikse, W. F., Rekaya, R. (2003). Application of a multiple-trait herd cluster model for genetic evaluation of dairy sires from seventeen countries. *Journal of Dairy Science*, 86: 376-382.

In: Milk Production
Editor: Boulbaba Rekik, pp. 171-184

ISBN 978-1-62100-061-7

Chapter IX

Fresh Ewe Milk Production and Cereal Nutrition: A Peripartal Interactive Model of Grain Choice and Level

A. Nikkhah[*] and M. Karam-Babaei
Department of Animal Sciences, Faculty of Agricultural Sciences,
University of Zanjan, Zanjan 313-45195 Iran

Abstract

Periparturient metabolic imbalances continue to challenge profitable ruminant farming. Sheep as a docile, cooperative, economical and highly adaptable ruminant is increasingly regarded as an experimental model for development of strategies that can attenuate physiological shifts in nutrient metabolism. The objective was to determine effects of peripartal dietary cereal choice, level, and their interaction on sheep production and metabolism. Twenty Afshari×Merino ewes (80.3 ± 2.0 kg body weight) were used in a completely randomized design experiment from 24 days prepartum through 21 days postpartum. Ewes were kept indoor in individual boxes (1.5 × 2.5 m) and received once daily at 0900 h total mixed rations (TMR) containing either 1) higher or 2) lower levels of concentrates based on either 1) corn grain (CO) or 2) 50 to 50 combination of wheat + barley grains (WB). Dietary concentrate level and cereal choice were studied in a 2 × 2 factorial arrangement. Ewes were stepped into the postpartum diet after feeding two prepartum diets, one from 24 to 10 days followed by another fed until lambing. Dry matter based dietary forages included 3 parts of chopped alfalfa hay and 1 part of corn silage. Prepartal DMI numerically increased by feeding CO vs. WB. Feeding CO and not WB at higher vs. lower level improved postpartum DMI. Lambing DMI tended to increase with the higher vs. lower WB (1.59 vs. 1.37 kg/d). Feeding CO instead of WB and feeding both at lower vs. higher level increased fecal pH. Postpartal rumen pH decreased when the inclusion rate of WB (5.7 vs. 6.2, $P<0.05$) and not CO (6.3 vs. 6.3) was increased. Rumen propionate concentrations were decreased (20.4 vs. 18.9 mmol/L), and acetate levels (67 vs. 70 mmol/L) as well as acetate to propionate ration (3.3 vs. 3.7)

[*] Email: nikkhah@znu.ac.ir

were decreased by feeding higher amount of both grain choices. Colostrum yield and composition, peripartal urine pH, lamb weight at lambing and 21 days of age, and placental weight and expulsion time were unaffected. Milk yield (1.64 vs. 1.27 kg/d) and milk fat yield (99 vs. 81 g/d) were increased by increasing dietary levels of both grains. Plasma glucose increased by feeding higher vs. lower WB (57.6 vs. 52.2 mg/dL). Feeding CO instead of WB tended to reduce peripartal plasma NEFA concentration (0.25 vs. 0.28 mmol/L) and significantly increased insulin to NEFA ratios (2.47 vs. 1.77). Data for provide novel nutritional insights into independent and interactive effects of dietary cereal choice and level on periparturient sheep productive and metabolic responses. These serve as a peripartal model for high-producing ruminants (e.g., dairy cows) and non-ruminants.

Keywords: Ewe, Periparturient, Milk, Nutrition, Model, Cereal.

Introduction

Physiological shifts from a pregnant, non-lactating to a non-pregnant, lactating status impose metabolic challenges to periparturient ruminants. The transition is concurrent with major and minor nutrients imbalances (NRC, 2001; NRC, 2007). The increased research on transition biology has quantified some facets of the challenge in feeding periparturient dairy cows at animal, organ, cell and gene levels (Drackley, 1999; Grummer et al., 2004; Janovick-Guretzky et al., 2007; Nikkhah et al., 2009). However, due in part to on-farm difficulties of conducting large ruminant studies, transition biology continues to generously host many unanswered or partly resolved questions. Thus, sheep as a docile and small experimental unit is an inexpensive nutritional model for studying transition metabolism in ruminants and non-ruminants.

Effects of late pregnancy nutrition planes on periparturient ewe metabolism and production have long been recognized (Banchero et al., 2006; Charismiadou et al., 2000; Hall et al., 1992; Treacher, 1970). However, global and regional clear-cut guidelines especially on energy nutrition for optimal health and productivity are yet to be developed (NRC, 2007). As for cereals, corn grain is much more slowly fermentable (40-70%) than wheat and barley grains (> 80%; Huntington, 1997; Nikkhah et al., 2010). Wheat grain is also on average somewhat more rapidly degradable than barley grain. Processing alters physical and chemical grain properties and their effects on rumen fermentation patterns and yields (Nikkhah, 2010a,b; Yang et al., 2000). Prepartal dietary starch source affects peripartal metabolism and performance (Banchero et al., 2006; Dann et al., 1999). Partial or full prepartal substitution of ground wheat grain for ground barley grain improved energy and calcium balances of periparturuient cows (Amanlou et al., 2008) and heifers (Nikkhah et al., 2010). However, performance of dairy cows fed either ground corn or ground barley grain based periparturient diets was similar (Sadri et al., 2009). Replacing steam-flaked corn for cracked corn lowered blood urea and NEFA, and increased milk yield in periparturient cows (Dann et al., 1999). Feeding more rapidly vs. more slowly degradable cereal grains (e.g., barley vs. corn) alters rumen fermentation patterns and effective fiber requirements for optimal productive response and health (Beauchemin and Rode, 1997). To gain positive wheat grain effects on DMI and calcium metabolism (Amanlou et al., 2008; Nikkhah et al., 2010) whilst minimizing the risk of its unfavorably high starch fermentability, and thus, periods of subacute rumen acidosis(Stone, 2004), the combined wheat and barley in peripartal diets may be a more desirable choice than each fed alone. A paucity of data exists on how dietary starch source interact with starch level on periparturient sheep performance. The objective was to determine independent and interactive effects of feeding peripartal diets with either corn grain (CO) or a combination of wheat and barley grains (WB) at higher or lower levels on feed intake, rumen fermentation, colostrum and milk properties, and blood metabolites of Afshari×Merino ewes as a model for high-producing ruminants.

Materials and Methods

Ewes and Diets Management

Sixteen late pregnant Afshari×Merino ewes (80.3 ± 2.0 kg body weight; 3.0 ± 0.45 BCS; mean ± SD) were used in a completely randomized design experiment from 24 ± 3 days prepartum through 21 days postpartum. Ewes were kept indoor in individual boxes (1.5 × 2.5 m) and fed once daily at 0900 h total mixed rations (TMR) containing an either 1) higher or 2) lower level of concentrate based on either 1) corn grain (CO) or 2) a 50:50 ratio of wheat grain : barley grain (WB). Ewes were stepped into the postpartum diet after feeding two prepartum diets (Table 1). The first prepartal diet (Pr1) was fed from 24 until 10 days prepartum, and had a forage to concentrate ratio (F:C) of 75:25 for the higher grain, and 65:35 for the lower grain diet. The second prepartal diet (Pr2) was fed from 10 days prepartum until lambing, and had a F:C of 70:30 for the lower grain and 60:40 for the higher grain diet. The postpartum ration (Po) was fed for 21 days starting from lambing, with a F:C of 65:35 for the lower grain and 50:50 for the higher grain diet (Table 1). Dietary concentrate level and cereal choice were studied in a 2 × 2 factorial model. Rations were offered to permit 5-10% orts. The peripartum ewes were randomly assigned to one of four treatments at the commencement of the study. Diets were formulated using the Cornell Net Carbohydrate and Protein System program (CNCPS, v.5). Within each grain level, diets were balanced to have similar energy, CP, and NDF densities (Table 1). Each box was equipped with a concrete feed bunker and a metal waterer. Manure was collected and bedding was cleaned daily. Prior to the experiment, all ewes were adapted to the experimental boxes and diets during a 4-day period. The experiment was conducted at the Sheep Production Facilities of the University of Zanjan's Research Farm (Zanjan, Iran) under the guidelines of the Iranian Council of Animal Care (1995).

Grain and Forage Properties and Processing

Wheat (*Triticum Spp.*), barley (*Hordeum Spp.*) and corn (*Zea mays L. ssp.*) grains were obtained from bulked sources fed in broad Iranian regions. Corn, wheat, and barley grains contained respectively 70, 75 and 58% starch; 8.5, 13 and 11% CP; 11, 10 and 20% NDF; 3.2, 3 and 7% ADF; and 4, 2.3 and 2.2% ether extract (DM-based). Whole corn, barley and wheat grains densities were 540, 500 and 550 g/L, respectively. All grains were ground by a commercial hammer mill commonly used by almost all ruminant farmers in Iran (Nikkhah et al., 2004, 2010).

Table 1. Periparturient dietary ingredients (%) and nutrient composition (DM based)

Diet	Higher corn			Lower corn			Higher WB			Lower WB		
Period	Pr1	Pr2	Po	Pr1	Pr2	Po	Pr1	Pr2	Po	Pr1	Pr2	Po
Alfalfa hay	55	40	34	60	47	44	55	40	34	60	47	44
Corn silage	10	20	16	15	23	21	10	20	16	15	23	21
Corn grain	24.5	28	30	17	19	24.5	-	-	-	-	-	-
Barley grain	-	-	-	-	-	-	8	16	18	6	11	13.5
Wheat grain	-	-	-	-	-	-	18	17	18	12	11	13.5
Wheat bran	4	5.5	6.5	2	3.5	3.5	5	2	2	3	1.5	2.5
Soybean meal	5.5	5.5	12.5	5	6.5	6	3	4	11	3	5.5	4.5
Vit & Min Suppl.[1]	1	1	1	1	1	1	1	1	1	1	1	1
Nutrients												
NEL, Mcal/kg	1.52	1.57	1.62	1.47	1.51	1.54	1.51	1.57	1.62	1.46	1.51	1.54
ME, Mcal/kg	2.44	2.51	2.59	2.36	2.43	2.47	2.42	2.51	2.59	2.34	2.42	2.46
CP %	13.9	13.2	15.6	13.8	13.7	13.4	13.8	13.2	15.6	13.6	13.7	13.4
NDF %	38.4	35.9	32.6	40.7	39.3	37.2	39.0	36.8	33.3	42.0	40.0	38.6
EE %	2.9	3.0	3.0	2.7	2.8	2.9	2.4	2.4	2.3	2.4	2.4	2.4

Pr1 = the first prepartum diet; Pr2 = the second prepatum diet; Po = postpartum diet; WB = the combination of wheat + barley grains.

[1]Contained 500000 IU Vitamin A, 100000 IU Vitamin D, 0.1 mg Vitamin E, 180 g Ca, 90 g P, 20 g Mg, 60 g Na, 2 g Mn, 3 g Fe, 0.3 g Cu, 0.1 g CO, 0.1 g I, 0.001 g Se, and 3 g antioxidants per kg.

Alfalfa hay contained 13.1% CP and 45.6% NDF, and was chopped using an on-farm chopper machine for an average theoretical chop length of 4 cm before mixing with concentrate. Corn grain had 8% CP and 50% NDF. Chopped alfalfa hay, processed corn silage, and concentrate for individual cows were weighed daily and mixed thoroughly to prepare TMR just before each feed delivery.

Feed Intake and Analyses, and Total Tract DM Digestibility

The amount of TMR offered and orts were measured daily for individual ewes to calculate DMI. Samples of TMR and orts were taken daily to obtain DM after oven-drying at 100°C for 24-h. Weekly feed and fecal samples were dried and ground to pass through 1-mm screen using a Wiley mill (Arthur H. Thomas Co., Philadelphia) to be stored at -20°C for later nutrient analysis.

Feed samples were analyzed for CP (method 984.13; AOAC, 1990), NDF (Van Soest et al., 1991; using heat-stable α-amylase and sodium sulfite) and ADF (method 973.18; AOAC, 1990). The acid insoluble ash (AIA; Nikkhah et al., 2004) was measured in feed and fecal samples as an internal marker to determine coefficient of DM total tract apparent digestibility.

Colostrum and Milk Production and Composition

Colostrum was obtained shortly after lambing and following an intramuscular 10 IU dose of oxytocin. Lambs received colostrums at 10% of body weight. Colostrum from individual ewes was sampled and frozen at -20C until analyzed for fat and protein by Milk-O-Scan (134 BN Foss Electric, Hillerd, Denmark) at the Jahad-Agriculture laboratory of the Zanjan province (Zanjan, Iran). Milk production was measured at days 5, 12 and 19 postpartum after a subcutaneous 10 IU oxytocin injection. Milking was performed by hand. After thorough mixing, 30 ml of milk from all ewes was sampled, added with potassium dichromate, stored at 4°C, and analyzed for fat and protein within 2 days by the Milk-O-Scan.

Blood Sampling and Analysis

Blood from individual ewes was obtained weekly from jugular vein and at 4 h post-feeding. Blood was collected into EDTA-containing 5-ml tubes, put in the ice, and centrifuged within 30 min at 3000 *g* for 15 min to harvest plasma. Plasma was stored at -20°C until analyzed for metabolites. Glucose (GOD-PAR enzymatic method), urea and albumin (Bromcresol Green method) were measured using spectrophotometer (Perkin- Elmer, Colemen Instruments Division, Oak Brook, IL, USA) and commercial kits (Pars Azmun Laboratory, Tehran, Iran) at the Nutrition Laboratory of the Animal Science Department (University of Zanjan, Zanjan, Iran). Plasma levels of BHBA, NEFA and insulin were measured at the Veterinary Clinical Laboratory of Mabna in Karaj (Karaj, Iran).

Fecal and Urine pH

For 2 days during the week before lambing, fecal samples were collected from rectum twice in morning and evening. Each sample was mixed with an equal portion of distilled waterand stirred for sufficient uniformity before immediate pH measurement using a portable pH meter (HI 8314 membrane pH meter, Hanna Instruments, Villafranca, Italy).

The remaining fecal samples were stored at -20°C for later DM digestibility analysis (Nikkhah et al., 2004). At week 1 prepartum and week 2 postpartum, urine was collected by manual stimulation of the lower vulva into plastic vials twice in morning and evening. Urine samples were stirred and analyzed for pH immediately upon sampling.

Body Weight and Body Condition Score

Body weight was measured weekly and shortly before feeding. Body condition score (BCS) was recorded by three trained individuals at the commencement and the end of the study as well as at lambing. A five-point scoring scale, with 1 being an emaciated or extremely thin ewe and 5 describing an obese or extremely fat ewe, was used.

Rumen Fluid Sampling and Analysis

Rumen fluid was sampled in the last week postpartum at 4-h post-feeding using a 1.5 m-length stomach tube. To minimize saliva contamination, the first 20 ml aspirated was removed. Rumen fluid samples were passed through 4-layer cheese cloths and analyzed immediately for pH (HI 8314 membrane pH meter, Hanna Instruments, Villafranca, Italy). A rumen fluid sample of 10 ml mixed with 2 ml of 25% metaphosphoric acid was stored at -20 C for later volatile fatty acids analysis using Gas Chromatography (Hewlett Packard Series 2 5890). Another rumen fluid sample of 10 ml added with 1 ml of 50% sulfuric acid was stored for ammonia analysis.

Statistical Analysis

Data were analyzed using the MIXED MODEL procedure of the SAS program (SAS Institute, 2003). The REML method was set to estimate least squares means, and the Kenward-Roger method was used to calculate denominator degrees of freedom. The model for repeated blood parameters included fixed effects of grain source, grain level and their interaction, week, and week interactions with grain source and level. Random effects included ewe within week by treatment, plus residuals. Due to no significance, the three-way interaction of 'week × grain source × grain level' was excluded from the model. For unrepeated parameters, the final model contained fixed effects of grain source, grain level and their interaction, plus random effects of ewe within treatment, and residuals. Normality of distribution and homogeneity of variance for residuals were tested and ensured using PROC

UNIVARIATE (SAS Institute, 2003). The P-values < 0.05 were declared as significant and those ≤ 0.10 were declared as trends for significance.

Results

Lamb and Ewe Body Weights, and Placenta Weight

Placenta weight did not differ for CO vs. WB (0.66 vs. 0.69 kg) or for the higher vs. lower grain diets (0.71 vs. 0.65 kg), respectively (Table 2). Lamb birth weight was not significantly affected by treatments, being on average 6.2 kg for CO and 7.4 kg for WB. Lamb weight at 3 weeks of age was similar among treatments, being 14.4 kg for CO and 14.8 kg for WB. Ewe BW changes prepartum and postpartum were comparable among treatments (Table 2). The average across-diet BW change was 2.2 kg from 24 days prepartum until lambing, and -0.69 kg from lambing until 21 days postpartum.

Ewe DMI and DM Digestibility

Postpartum, an interaction of dietary grain choice and level existed (P=0.05), such that feeding respectively the higher vs. lower level of CO (2.30 vs. 1.98 kg/d) and not of WB (2.03 vs. 2.00 kg/d) improved DMI. Prepartal DMI was numerically greater (P=0.11) for CO than for WB (1.67 vs. 1.52 kg/d). However, dietary grain level and its interaction with grain choice did not affect prepartal DMI (Table 2). Total tract apparent DM digestibility was similar among diets, both prepartum and postpartum. The average across-treatment DM digestibility coefficients were 63.2% and 67.1% for prepartal and postpartal diets, respectively.

Fecal and Urine pH

Prepartum, fecal pH was higher for CO than for WB (6.7 vs. 6.4, P=0.02). The higher than lower grain diets tended to decrease fecal pH peripartum (6.5 vs. 6.7, P=0.10). A numerical interaction between dietary grain choice and level was found on prepartal urine pH (P=0.11), such that increased dietary grain inclusion reduced urine pH for WB (7.24 vs. 7.83), but not for CO (7.63 vs. 7.52). Postpartal urine pH did not significantly differ among treatments (Table 2).

Milk and Colostrum Production and Composition

Colostrum volume, and content and yield of fat and protein were not affected (Table 3). The average colostrum yield was 374 g, and the average colostrum fat and protein contents were 11.6% and 12.9%, respectively (Table 3). Dietary grain level (P=0.02) and not grain choice (P=0.32) affected milk yield. Milk yield was 1.64 kg/d for the higher grain diet and

1.27 kg/d for the lower grain diet. Inversely, milk fat content was decreased by feeding the higher vs. lower grain diet (5.9 vs. 6.3% vs. P=0.03). As a result, milk fat yield tended to increase by the higher vs. lower grain diet (95 vs. 80 g/d; P=0.06). Milk content and yield of protein was comparable among treatments, being on average 5.4% and 76 g/d across diets, respectively (Table 3).

Rumen Fermentation Parameters

Dietary grain source and level interacted on rumen fluid pH (P=0.002), such that increasing grain inclusion rate did not affect rumen pH for CO (6.28 vs. 6.25), but reduced it for WB (5.70 vs. 6.23)(Table 4). Rumen fluid ammonia and total VFA concentrations were unaffected, being on average 17.6 mg/dL and 106 mmol/L, respectively. Rumen fluid propionate levels were higher (20.5 vs. 18.9 mmol/L) and acetate levels were lower (67.1 vs. 70.0 mmol/L) for the higher grain diet than for the lower grain diet (P<0.05; Table 4). Rumen fluid acetate to propionate ratio was lower for the higher grain diet (3.3 vs. 3.7; P=0.02).

Blood Metabolites and Insulin

Increasing dietary grain level increased plasma glucose for WB (57.6 vs. 52.2 mg/dL) and not for CO (56.0 vs. 54.5 mg/dL). Ewes fed CO tended to have lower plasma NEFA concentrations (0.25 vs. 0.28 mmol/L; P=0.06) and had significantly higher plasma ratio of insulin to NEFA (2.4 vs. 1.8; P=0.009) than ewes fed WB, while similar plasma insulin (0.38 vs. 0.35 ng/L) (Table 5). Plasma levels of urea, BHBA and albumin were not different among diets (Table 5).

Discussion

This study provides novel findings that feeding periparturient ewes ground corn grain (CO) vs. combined ground wheat + barley grain (WB) based total mixed rations improves feed intake and insulin efficiency, and attenuates negative energy balance. These data are of health and metabolic importance, considering that the ewes were utilized as a peripartal model for other species of mainly high-producing cows. The interaction of dietary grain choice and level was for the first time established to affect peripartal ewe performance. The tendency for greater prepartum DMI of ewes fed CO instead of WB provides evidence that ground corn grain could be a superior choice over combined wheat and barley grains in improving ration palatability and stimulating feed intake. Recently, ground wheat grain proved to be more effective than ground barley grain in improving energy intake of periparturient cows and heifers (Amanlou et al., 2008; Nikkhah et al., 2010). The positive effect of CO on feed intake became more noticeable postpartum when increasing dietary grain improved DMI for CO by about 300 g/d.

Table 2. Peripartum ewe performance and lamb weight

Grain source (GS)	Corn		Wheat + Barley			*P*-value		
Grain level (GL)	Higher	Lower	Higher	Lower	SEM	GS	GL	GS × GL
Prepartum								
DMI, kg/d	1.67	1.66	1.60	1.45	0.08	0.11	0.34	0.37
Total tract DM digestibility %	64.4	61.9	63.8	62.6	2.03	0.96	0.38	0.76
BW changes, kg	2.44	2.42	2.28	1.64	0.52	0.37	0.52	0.55
Fecal pH	6.6	6.9	6.4	6.5	0.09	0.02	0.10	0.21
Urine pH	7.6	7.5	7.2	7.8	0.20	0.85	0.25	0.11
Postpartum								
DMI, kg/d	2.30[a]	1.98[b]	2.03[b]	2.0[b]	0.01	0.10	0.02	0.05
Total tract DM digestibility %	69.2	65.7	68.8	64.6	2.38	0.77	0.13	0.89
BW changes	-0.26	-0.63	-1.0	-0.88	0.52	0.37	0.82	0.65
Urine pH	7.7	7.9	7.5	7.8	0.12	0.30	0.14	0.61
Lamb birth weight, kg	6.5	6.0	6.6	8.2	1.13	0.33	0.65	0.37
Lamb weight at week 3	13.9	14.8	15.7	14.1	1.92	0.79	0.83	0.53
Placenta weight, kg	0.72	0.61	0.70	0.68	0.10	0.80	0.52	0.66

Table 3. Ewe milk and colostrum production and composition

Grain source (GS)	Corn		Wheat + Barley			*P*-value		
Grain level (GL)	Higher	Lower	Higher	Lower	SEM	GS	GL	GS × GL
Colostrum yield, kg/d	0.37	0.39	0.37	0.37	0.06	0.84	0.96	0.86
Fat %	11.6	11.8	11.3	11.6	0.29	0.45	0.38	0.97
Protein %	13.0	12.7	13.2	12.8	0.69	0.74	0.62	0.95
Fat yield, g/d	43	45	42	42	6	0.75	0.82	0.91
Protein yield, g/d	43	45	42	47	8	0.96	0.90	0.88
Milk yield, kg/d	1.72	1.31	1.55	1.22	0.15	0.32	0.02	0.80
Fat %	5.9	6.3	5.9	6.2	0.16	0.63	0.03	0.94
Protein %	5.5	5.2	5.4	5.4	0.18	0.79	0.49	0.57
Fat yield, g/d	100	80	87	79	9	0.29	0.06	0.32
Protein yield, g/d	78	67	84	73	8	0.45	0.16	0.99

Table 4. Postpartum rumen fermentation indicators

Grain source (GS)	Corn		Wheat + Barley			*P*-value		
Grain level (GL)	Higher	Lower	Higher	Lower	SEM	GS	GL	GS × GL
Rumen pH	6.28	6.25	5.70	6.23	0.07	0.001	0.004	0.002
Ammonia, mg/dL	17.2	17.7	17.5	17.9	1.33	0.87	0.77	0.98
Acetate, mmol/L	67.0	69.6	67.2	70.3	1.15	0.70	0.03	0.77
Propionate, mmol/L	20.2	19.0	20.7	18.9	0.64	0.79	0.04	0.66
Butyrate, mmol/L	14.3	14.1	13.7	13.6	0.46	0.25	0.69	0.94
Total VFA, mmol/L	105.7	106.5	106.0	107.0	0.96	0.68	0.38	0.90
Acetate : propionate	3.34	3.69	3.25	3.74	0.15	0.92	0.02	0.68

VFA = volatile fatty acids.

Table 5. Periparturient blood metabolites and insulin concentrations

Grain source (GS)	Corn		Wheat + Barley			*P*-value		
Grain level (GL)	Higher	Lower	Higher	Lower	SEM	GS	GL	GS × GL
Glucose, mg/dL	56.0	54.5	57.6	52.2	1.08	0.75	0.002	0.07
NEFA, mmol/L	0.24	0.26	0.29	0.27	0.01	0.06	0.72	0.19
BHBA, mmol/L	0.58	0.65	0.66	0.63	0.05	0.51	0.72	0.25
Insulin, ng/L	0.39	0.37	0.38	0.33	0.02	0.34	0.15	0.62
Insulin/NEFA	2.3	2.6	1.8	1.7	0.10	0.01	0.80	0.39
Albumin, g/dL	3.45	3.41	3.37	3.30	0.08	0.30	0.57	0.87
Urea, mg/dL	13.3	13.6	12.9	13.6	0.32	0.50	0.16	0.61

The lower prepartal fecal pH in WB than in CO ewes provides another evidence for more effective nutrient use along the gut, since a lower fecal pH value would suggest a more extensive hindgut fermentation (Nikkhah et al., 2004). The hindgut digestion is certainly inferior to nutrient fermentation in the rumen and upper intestine, where microbial mass, VFA and direct nutrient absorption can occur; whereas lower gut nutrient digestion provides minor VFA and leads to microbial and non-microbial nutrient excretion via feces (Nocek and Tamminga, 1991).

Feeding cows ground corn vs. ground barley based prepartuma (21% grain) and postpartum (28% grain) diets did not affect DMI, milk production and energy balance (Sadri et al., 2009). Corn and barley based diets in that study had exactly the same portion of grains. As a result, contents of NEL, CP, NDF and NFC between corn and barley based diets were not similar because corn grain has more starch, fat and energy, and less fiber and protein than barley grain. It is thus probable that cow response to dietary grain choice was confounded with altered dietary nutrient composition. For this reason plus the sole use of barley and not barley + wheat, the results of Sadri et al. (2009) cannot be accurately compared with the current findings.

The lower rumen pH at the higher WB level compared with other treatments suggests an occurrence of subacute rumen acidosis (SARA; Krause and Oetzel, 2006) and supports the lower fecal pH. These results suggest that increased rumen fermentation for WB greatly reduced rumen pH that may in consequence have reduced rumen nutrient assimilation, thus increasing partially-digested starch inflow to the lower gut. As a result, more nutrients were available for the lower gut fermentation that would expectedly lower fecal pH, as observed for WB than for CO. These findings are consistent with the proved fact that corn grain possesses lower rumen digestibility than wheat and barley grains (e.g., 40-50% vs. 80-90%; Huntington, 1997). The tendency for reduced urine pH of ewes fed the higher vs. lower WB suggests an increased acidity of extracellular body fluids. Altered urine pH has been used as a reliable indicator of altered electrolyte metabolism and body fluids acid-base status (Tucker et al., 1988). Recently, wheat grain was fed to prepartal dairy cows to reduce extracellular alkalinity and improve gradual bone calcium release and intestinal absorption (Amanlou et al., 2008; Nikkhah et al., 2010). Reduced urine pH concurred with increased circulating calcium levels, thus improving milk production. As such, increasing WB feeding level in the present study may have similarly reduced extracellular alkalinity, thereby reducing urine pH, which was additionally supported by the reduced rumen pH.

Increased postpartum rumen propionate and reduced acetate concentrations with increased dietary grain level for both WB and CO suggest more suitable conditions for amylolytic bacteria. However, the total VFA levels remained unchanged. It should be noted that rumen VFA concentrations do not essentially represent rumen VFA production because a multitude of factors (pH, saliva flow, rumen fluid and solids passage rate, buffering capacity and absorption rate) affect concentrations (France and Dijkstra, 2005). The data suggest that corn starch may be increased from 24.5% up to 30% of the postpartal diet DM without compromising rumen pH; but similar increases in starch supply from wheat + barley grains extremely lowers rumen pH. These translate into about 17% and 22% starch from cereal grains from lower and higher grain diets, respectively.

The finding that increasing dietary grain increased milk fat output by 20 g for CO and by 8 g for WB ewes suggests that mammary fat precursors availability was greater for the higher grain diets. This concurs with the increased DMI of about 300 g by feeding the higher vs.

lower CO diet (2.3 vs. 2.0 kg/d). But, the increased DMI was only about 30 g for the WB diet, yet milk fat yield increased by 8 g. These changes suggest either 1) more efficient nutrient use for milk secretion, or 2) more reliance on body reserves, or a combination of both for WB than for CO. The increased blood glucose from 52.2 to 57.6 mg/dL by feeding the higher vs. lower WB lends support to the first likelihood. However, due to lower plasma NEFA for CO vs. WB, it can be suggested that the ewes on WB mobilized more fat from peripheral and visceral adipots than did the CO ewes. Hence, despite only 30 g more DMI, increased milk fat yield by feeding the higher vs. lower WB may be partly justified by an increased mammary NEFA inflow. Internal body fats may supply considerable amount of long-chain fatty acids for milk fat secretion in fresh ruminants (Bauman and Griinari, 2003). In light of the similar plasma insulin levels, the higher insulin to NEFA ratio for CO than for WG strengthens a suggestion that insulin efficiency was improved by feeding CO (Nikkhah et al., 2009). In other words, feeding CO instead of WB decreased the demand for insulin to maintain milk production, caused by the greater DMI. Increased energy intake may have reduced dependence on body fat stores (Vernon et al., 1990).

Unaltered plasma albumin and urea levels are inter-supportive, suggesting little treatment effects on liver nitrogen metabolism and protein synthesis (Nikkhah et al., 2010). Altogether, the productive and metabolic data favor ground corn grain as a peripartal cereal choice over wheat + barley grains, particularly in high grain diets. While being more energy dense, the corn based rations improve nutrient intake and attenuate negative energy balance without increasing the risk from disturbed rumen fermentation. Where inexpensive and more accessible than corn grain, the mixture of wheat and barley grains may be effectively fed to periparturient ewes rather at lower levels.

Conclusion

Prepartal DMI numerically increased by feeding ground corn based mixed rations (CO) vs. wheat + barley grain based mixed rations (WB). Feeding CO and not WB at a higher vs. a lower level improved postpartum DMI. Lambing DMI tended to increase when WB was fed at higher than lower level. Feeding CO instead of WB, and feeding higher grain diets increased fecal pH. Postpartum, rumen pH decreased by feeding more of WB and not more of CO. Rumen propionate levels were higher and acetate levels were lower at higher dietary grain levels. Increased feeding of WB and not CO reduced prepartal urine pH. Colostrum properties, postpartal urine pH, lamb weight at birth and 21 days of age, and placenta weight were not affected. Milk yield (1.64 vs. 1.27 kg/d) and milk fat yield (99 vs. 81 g/d) were increased by increasing dietary levels of WB and CO.

Plasma glucose increased by feeding the higher vs. lower WB level. Feeding CO instead of WB tended to reduce peripartal plasma NEFA, and increased insulin to NEFA ratios. Data provide novel nutritional insights into independent and interactive effects of dietary cereal choice and level on periparturient sheep biology. These serve as a nutritional model for high-producing ruminants (e.g., dairy cows) and non-ruminants.

Acknowledgments

The Iranian Ministry of Science, Research and Technology is acknowledged for supporting the programs of A. Nikkhah for improving science education in the third millennium. The University of Zanjan (Zanjan, Iran) is thankfully acknowledged for research facilities and funding. The employees of the Sheep Production Complex, especially D. Aliyari the Farm Manager, are thanked for their cooperation.

References

Amanlou, H., Zahmatkesh, D. & Nikkhah, A. (2008). Wheat grain as a prepartum cereal choice for periparturient cows. *J. Anim. Physiol. Anim. Nutr.* 92, 605-613.

Banchero, G. E., Quintans, G., Vazquez, F., Gigena, A., Manna, La., Lindsay, D. R. & Milton, J. T. B. (2006). Effect of supplementation of ewes with barley or maize during the last week of pregnancy on colostrum production. *Br. J. Nutr.* , 625-630.

Bauman, D. E. & Griinari, J. M. (2003). Nutritional regulation of milk fat synthesis. *Annu. Rev. Nutr.* 23, 203-227.

Beauchemin, K. A. & Rode, L. M. (1997). Minimum versus optimum concentrations of fiber in dairy cow diets based on barley silage and concentrates of barley or corn. *J. Dairy Sci.* 80, 1629-1639.

Charismiadou, M. A., Bizelis, J. A. & Rogdakis, E. (2000). Metabolic changes during the perinatal period in dairy sheep in relation to level of nutrition and breed. I. late pregnancy. *J. Anim. Phys. Anim. Nutr.* 84, 61–72.

Dann, H. M., Varga, G. A. & Putnam, D. E. (1999). Improving energy supply to late gestation and early postpartum dairy cows. *J. Dairy Sci.* 82, 1765–1778.

France, J. & Dijkstra, J. (2005). Volatile fatty acid production. In: Dijkstra, J., J. M. Forbes and J. France. Quantitative aspects of ruminant digestion and metabolism. 6, 157-176. 2nd Edition. CABI Publishing. Wallingford, UK.

Grummer, R. R., Mashek, G. D. & Hayirli, A. (2004). Dry matter intake and energy balance in the transition period. *Vet. Clin. Food. Anim.* 20, 447–470.

Hall, D. G., Holst, P. J. & Shutt, D. A. (1992). The effect of nutritional supplements in late pregnancy on ewe colostrum production plasma progesterone and IGF-1 concentrations. *Aust. J. Agric. Res.* 43, 325–337.

Huntington, G. B. (1997). Starch utilization by ruminants: From basics to the bunk. *J. Anim. Sci.* 75:852–867.

Janovick-Guretzky, N. A., Dann, H. M., Carlson, D. B., Murphy, M. R., Loor, J. J. & J. K. Drackley. (2007). Housekeeping gene expression in bovine liver is affected by physiological state, feed intake, and dietary treatment. J. Dairy Sci. 90, 2246-2252.

Krause, K. M. &Oetzel, G. R. (2006). Understanding and preventing subacute ruminal acidosis in dairy herds: A review. Anim. Feed Sci. Technol. 126, 215-236.National Research Council. (2001). Nutrient Requirements of Dairy Cattle. 7th rev. ed. National Academy Press, Washington, DC.

National Research Council. (2007). Nutrient Requirements of Small Ruminants. 5th rev. ed. National Academy Press, Washington, DC.

Nikkhah, A. (2010a). Barley grain for rumen and ruminants: over-modernized uses of an inimitable fuel. In Barley. Nova Science Publishers, Inc, NY, USA.

Nikkhah, A. (2010b). Optimizing barley grain use by dairy cows: a betterment of current perceptions. In Grain Production. Nova Science Publishers, Inc, NY, USA.

Nikkhah, A., Ehsanbakhsh, F., Zahmatkesh, D. & Amanlou, H. (2010). Prepartal provision of wheat grain for easier metabolic transition in periparturient Holstein heifers. Animal. In press. doi:10.1017/S1751731110002065.

Nikkhah, A. J. Loor, R. Wallace, D. Graugnard, J. Vasquez, B. Richards, and J. Drackley. (2009). Free-choice access to a moderate-energy diet increases internal body fat in dry cows. Illinois Dairy Report, IL, USA.

Nocek, J. E. & S. Tamminga. (1990). Site of digestion of starch in the gastrointestinal tract of dairy cows and its effects milk yield and composition. *J. Dairy Sci.* 74, 3598-3629.

Sadri , H., Ghorbani, G. R., Rahmani, H. R., Samie, A. H., Khorvash, M. & Bruckmaier. (2009). Chromium supplementation and substitution of barley grain with corn: effects on performance and lactation in periparturient dairy cows. *J. Dairy Sci.* 92, 5411–5418.

Stone, W. C. (2004). Nutritional approaches to minimize subacute ruminal acidosis and laminitis in dairy cattle. *J. Dairy Sci.* 87, E13-E26.

Treacher, T. T. 1970. Effects of nutrition in late pregnancy on subsequent milk production in ewes. *Anim. Prod.* 12, 23-36.

Tucker, W. B., Harrison, G. A. & Hemken, R. W. (1988). Influence of dietary cation–anion balance on milk, blood, urine, and rumen fluid in lactating dairy cattle. *J. Dairy Sci.* 71, 346–354.

Van Soest, P. J. (1994). Nutritional Ecology of the Ruminant. 2nd ed. Cornell Univ. Press, Ithaca, NY.

Vernon, R. G., Faulkner, A., Hay, W. W., Calvert, D. T. & Flint, D. J. (1990). Insulin resistance of hind-limb tissues in vivo in lactating sheep. *Bio J.* 270, 783–786.

Yang, W. Z., Beauchemin, K. A. & Rode, L. M. (2000). Effects of barley grain processing on extent of digestion and milk production of lactating cows. *J. Dairy Sci.* 83, 554-568.

In: Milk Production
Editor: Boulbaba Rekik, pp. 185-241

ISBN 978-1-62100-061-7

Chapter X

Fatty Acid Composition of Milk Lipids in Response to Dietary Fish and Safflower Oils

K. L. Jacobsen, S. Harris, E. J. DePeters* and S. J. Taylor
Department of Animal Science, One Shields Avenue,
University of California, Davis, CA 95616

Abstract

The effects of dietary fish and safflower oils were studied using eight lactating Holstein cows, 4 primiparous (P) and 4 multiparous (M), in a split-plot Latin square design. Cows were fed a basal diet composed of 41% sliced alfalfa hay and 58% concentrate ingredients to evaluate the effects of type and level of oil supplementation. Type of dietary oil was either 100% safflower oil (S) or 75% safflower oil + 25% fish oil (FS), fed at either 400 g/cow (L) or 800 g/cow daily (H). Yields of milk and milk components and dry matter intake were not affected by type or level of oil supplemented. Concentration of total *trans* fatty acids (FA) in milk fat triacylglycerol increased with the addition of fish oil and the higher level of oil. The concentration of C18:3 n3 decreased as the level of oil in the diet increased. Fish oil supplementation, particularly at 800g/day, increased concentrations of *cis*-9 *trans*-11 C18:2 and *trans*-11 C18:1. The high oil level increased concentrations of C20:5 and C22:5, but not C22:6, in milk fat. The phospholipid (PL) fraction was a rich source of polyunsaturated FA, and feeding diets rich in linoleic acid in conjunction with a source of marine oil increased the *cis*-9 *trans*-11 C18:2 (rumenic acid, RA) and *trans*-11 C18:1 (*trans* vaccenic acid, TVA) concentrations in PL. The FA compositions of plasma triglyceride (TG) and PL were not affected by parity, and few FA were altered by type or level of oil in TG. Concentration of C20:5 was not impacted by type or level of oil in the diet, and C22:6 was not measurable. Total TG in plasma was not affected by treatments. Plasma PL FA were affected by oil type; *trans*-11 C18:1, C20:5, and C22:6 increased, and C18:3 n3

* Corresponding author: ejdepeters@ucdavis.edu

decreased with addition of fish oil, although oil level did not influence FA profile. There was no significant effect on plasma concentration of TG and non-esterified FA for either type or level of oil supplementation, possibly due to the high removal rate of TG by the mammary gland. Although a source of C20:5 and C22:6, dietary fish oil primarily increased the TVA and RA concentrations in milk fat, particularly in the PL fraction.

Keywords: milk fatty acids, fish oil, safflower.

Introduction

Fish lipids are a rich source of long-chain n3 fatty acids. Feeding fish oil, either as oil or as a component of fish meal, to alter the fatty acid (FA) composition of milk fat of lactating cows has been studied by numerous researchers. But, the success in changing the long-chain n3 FA has not been promising (AbuGhazaleh et al., 2001, 2002, 2003; Donovan et al., 2000), and in some cases compromised production performance as determined by decreased dry matter intake, milk yield and/or milk fat concentration (AbuGhazaleh et al., 2001; Donovan et al., 2000; Rego et al., 2005). In addition to modifying FA composition of milk fat, the n3 FA present in fish oil when fed could have beneficial systemic effects in the animal. α-linolenic acid (C18:3 n3) influenced immune response (Lessard et al., 2003) and reproduction (Petit et al., 2001, 2002; Robinson et al., 2002) in dairy cows with broad implications across mammals (Abayasekara and Wathes, 1999). Therefore, feeding fish oil offers area for study.

Doreau and Chilliard (1997) infused 276 g of menhaden type oil into the rumen only to find little eicosapentaenoic acid (C20:5 n3) and docosahexaenoic acid (C22:6 n3) in duodenal contents, which suggested that little of these unsaturated FA escaped rumen biohydrogenation. Gulati et al. (2002) noted that the amount of fish oil fed might influence the escape of unsaturated FA from the rumen. As the proportion of fish oil in *in vitro*rumen incubations increased, there was a corresponding increase in C18:1 *trans* 11 (Gulati et al., 1999), an intermediate indicating incomplete biohydrogenation (Harfoot, 1978). Gulati et al. (1999) proposed that fiber type and level might play a role in biohydrogenation with alfalfa fiber more likely to bind C20:5 and C22:6 than corn silage fiber thus reducing their biohydrogenation. Gulati et al. (2002) also suggested that higher levels of fish oil supplementation might reduce rumen biohydrogenation and enhance incorporation of n3 FA into milk fat.

Rumen-unprotected dietary fish oil increased levels of C20:5 and C22:6 in milk fat, although transfer efficiency of these FA into milk was low, with coefficients ranging from 0.007 to 0.34 (Rego et al., 2005; Palmquist and Griinari, 2006). Rumen infusion of fish oil increased both C20:5 and C22:6 in milk fat above the concentrations observed for control animals receiving no fish oil (0.36 versus 0.08 g/100g total FA and 0.17 versus 0.04 g/100g total FA, respectively), although the effect was more pronounced if the fish oil was infused duodenally (1.47 and 0.47 g/100g total FA). Despite increased levels, transfer efficiency nonetheless was low, with coefficients ranging from 0.031 to 0.165 (Loor *et al.*, 2005).

Rumen-protection of dietary fish oil enhanced transfer efficiencies of C20:5 and C22:6 (0.211 and 0.078. respectively) (Gulati et al., 2003), with greater efficiencies achieved when providing a mixture of soybean and fish oils (0.240 and 0.144, respectively) (Gulati et al., 2002). Data for transfer of unprotected FA vary among goats; Kitessa et al. (2001b) reported coefficients approximating 0.035 for both C20:5 and C22:6, whereas Gulati et al. (1999) reported 0.06 and 0.05, respectively. Transfer rates using protected oil are similar between the two reports, with 0.08 for C20:5 and 0.05-0.07 for C22:6.

Plasma sources of FA for mammary synthesis of lipids include TAG, cholesterol esters, and non-esterified FA (Shennan and Peaker, 2000). Although FA uptake by the mammary epithelial cell is not well understood (Shennan and Peaker, 2000), in their review of lipid metabolism in the mammary gland Barber et al. (1997) proposed that based on the limited data available, FA binding proteins were likely responsible for the transport of FA across the plasma membrane. Subsequent work in this laboratory identified the presence of FA translocator (FAT) in lactating mammary tissue of the rat (Sumathipala et al., 1998). Feeding fish oil increased the concentration of C20:5 and C22:6 in the cholesterol ester and phospholipid of high-density lipoprotein in plasma (Offer et al., 2001). These researchers postulated that based on the 16-fold difference between concentration of C22:6 in plasma compared with milk, there was a poor transfer of C22:6 from blood lipoproteins into milk lipid.

The objective of this study was to evaluate the effect of increasing the amount of fish oil in the dietary lipid supplement of a diet based on alfalfa hay as a method to increase the unsaturated FA composition of milk lipids.

Materials and Methods

Animals, Diets, Treatments, and Experimental Design

Lactating Holstein cows were randomly selected within parity between 60 and 100 days in milk (DIM) from the general herd population at the University of California, Davis Dairy Teaching and Research Facility. Eight lactating Holstein cows, 4 primiparous (P) and 4 multiparous (M), were used in a split-plot Latin square design. The factor associated with the main plots was parity. The error term used to test the effects of parity was cow within parity. The factors associated with the sub plots were type of oil supplemented and level of lipid supplementation.

Periods were 21 d in length with the last 7 d of each period used for collection of data. At the start of the study, primiparous cows averaged 99 ± 10 days-in-milk and had a mean initial body weight of 458 ± 25 kg. All multiparous cows were in their second lactation and averaged 94.5 ± 29 days-in-milk at the start of the trial with a mean initial body weight of 542 ± 52 kg. The same two individuals recorded weights and body condition scores (Wildman et al., 1982) of the cows weekly. All cows received 500 mg of exogenous bST (Posilac®, Monsanto Co., St. Louis, MO) once every 14 d for the duration of the trial. The experiment was conducted from November 2003 to January 2004. The Institutional Animal Care and Use Committee of the University of California at Davis approved care of the animals.

The basal diet was composed of 41% sliced alfalfa-hay and 59% concentrate ingredients (Table 1). The concentrate ingredients were mixed at the Feed Mill at the University of California at Davis. The alfalfa was passed through a bale slicer (Montano Inc., Merced, CA) and put into a mixer wagon (Laird Welding and Manufacturing Co., Merced, CA). The concentrate was added to the alfalfa hay in the mixer wagon and the ingredients were mixed for approximately 20 min.

Table 1. Ingredient and chemical composition of diets (DM basis)

Ingredient composition (g/kg):		
Alfalfa hay, chopped	410	
Beet pulp, shredded	135	
Corn grain, flaked	100	
Barley grain, rolled	100	
Molasses liquid, cane	50	
Almond, hulls	40	
Soybean, meal	30	
Mineral-vitamin supplement	20	
Upland cottonseed, whole	100	
Chemical composition (g/kg):		SD
Crude protein	177.8	8.51
Ether extract	28.7	3.72
Neutral detergent fibre	319.5	19.03
Acid detergent fibre	219.7	9.56
Organic matter	830.0	4.03
Ash	91.2	2.39
Calcium	8.5	0.49
Phosphorus	4.0	0.21
Sodium	9.8	1.20
Zinc	0.3	0.04
Magnesium	10.1	0.26

Dietary treatments were (1) type of lipid supplementation and (2) level of lipid supplementation. Type of lipids was safflower oil (S) and fish oil (F) with the lipid supplement either 100% S or a 75% S + 25% F mixture (w:w). Level of lipid supplementation was either 400 (L, low) or 800 (H, high) g oil/cow daily fed in two equal feedings. The four dietary treatments were:

LS (400 g oil/d; 100% safflower oil),
HS (800 g oil/d; 100% safflower oil),
LFS (400 g oil/d; 75% safflower oil + 25% fish oil)
HFS (800 g oil/d; 75% safflower oil + 25% fish oil)

Each treatment was fed as a total mixed ration. Natural Oils International (Simi Valley, CA) provided both the safflower and fish oils. Wheat mill run (453.6 g/cow/feeding) was used as a carrier for the oils. The wheat mill run and oil mixtures were prepared twice weekly using a Hobart mixer (Hobart Corp. Mixer Model M802, Troy, OH). Oil/wheat mill run mixtures were added to the basal diet and mixed by hand in the feed manger to create each total mixed-ration twice daily.Cows were housed in an open corral with access to a concrete exercise area and sand-bedded free-stalls and fed using Calan Broadbent feeding doors (American Calan Inc., Northwood, NH). Cows were fed twice daily at approximately 0730 and 2030 h following milking for *ad libitum* intake with the amounts of feed offered and refused recorded daily.

Samples of the oil and basal diet were collected weekly to assure no change in composition. Samples of oil were taken at the beginning of every week and frozen. At the end of each period, oil samples were pooled and analyzed for FA composition. The basal diet was sampled daily during the last 7 d of each period with daily samples composited by week for chemical analysis. A portion of the week 3 composite sample was sub-sampled and oven dried at 100˚C for 16 h to determine its dry matter content. The remaining portion was air dried at room temperature (25˚C) for approximately 96 h and then ground through a 1 mm screen using a Wiley mill (Arthur A. Thomas, Philadelphia, PA). The ground samples were stored in sealed plastic containers at room temperature(25˚C) until analyzed.

Cows were milked daily at 0630 and 1930 h, and milk yield was recorded at each milking. Milk samples were collected twice during week 3 of each period. Milk collection consisted of sampling from consecutive evening and morning milkings. Westfalia milk meters and samplers (Westfalia, Naperville, IL) were used to measure milk yield and collect milk samples. Milk samples were preserved in 2-bromo-2-nitro-propane-1, 3-diol (Siliker Labs, Inc., Modesto, CA), and kept refrigerated (5°C) until analyzed immediately following the morning collection. Prior to analyses, milk samples were placed in a water bath (40°C) and warmed, and the a.m. and p.m. samples from each cow were gently mixed to create a daily composite weighted to the a.m. and p.m. milk yields. Milk samples were analyzed for their content of fat, protein, solids-not-fat (SNF), lactose, urea N, and FA composition.

Blood Sampling

Blood was collected during week 3 from a coccygeal vessel of each cow prior to feeding and at 2 h post feeding. Blood was collected into 10 ml evacuated tubes containing sodium heparin (Vacutainer, Becton Dickinson, and Rutherford, NJ). Blood samples were placed in ice immediately after collection, and transported to the laboratory within 15 min where the samples were centrifuged at 3000 x *g* for 10 min. Plasma was recovered and pipetted into 14 ml snap-cap plastic test tubes (Starstedt, Inc, Newton, NC) and stored at -20˚ C for subsequent analysis of glucose, urea N, triglycerides and non-esterified FA (NEFA). Blood plasma for the 2 h post feeding was analyzed for FA composition.

Analytical Procedures

Composite ground samples of diets were analyzed for DM (A.O.A.C., 1995; Method #925.40), total N (A.O.A.C., 1995; Method #984.13) with crude protein (CP) calculated as the = percentage of N * 6.25, EE (A.O.A.C., 1995; Method #920.39C), ash (A.O.A.C., 1995, Method #923.03), NDF (Van Soest et al., 1991), and ADF, cellulose, and lignin (Robertson et al., 1981). Calcium and Mg in feeds were determined (A.O.A.C., 1995; Method #968.08) using an Atomic Absorption Spectrometer (Analyst 300, Perkin-Elmer Instruments), and P in feed was determined according to a Technicon autoanalyzer method N-4C (Kraml, 1966). Samples of feed refusals remaining during week 3 of each period were analyzed for DM

(A.O.A.C., 1995; Method #925.40), total N (A.O.A.C., 1995; Method #984.13), ether extract (A.O.A.C., 1995; Method #920.39C), ash (A.O.A.C., 1995, Method #923.03), NDF (Van Soest et al., 1991), and ADF (Robertson et al., 1981).

Pooled milk samples from each sample day were analyzed for fat, protein, lactose and SNF (A.O.A.C., 1995, Method #972.16) with an infrared analyzer and for urea N by Technicon Autoanalyzer method N-10a (Marsh et al., 1957). Milk samples from sample day one during week 3 were analyzed for total FA by GLC (Crocker et al., 1998; DePeters et al., 2001); for total N (A.O.A.C., 1995; Method #991.20); for total solids by drying in a 100°C oven for 4 h (A.O.A.C., 1995, Method #990.20); and for ash (A.O.A.C., 1995; Method #925.21) using a muffle furnace at 575°C for 3 h then allowed to cool in desiccators before weighing to determine total ash content. A 10-ml aliquot was frozen (-20°C) for subsequent analysis of the FA composition of milk lipids including the triacylglycerol (TG) and phospholipid (PL) fractions. Milk TG was isolated and FA according to Crocker et al. (1998) and DePeters et al. (2001). The PL fraction was extracted by methods of Bitman et al. (1984) with subsequent methylation and GC analysis according to DePeters et al. (2001). Methyl esters of FA were separated and quantified by GLC (Hewlett-Packard model 5890, equipped with flame ionization detector and model 7673A auto injector, Palo Alto, CA) with a Supelco SP-2560 fused silica capillary column (100 m x 0.25 mm i.d., 0.25 μm film thickness; Supelco, Inc., Bellefonte, PA). The carrier gas was H_2 with 0.77 ml/min flow rate (linear flow rate, 27 cm/sec), 220° C injector temperature, 1:88 split ratio, and constant 175° C column temperature. Fatty acid peaks were identified by comparison with a standard mixture containing known FA.

Concentrations of NEFA in plasma were quantified using a commercial assay (NEFA-C, Wako Pure Chemicals Industries Ltd. Osaka, Japan). Plasma glucose was analyzed according to Technicon Autoanalyzer method N-2b (Hoffman, 1937), and plasma urea N was determined by Technicon Autoanalyzer method N-10a (Marsh et al., 1957). Triglyceride concentration in all plasma samples was analyzed using a colorimetric assay kit (Trr22421, Thermo Electron Clinical Chemistry, Noble Park, Australia).

Statistical Analyses

Milk and blood data were analyzed in accordance with a split-plot Latin square model (Gill and Hafs, 1971). In addition, data were analyzed across levels of supplementation (H, L) and type of lipid supplemented (F, FS) where each cow was selected from one of two parity classifications (P, M). Accordingly, our model included terms for period, level of supplementation, and type of lipid supplemented with cow nested within a parity class. Fixed effects were parity, type, level, parity x type, parity x level, type x level, and parity x type x level. In the analyses, cow was included as a random effect such that the appropriate model was a mixed model. Initially we included covariates for dry matter intake (DMI) or fat intake, but these effects were found to be not significant and, therefore, were removed from the analyses presented. Determination of significance for the main effects and interactions was accomplished through PROC MIXED of the Statistical Analysis System (SAS version 8.03). Significance was declared at $P < 0.05$.

The analysis of the FA present in oils followed that of a two-way classification without interaction. The factors in this analysis of variance were period and type of lipid supplemented. Determination of significance was accomplished through PROC GLM of the

Statistical Analysis System (SAS version 8.03). Because of mastitis, the data from period 4 of one cow were deleted as it was determined that her mastitis infection was not related to treatment. Statistical analyses were performed with this cow represented as missing data for the last period.

Results

Diet

A basal diet prepared as a total mixed ration was fed to all cows (Table 1). Oils added to the basal diet included safflower oil and fish oil (Table 2). Safflower oil was high in linoleic acid (77.11g/100 g FA) and total unsaturated FA (90.59 g/100 g FA). Fish oil was characterized by its content of long-chain FA greater than 20 carbons, particularly rich in C20:5 n3 (eicosapentaenoic acid; EPA; 14.54 g/100 g FA) and C22:6 n3 (docosahexaenoic acid; DHA; 15.05g/100 g FA). Neither EPA nor DHA was detectable in the basal diet or the safflower oil. Levels of C22:5 n6 (docosapentaenoic acid; DPA) were low in the basal diet and fish oil (0.005g/100g FA and 2.89 g/100g FA, respectively), and not detectable in the safflower oil.

Table 2. Average fatty acid composition of basal diet (g/100 g of DM), safflower oil and fish oil (g/100 g FA)

Fatty acid	Basal diet	SD	Safflower oil	SD	Fish oil	SD
C4:0	-	-	-	-	-	-
C6:0	-	-	-	-	-	-
C8:0	-	-	-	-	-	-
C10:0	-	-	-	-	-	-
C12:0	0.009	0.001	0.023	0.006	0.094	0.002
C13:0	-	-	-	-	0.047	0.002
C14:0	0.018	0.002	0.111	0.003	9.621	0.035
C14:1 *cis*	-	0.003	-	-	0.048	0.010
C15:0	0.004	0.000	0.018	0.001	0.833	0.022
C16:0	0.633	0.055	6.710	0.020	21.316	0.494
C16:1 *trans*	-	-	-	-	0.337	0.005
C16:1 *cis*	0.010	0.001	0.088	0.002	12.241	0.173
C17:0	0.010	0.001	0.038	0.012	0.680	0.015
C18:0	0.082	0.006	2.515	0.009	4.040	0.080
C18:1 *trans* 5	0.011	0.005	-	-	-	-
C18:1 *trans* 7	-	-	-	-	-	-
C18:1 *trans* 6&8	0.001	0.001	-	-	-	-
C18:1 *trans* 9	0.002	0.001	-	-	0.130	0.010
C18:1 *trans* 10	0.002	0.001	-	-	-	-
C18:1 *trans* 11	0.001	0.001	-	-	0.012	0.001

Table 2. (Continued)

Fatty acid	Basal diet	SD	Safflower oil	SD	Fish oil	SD
C18:1 *trans* 12	0.001	0.000	-	-	1.635	0.042
C18:1 *trans* 13&14	0.002	0.000	-		0.000	0.000
C18:1 *cis* 9&10	0.342	0.047	12.615	0.011	8.688	0.020
C18:1 *cis* 11	0.019	0.002	0.687	0.002	3.638	0.017
C18:1 *cis* 12	0.002	0.001	-	-	0.030	0.001
C18:1 *cis* 13	0.003	0.001	-	-	0.104	0.010
C18:1 *trans* 16	-	-	-	-	0.017	0.002
C18:2	0.984	0.209	77.109	0.024	1.523	0.015
C18:3	0.143	0.028	0.086	0.008	1.359	0.023
RA[1]	-	-	-	-	-	-
C20:4	-	-	-	-	1.134	0.008
C20:5	-	-	-	-	14.536	0.129
C22:5	0.005	0.000	-	-	2.886	0.018
C22:6	-	-	-	-	15.048	0.239
SCFA[2]	-	-	-	-	-	-
MCFA[3]	0.027	0.002	0.134	0.009	9.716	0.034
LCFA[4]	1.551	0.284	92.325	0.020	15.611	0.084
Total *trans*[5]	0.021	0.006	-	-	2.132	0.041
Total Unsat *cis*[6]	1.505	0.282	90.585	0.020	61.207	0.575
PUFA[7]	1.133	0.233	77.195	0.031	36.487	0.405

[1]RA = Rumenic acid, *cis*-9, *trans*-11 C18:2
[2]Short chain fatty acids = sum of C4:0 to C8:0
[3]Medium chain fatty acids = sum of C10:0 to C17:0
[4]Long chain fatty acids = sum of C18:0 to C20:4.
[5]Sum of *trans* fatty acids.
[6]Sum of unsaturated *cis* fatty acids.
[7]Polyunsaturated fatty acids.

Production Performance: Milk Yield and Composition

Multiparous cows produced significantly (P=0.03) more milk than primiparous cows (Table 3). Milk production for multiparous and primiparous cows was not affected by type or level of oil supplemented and averaged 41.02 and 48.4 kg/d, respectively. Yield of 4% fat-corrected milk (FCM) was not different for type or level of oil supplemented. There was a significant interaction (P<0.03) between parity and level of oil supplemented for FCM with primiparous cows producing less FCM on high oil diets while the multiparous cows produced more FCM. This response is reflected in the milk fat percentages and yields, which were not different for type or level of oil but there was a parity by level effect for percentage (P=0.06) and yield (P<0.03) for milk fat. Milk fat content and yield were reduced with the higher oil intake for primiparous cows, in contrast to the multiparous cows which had increased milk fat content and yield with the higher oil intake.

Table 3A. Yield and composition of milk from primiparous and multiparous cows supplemented with either safflower oil or a combination of safflower and fish oils

		Safflower Oil		Safflower/Fish Oil								
	Parity	400	800	400	800	SE	PE[1]	TE[2]	LE[3]	PE X TE	PE X LE	TE X LE
							P					
Milk (kg/d)	P[4]	41.3	40.6	41.3	40.8	2.24	0.03	0.61	0.51	0.67	0.25	0.96
	M[5]	46.8	48.9	47.9	50.0							
4%FCM[6] (kg/d)	P	44.3	38.6	41.8	40.9	3.28	0.12	0.55	0.80	0.52	0.03	0.41
	M	45.6	49.5	47.3	51.6							
FCM/DMI[7]	P	1.80	1.61	1.77	1.71	0.094	0.10	0.47	0.77	0.82	0.05	0.55
	M	1.48	1.62	1.52	1.70							
Fat (g/kg)	P	44.6	36.6	40.6	39.9	3.01	0.93	0.73	0.61	0.58	0.06	0.33
	M	38.0	40.7	39.7	42.1							
Fat (kg/d)	P	1.85	1.49	1.68	1.63	0.177	0.23	0.59	0.94	0.53	0.03	0.34
	M	1.79	2.00	1.88	2.11							
Crude protein (g/kg)	P	33.3	33.9	32.5	34.0	1.12	0.81	0.76	0.07	0.71	0.98	0.94
	M	33.1	34.4	33.5	34.1							
Crude protein (kg/d)	P	1.38	1.37	1.34	1.37	0.075	0.03	0.87	0.07	0.50	0.16	0.96
	M	1.55	1.68	1.59	1.69							

Table 3A. (Continued)

	Parity	Safflower Oil 400	Safflower Oil 800	Safflower/Fish Oil 400	Safflower/Fish Oil 800	SE	PE[1]	TE[2]	LE[3]	PE X TE	PE X LE	TE X LE
							P					
Lactose (g/kg)	P	52.4	50.1	51.4	51.3	1.02	0.15	0.90	0.89	0.97	0.78	0.22
	M	49.4	50.1	48.8	50.9							

[1]PE = Parity effects; [2]TE = Type effects, [3]LE = Level effects, PE x TE = Parity x Type, PE x LE = Parity x Level, TE x LE = Type x Level
[4]P = Primiparous cows.
[5]M = Multiparous cows.
[6]FCM = Fat-corrected milk
[7]FCM/DMI = Fat corrected milk / Dry matter intake

Table 3B. Yield and composition of milk from primiparous and multiparous cows supplemented with either safflower oilor a combination of safflower and fish oils

	Parity	Safflower Oil 400	Safflower Oil 800	Safflower/Fish Oil 400	Safflower/Fish Oil 800	SE	PE[1]	TE[2]	LE[3]	PE X TE	PE X LE	TE X LE
							P					
Lactose, kg/d	P[4]	1.08	1.02	1.06	1.05	0.062	0.08	0.70	0.48	0.86	0.08	0.46
	M[5]	1.16	1.23	1.16	1.26							
SNF[8] (g/kg)	P	91.0	89.8	89.9	90.9	0.84	0.45	0.94	0.28	0.97	0.23	0.26
	M	89.1	90.2	89.1	90.1							
SNF (kg/d)	P	3.76	3.65	3.71	3.71	0.199	0.04	0.65	0.36	0.71	0.15	0.75
	M	4.18	4.41	4.25	4.50							

Table 3B. Yield and composition of milk from primiparous and multiparous cows supplemented with either safflower oilor a combination of safflower and fish oils

	Parity	Safflower Oil		Safflower/Fish Oil		SE	PE[1]	TE[2]	LE[3]	PE X TE	PE X LE	TE X LE
		400	800	400	800							
							--------	--------	--------	P	--------	--------
Total solids (g/kg)	P	122.9	124.5	127.1	123.2	3.41	0.94	0.72	0.72	0.43	0.56	0.99
	M	126.3	123.8	123.0	126.1							
Total solids (kg/d)	P	5.08	5.06	5.25	5.03	0.337	0.06	0.56	0.59	0.90	0.18	0.83
	M	5.93	6.06	5.88	6.32							
Ash (g/kg)	P	6.9	7.0	6.8	7.1	0.14	0.56	0.84	0.41	0.79	0.04	0.79
	M	7.0	7.1	7.2	6.9							
Ash (kg/d)	P	0.28	0.28	0.28	0.29	0.016	0.02	0.58	0.36	0.73	0.73	0.92
	M	0.33	0.34	0.34	0.34							

[1]PE = Parity effects; [2]TE = Type effects, [3]LE = Level effects, PE x TE = Parity x Type, PE x LE = Parity x Level, TE x LE = Type x Level
[4]P = Primiparous cows.
[5]M = Multiparous cows.
[6]FCM = Fat-corrected milk
[7]SNF = Solids not fat

Table 4. Nutrient intake of primiparous and multiparous cows supplemented with either safflower oil or a combination of safflower and fish oils

Intake		Safflower Oil		Safflower/Fish Oil								
kg/day	Parity	400	800	400	800	SE	PE[1]	TE[2]	LE[3]	PE X TE	PE X LE	TE X LE
							P					
DM	P[4]	24.7	24.0	23.5	23.7	1.15	0.001	0.730	0.702	0.514	0.992	0.756
	M[5]	30.8	30.6	31.0	30.8							
DMI (% BW)	P	4.45	4.27	4.18	4.21	0.570	0.298	0.811	0.641	0.306	0.912	0.783
	M	4.57	4.56	4.71	4.63							
OM[6]	P	19.5	18.6	18.6	18.4	0.95	0.001	0.759	0.352	0.507	0.994	0.748
	M	24.6	24.1	24.9	24.3							
Ash	P	2.18	2.08	2.09	2.09	0.098	0.0005	0.863	0.408	0.627	0.950	0.631
	M	2.77	2.71	2.79	2.74							
CP[7]	P	4.41	4.20	4.20	4.20	0.193	0.0007	0.727	0.375	0.573	0.982	0.595
	M	5.54	5.41	5.55	5.46							
NDF[8]	P	7.71	7.22	7.28	7.20	0.467	0.004	0.758	0.415	0.634	0.842	0.843
	M	9.53	9.45	9.67	9.40							
ADF[9]	P	4.72	4.31	4.46	4.47	0.343	0.009	0.825	0.435	0.985	0.871	0.722
	M	5.85	5.79	5.88	5.68							
EE[10]	P	1.11	1.46	1.07	1.45	0.031	0.0008	0.619	<.0001	0.408	0.677	0.409
	M	1.29	1.65	1.28	1.67							
% Oil	P	1.63	3.38	1.70	3.41	0.124	0.0011	0.941	<.0001	0.632	0.022	0.836
	M	1.31	2.62	1.29	2.57							

Intake		Safflower Oil		Safflower/Fish Oil								
kg/day	Parity	400	800	400	800	SE	PE[1]	TE[2]	LE[3]	PE X TE	PE X LE	TE X LE
							P					
BCS	P	3.13	2.88	2.94	3.00	0.196	0.27	0.669	0.669	0.940	0.529	0.529
	M	2.69	2.75	2.69	2.66							

[1]PE = Parity effects; [2]TE = Type effects, [3]LE = Level effects, PE x TE = Parity x Type, PE x LE = Parity x Level, TE x LE = Type x Level
[4]P = Primiparous cows.
[5]M = Multiparous cows.
[6]OM = Organic matter
[7]CP = Crude protein
[8]NDF = Neutral detergent fiber
[9]ADF = Acid detergent fiber
[10]EE = Ether extract

Table 5. Mineral composition of milk (mg/100g of milk) from primiparous and multiparous cows supplemented with either safflower oil or a combination of safflower and fish oils

		Safflower Oil		Safflower/Fish Oil								
	Parity	400	800	400	800	SE	PE[1]	TE[2]	LE[3]	PE X TE	PE X LE	TE X LE
							P					
Calcium	P[4]	82.52	84.67	86.96	87.23	5.984	0.50	0.31	0.45	0.67	0.95	0.10
	M[5]	70.97	87.04	86.68	79.31							
Potassium	P	161.68	158.28	159.03	159.81	8.464	0.96	0.68	0.51	0.62	0.38	0.56
	M	148.72	166.32	162.67	162.80							

Table 5. (Continued)

	Parity	Safflower Oil 400	Safflower Oil 800	Safflower/Fish Oil 400	Safflower/Fish Oil 800	SE	PE[1]	TE[2]	LE[3]	PE X TE	PE X LE	TE X LE
							---P					-----------------------------------
Phosphorus	P	108.60	107.90	103.70	106.90	4.716	0.78	0.70	0.28	0.30	0.61	0.99
	M	101.80	107.10	105.10	103.40							
Magnesium	P	12.18	11.58	11.11	11.02	0.804	0.24	0.65	0.76	0.27	0.16	0.65
	M	9.46	10.72	10.53	10.50							
Sodium	P	60.02	58.46	57.12	60.99	6.871	0.11	0.18	0.71	0.17	0.90	0.58
	M	58.29	68.62	79.46	73.78							

[1]PE = Parity effects; [2]TE = Type effects, [3]LE = Level effects, PE x TE = Parity x Type, PE x LE = Parity x Level, TE x LE = Type x Level
[4]P = Primiparous cows.
[5]M = Multiparous cows.

Primiparous cows showed the greatest decrease in milk fat yield (24%) on the S diet and decreased only 3% on the FS diet when dietary oil intake was increased. Milk fat yield increased approximately 12% on both diets when the level of dietary oil increased for multiparous cows. Milk composition of total protein, lactose, SNF, total solids, and ash was similar for all treatments (Tables 3A and 3B). Ash concentration showed a significant interaction (P=0.04) between parity and level of oil supplementation. Yields for milk protein (P=0.03), SNF (P=0.04), and ash (P=0.02) were significantly higher for multiparous cows compared to primiparous cows as expected. Milk protein yield tended (P=0.07) to be higher with the higher level of oil supplementation, and multiparous cows tended (P=0.06) to yield more total solids compared to primiparous cows.

Multiparous cows had higher (P=0.001) dry matter intake (DMI) than primiparous cows, averaging 30.8 and 24.0 kg/d, respectively (Table 4). Multiparous cows' DMI averaged 4.6% of body weight (BW) while primiparous cows' DMI averaged 4.3% of their body weight calculated as (DMI/body weight) *100%. Organic matter, ash, CP, NDF, ADF, EE, and percentage oil intake were higher (P≤0.004 for all) for multiparous than primiparous cows. Percentage oil intake was calculated as the proportion of the daily DMI that was represented by either 400 or 800 g oil/cow daily. Because primiparous cows consumed less dry matter, the supplemental oil, either 400 or 800 g/d, contributed a higher proportion of the diet. Ether extract and percent oil intake were higher (P≤0.001) in those diets containing higher levels of oil supplementation. A significant (P=0.022) interaction between parity and level of supplemental oil was observed for percentage of oil intake. The level of oil supplementation in the present study was adequate to increase daily intake of total FA without negatively affecting DMI. The mineral composition of milk was not affected by type or level of lipid supplementation, and concentrations agree with previous observations (Johnson, 1974; McCaughey et al., 2005).

Fatty Acid Composition and Yield of Milk Triacylglycerol

Level Effects

The FA composition of milk fat was altered by level of dietary lipid (Table 6). In the current study, the total concentration of short-chain (SCFA) and medium-chain (MCFA) FA decreased (P<0.0001) as the level of oil supplemented increased. Most of the individual SCFA and MCFA significantly decreased in milk fat, except for *cis*-C14:1, *trans*-C16:1, and *cis*-C16:1. Concentrations of long-chain FA (LCFA) increased with increasing level of lipid supplementation. Stearic acid (C18:0) was not affected by level of lipid supplementation, however there was a significant level by type effect. Concentration of C18:0 increased in milk fat with increasing amount of S fed but decreased with increasing amount of FS fed. The concentration of all *trans* monoenoic isomers of C18:1 significantly increased with the higher level (HS and HFS) of oil supplementation. There was an approximate increase of 44% of the *trans*-10 C18:1 isomer for both parities and treatments even though there was no detrimental effect on milk fat concentration (Table 3). The *trans*-16 C18:1 isomer also had a significant type by level interaction where concentration increased with the higher level of S but not the higher level of FS. The concentration of total *trans* FA (TFA) significantly increased (P<0.001) from 7.06 to 11.43 g/100g lipid as the level of oil supplemented increased. As the level of oil in the diet increased the concentration of C18:3 n3 decreased (P=0.0001). Rumenic acid (*cis*-9 *trans*-11 C18:2) and TVA (*trans*-11C18:1; *trans* vaccenic acid) concentrations increased as the level of dietary oil increased. Rumenic acid and TVA concentrations increased, on average, 67.5% and 77%, respectively, for the S diet and 145% and 90%, respectively, for the FS diet as dietary oils increased from low to high. An interaction between parity and level of supplementation approached significance (P=0.08) for concentration of rumenic acid in milk fat, likely due to the fact that fixed amounts of oil were fed to both primiparous and multiparous, such that the lower DMI of primiparous cows increased the concentration of FA in their diets.

Oleic acid (*cis*-9 C18:1) could not be separated from *cis*-10 C18:1, but the latter is in small concentrations so *cis*-9 and -10 C18:1 together are referred to as oleic acid. The concentration of *cis*- 9 and10 C18:1 was not affected by level of oil supplementation, but a tendency (P=0.06) for interaction between type and level of supplemental fat was observed for its concentration, where oleic acid increased with higher level of S feeding, but not with a higher level of FS feeding. Increasing lipid supplementation from 400 to 800 g increased the concentration of *cis*-12 C18:1 in milk fat, but the increase was greater for S than SF (P=0.0002; TE x LE). The *cis*-13 C18:1 isomer concentration increased (P<0.0001) with level of oil supplementation, but the change was small. The concentration of C18:2 increased (P<0.004) with the level of oil feeding, but C18:2 concentrations for primiparous and multiparous cows responded differently with type of oil fed (P=0.013; TE x LE). Changes in C18:3 n3 and C20:4 n6 were small, but increasing level of oil supplementation decreased (P<0.0001) their concentrations in milk fat. Changes in C20:5 n3 and C22:5 n6 were significant but biologically small. There was no change in C22:6 n3 concentration due to level of oil supplemented. Concentration of total unsaturated *cis* FA and PUFA were not different for level of oil supplementation, but there were type by level interactions where both tended to increase with higher S in the diet but decrease with higher FS in the diet.

Yields of SCFA (P=0.011) and MCFA (P=0.011) decreased as the level of oil supplementation increased (Table 8). Yield of FA was calculated without consideration for the glycerol content of the lipid. In contrast, daily yields of LCFA (P=0.0002) and PUFA (P<0.0001) increased with additional supplemental oil. Yields of individual FA reflect changes in concentration. Yields of *cis*-C14:1, *cis*-C14:1, *trans*-C16:1, C18:0, and *cis*-9 and -10 C18:1 were not different for level of oil supplementation. The daily yields of TVA and rumenic acid increased as the level of oil in the diet was increased. TVA daily yields increased, on average, 81% for the S diet and 149% on the FS diet as the dietary oil levels increased. Rumenic acid yields increased as dietary oils increased from low to high, on average, 71% for the S diet and 144% on the FS diet.

Type Effects

The composition of milk fat was altered by the type of dietary lipid supplemented (Tables 6A-6F). Concentration of SCFA was not affected by type of lipid (Table 6E). Concentration of MCFA was greater (P = 0.004) for FS than S as a result of the higher C16:0. Many of the individual LCFA showed type effects, as fish oil was included in the diet. Stearic acid (C18:0) decreased (P=0.003) in both primiparous and multiparous cows when fish oil was included in the lipid supplement. There was an overall decrease in both primiparous and multiparous cows in the concentration of *cis*-9 and -10 C18:1 with fish oil. Concentration of *cis*-9 and-10 C18:1 decreased by 15% for primiparous cows fed the HFS diet when compared to primiparous cows fed the HS diet. The inclusion of fish oil significantly (P<0.008) affected the concentration of C18:2 n6. Primiparous cows showed elevated concentrations of C18:2 n6, while multiparous cows remained unchanged (P=0.0003; PE x TE).

Cows receiving the S treatments showed a slight increase (P=0.005) in the concentration of LCFA as compared to those receiving the FS treatments. The concentration of PUFA tended (P<0.07) to be lower with the addition of fish oil. The concentration of total *trans* FA increased (P<0.02) with the addition of fish oil and tended (P<0.09; TE x LE) to be greater with the higher level of oil. Ratio 1 (ratio 1 = *cis*-9 and -10 C18:1/C18:0) significantly (P<0.02) increased by 11.4% for both primiparous and multiparous cows with the addition of fish oil in the diet. Ratio 2 (ratio 2 = rumenic acid/TVA) did not show significant effects for type of lipid supplemented. Fish oil supplementation increased the concentration of TVA (*trans*-11 C18:1, P=0.0001) and rumenic (*cis*-9 *trans*-11 C18:2, P=0.0002) acid in milk fat. Type by level interaction was significant (P < 0.004 for both) since the increases in TVA and rumenic acid were greater for increasing levels of fish oil than safflower oil. When concentration levels from cows receiving the HFS diet were compared to those receiving the HS diet, rumenic acid concentration increased approximately 89%, and TVA increased approximately 90%. Inclusion of fish oil significantly increased concentration of *cis*-11 C18:1 and decreased concentrations of *cis*-12 C18:1 and *trans*-16 C18:1. The importance of these changes is unknown. Fish oil increased (P<0.0001) the concentration of C20:5 n3 and C22:6 n3 in milk fat although the changes were small. There were no significant interaction effects for either FA. Daily yields of C18:0 were lower (P=0.003) with the addition of fish oil; C18:0 decreased 26% on average for cows of both parities. Yields of TVA and rumenic acid increased (P<0.0001) by 90% and 91%, respectively, when compared to cows (within parity) on the FS diet. Yields of C20:5 n3 and C22:6 n3 increased (P<0.0001) two-fold but remained biologically small. Yields of PUFA tended (P<0.06) to be higher in those cows fed the FS diets.

Fatty Acid Composition of Milk Phospholipid

Level Effects:

Short-chain FA were not incorporated into the phospholipid (PL) fraction of milk fat (Table 8). Proportion of MCFA decreased (P=0.001) with level of oil feeding since C14:0, C15:0, and C16:0 were reduced with increasing level of oil. Proportion of LCFA increased (P=0.001) with 800g versus 400g of oil. The proportion of most *trans* FA significantly increased with increased oil feeding except *trans*-10 C18:1. The proportion of *cis*-9 and -10 C18:1 was decreased (P < 0.01) with increased oil feeding whereas concentrations of *cis*-12 and *cis*-13 were increased (Table 8). Both parities showed a negative effect (P<0.0001) for C18:3 n3 when the level of oil supplemented increased. The lowest concentration for both parities occurred on the HFS treatment (P=0.003; TE x LE). Linolenic, C20:4 n6, and DPA decreased and rumenic acid increased in milk PL with increased level of oil feeding. The concentration of total *trans*-FA in the PL fraction increased (P<0.0001) as the level of dietary oil increased.

Table 6A. Total fatty acid composition of milk fat (g/100g fat) from primiparous and multiparous cows supplemented with either safflower oil or a combination of safflower and fish oils

		Safflower Oil		Safflower/Fish Oil								
Fatty Acid	Parity	400	800	400	800	SE	PE[1]	TE[2]	LE[3]	PE X TE	PE X LE	TE X LE
							P					
C4:0	P[4]	4.19	3.79	4.37	3.45	0.250	0.195	0.817	0.004	0.751	0.194	0.765
	M[5]	4.57	4.11	4.41	4.29							
C6:0	P	1.77	1.48	1.78	1.40	0.144	0.048	0.987	0.0002	0.552	0.878	0.768
	M	2.22	1.78	2.22	1.91							
C8:0	P	0.86	0.69	0.85	0.67	0.098	0.058	0.771	0.0001	0.528	0.479	0.619
	M	1.18	0.90	1.17	0.98							
C10:0	P	1.71	1.33	1.67	1.36	0.234	0.072	0.632	<.0001	0.558	0.191	0.502
	M	2.45	1.84	2.46	1.99							
C12:0	P	1.88	1.52	1.84	1.62	0.228	0.057	0.359	<.0001	0.517	0.093	0.402
	M	2.65	2.04	2.72	2.22							
C13:0	P	0.08	0.07	0.08	0.07	0.014	0.065	0.850	0.0003	0.272	0.061	0.396
	M	0.13	0.10	0.13	0.11							

[1]PE = Parity effects; [2]TE = Type effects, [3]LE = Level effects, PE x TE = Parity x Type, PE x LE = Parity x Level, TE x LE = Type x Level
[4]P = Primiparous cows. [5]M = Multiparous cows.

Table 6B. Total fatty acid composition of milk fat (g/100g of fat) from primiparous and multiparous cows supplemented with either safflower oil or a combination of safflower and fish oils

		Safflower Oil		Safflower/Fish Oil								
Fatty Acid	Parity	400	800	400	800	SE	PE[1]	TE[2]	LE[3]	PE X TE	PE X LE	TE X LE
							P					
C14:0	P[4]	8.20	6.94	8.26	7.64	0.558	0.067	0.222	0.000	0.863	0.368	0.248
	M[5]	10.03	8.31	10.02	8.90							
C14:1 *cis*	P	0.50	0.47	0.51	0.53	0.082	0.142	0.164	0.236	0.775	0.305	0.269
	M	0.71	0.61	0.72	0.69							
C15:0	P	0.78	0.68	0.78	0.73	0.045	0.121	0.157	0.000	0.815	0.363	0.206
	M	0.90	0.76	0.90	0.82							
C16:0	P	25.35	22.50	26.17	25.56	0.678	0.355	0.005	<.0001	0.222	0.075	0.039
	M	26.86	23.42	27.53	24.91							
C16:1 *trans*	P	0.40	0.45	0.48	0.67	0.073	0.212	0.060	0.136	0.365	0.366	0.495
	M	0.67	0.40	0.45	0.45							
C16:1 *cis*	P	1.08	0.98	1.15	1.31	0.126	0.465	0.010	0.988	0.390	0.602	0.165
	M	0.98	0.93	1.07	1.07							

Table 6B. (Continued)

Fatty Acid	Parity	Safflower Oil		Safflower/Fish Oil		SE	PE[1]	TE[2]	LE[3]	PE X TE	PE X LE	TE X LE
		400	800	400	800							
							----------	----------	----P----	----------	----------	----------
C17:0	P	0.61	0.51	0.66	0.58	0.021	0.005	0.002	<.0001	0.689	0.527	0.467
	M	0.55	0.47	0.59	0.53							

1PE = Parity effects; 2TE = Type effects, 3LE = Level effects, PE x TE = Parity x Type, PE x LE = Parity x Level, TE x LE = Type x Level
4P = Primiparous cows
5M = Multiparous cows

Table 6C. Total fatty acid composition of milk fat (g/100g of fat) from primiparous and multiparous cows supplemented with either safflower oil or a combination of safflower and fish oils

Fatty Acid	Parity	Safflower Oil		Safflower/Fish Oil		SE	PE[1]	TE[2]	LE[3]	PE X TE	PE X LE	TE X LE
		400	800	400	800							
							----------	----------	----P----	----------	----------	----------
C18:0	P[4]	15.79	16.39	14.95	12.50	0.993	0.930	0.003	0.792	0.880	0.119	0.024
	M[5]	15.12	16.70	13.81	13.57							
C18:1 t5	P	0.03	0.04	0.03	0.04	0.037	0.616	0.360	0.001	0.997	0.532	0.849
	M	0.03	0.05	0.03	0.04							
C18:1 t7	P	0.03	0.04	0.03	0.04	0.003	0.735	0.420	0.001	0.957	0.056	0.489
	M	0.03	0.04	0.03	0.03							

Fatty Acid	Parity	Safflower Oil		Safflower/Fish Oil		SE	PE[1]	TE[2]	LE[3]	PE X TE	PE X LE	TE X LE
		400	800	400	800							
							P					
C18:1 *trans* 6&8	P	0.48	0.66	0.47	0.78	0.056	0.415	0.167	<.0001	0.957	0.677	0.291
	M	0.43	0.63	0.46	0.70							
C18:1 t9	P	0.47	0.66	0.46	0.69	0.035	0.396	0.306	<.0001	0.782	0.858	0.863
	M	0.41	0.64	0.45	0.65							
C18:1 t10	P	0.86	1.05	0.67	1.75	0.323	0.398	0.703	0.053	0.473	0.497	0.438
	M	0.70	1.09	0.69	0.94							
C18:1 t11	P	1.79	3.19	2.18	6.16	0.493	0.196	0.0001	<.0001	0.612	0.140	0.004
	M	1.39	2.45	2.06	4.54							
C18:1 t12	P	0.97	1.38	0.94	1.28	0.065	0.697	0.895	<.0001	0.129	0.123	0.416
	M	0.83	1.34	0.91	1.37							
C18:1 *trans* 13&14	P	1.59	2.19	1.48	1.88	0.122	0.315	0.209	<.0001	0.177	0.182	0.243
	M	1.52	2.30	1.60	2.23							

[1]PE = Parity effects; [2]TE = Type effects, [3]LE = Level effects, PE x TE = Parity x Type, PE x LE = Parity x Level, TE x LE = Type x Level
[4]P = Primiparous cows
[5]M = Multiparous cows

Table 6D. Total fatty acid composition of milk fat (g/100g of fat) from primiparous and multiparous cows supplemented with either safflower oil or a combination of safflower and fish oils

Fatty Acid	Parity	Safflower Oil		Safflower/Fish Oil		SE	PE[1]	TE[2]	LE[3]	PE X TE	PE X LE	TE X LE
		400	800	400	800							
							P					
C18:1 *cis* 9&10	P[4]	23.34	24.23	23.29	21.00	1.218	0.064	0.019	0.900	0.819	0.196	0.061
	M[5]	19.62	21.20	18.97	19.08							
C18:1 c11	P	0.48	0.44	0.50	0.52	0.027	<.00001	0.018	0.665	0.862	0.367	0.150
	M	0.37	0.37	0.39	0.44							
C18:1 c12	P	1.10	2.00	0.86	1.04	0.097	0.991	<.0001	<.0001	0.002	0.200	0.0002
	M	0.97	1.76	0.86	1.41							
C18:1 c13	P	0.10	0.12	0.09	0.13	0.007	0.087	0.540	<.0001	0.555	0.606	0.583
	M	0.08	0.12	0.08	0.11							
C18:1 t16	P	0.67	0.83	0.62	0.56	0.040	0.776	0.0002	0.003	0.328	0.008	0.003
	M	0.62	0.85	0.59	0.66							
C18:2	P	3.44	3.59	3.21	3.20	0.163	0.452	0.008	0.004	0.0003	0.113	0.013
	M	2.98	3.32	3.18	3.25							
C18:3	P	0.52	0.45	0.53	0.42	0.033	0.984	0.266	0.0001	0.143	0.529	0.624
	M	0.49	0.42	0.54	0.47							
RA[6]	P	0.62	1.07	0.78	2.16	0.182	0.180	0.0002	<.0001	0.302	0.085	0.004
	M	0.51	0.82	0.69	1.45							

[1]PE = Parity effects; [2]TE = Type effects, [3]LE = Level effects, PE x TE = Parity x Type, PE x LE = Parity x Level, TE x LE = Type x Level
[4]P = Primiparous cows.
[5]M = Multiparous cows.
[6]RA = Rumenic acid, *cis*-9, *trans*-11 C18:2

Table 6E. Total fatty acid composition of milk fat (g/100g of fat) from primiparous and multiparous cows supplemented with either safflower oil or a combination of safflower and fish oils

		Safflower Oil		Safflower/Fish Oil								
Fatty Acid	Parity	400	800	400	800	SE	PE[1]	TE[2]	LE[3]	PE X TE	PE X LE	TE X LE
							P					
C20:4	P[4]	0.13	0.11	0.11	0.08	0.007	0.136	0.001	<.0001	0.338	0.676	0.119
	M[5]	0.13	0.12	0.13	0.10							
C20:5	P	0.04	0.04	0.06	0.05	0.004	0.632	<.00001	0.068	0.577	0.613	0.578
	M	0.04	0.03	0.05	0.05							
C22:5	P	0.09	0.09	0.11	0.08	0.006	0.026	0.107	0.017	0.903	0.071	0.434
	M	0.07	0.07	0.08	0.08							
C22:6	P	0.02	0.02	0.03	0.03	0.003	0.397	<.00001	0.960	0.526	0.147	0.193
	M	0.02	0.01	0.03	0.04							
SCFA[6]	P	6.83	5.95	7.00	5.52	0.430	0.059	0.938	0.001	0.629	0.526	0.998
	M	7.97	6.78	7.76	7.18							
MCFA[7]	P	40.60	35.46	41.58	40.06	1.566	0.096	0.004	<.0001	0.515	0.089	0.059
	M	45.64	38.88	46.57	41.73							
LCFA[8]	P	52.40	58.43	51.22	54.25	1.651	0.052	0.005	<.0001	0.648	0.143	0.070
	M	46.21	54.20	45.46	50.91							

[1]PE = Parity effects; [2]TE = Type effects, [3]LE = Level effects, PE x TE = Parity x Type, PE x LE = Parity x Level, TE x LE = Type x Level
[4]P = Primiparous cows.
[5]M = Multiparous cows.
[6]Short chain fatty acids = sum of C4:0 to C8:0.
[7]Medium chain fatty acids = sum of C10:0 to C17:0. [8]Long chain fatty acids = sum of C18:0 to C20:4.

Table 6F. Total fatty acid composition of milk fat (g/100g of fat) from primiparous and multiparous cows supplemented with either safflower oil or a combination of safflower and fish oils

		Safflower Oil		Safflower/Fish Oil								
Fatty Acid	Parity	400	800	400	800	SE	PE[1]	TE[2]	LE[3]	PE X TE	PE X LE	TE X LE
							P					
Ratio 1[6]	P[4]	1.48	1.50	1.61	1.69	0.154	0.352	0.016	0.473	0.606	0.789	0.559
	M[5]	1.32	1.31	1.39	1.45							
Ratio 2[7]	P	0.35	0.34	0.36	0.36	0.035	0.881	0.920	0.491	0.502	0.909	0.598
	M	0.36	0.34	0.34	0.34							
Total *trans*[8]	P	7.29	10.49	7.38	13.86	0.945	0.322	0.016	<.0001	0.758	0.422	0.089
	M	6.32	9.79	7.25	11.59							
Total unsat *cis* FA[9]	P	29.73	30.51	29.60	27.36	1.331	0.074	0.036	0.876	0.609	0.177	0.069
	M	25.49	27.18	25.23	25.37							
Sat. fats	P	61.24	55.90	61.39	55.57	1.652	0.066	0.621	<.0001	0.727	0.724	0.999
	M	66.67	60.43	60.43	60.17							
PUFA[10]	P	4.23	4.28	4.06	3.86	0.182	0.477	0.063	0.765	0.0001	0.053	0.009
	M	3.74	3.97	4.00	3.97							

[1]PE = Parity effects; [2]TE = Type effects, [3]LE = Level effects, PE x TE = Parity x Type, PE x LE = Parity x Level, TE x LE = Type x Level
[4]P = Primiparous cows.
[5]M = Multiparous cows.
[6]Ratio 1 = *cis*-9&10 C18:1 / C18:0.
[7]Ratio 2 = RA / *trans*-11 C18:1
[8]Sum of *trans* fatty acids.
[9]Sum of unsaturated *cis* fatty acids. [10]Polyunsaturated fatty acids.

Table 7A. Total fatty acid yieldof milk fat (g/d) from primiparous and multiparous cows supplemented with either safflower oil or a combination of safflower and fish oils

Fatty Acid	Parity	Safflower Oil		Safflower/Fish Oil		SE	PE[1]	TE[2]	LE[3]	PE X TE	PE X LE	TE X LE
		400	800	400	800							
							--------------------	--------------------	P	--------------------	--------------------	--------------------
C4:0	P[4]	173.74	154.93	178.00	146.15	17.932	0.083	0.994	0.049	0.756	0.212	0.996
	M[5]	214.69	202.06	210.63	210.88							
C6:0	P	73.64	60.62	72.74	58.91	9.481	0.034	0.694	0.005	0.484	0.962	0.715
	M	104.46	87.17	105.58	95.22							
C8:0	P	36.02	28.18	34.84	28.31	5.733	0.036	0.593	0.003	0.452	0.593	0.581
	M	55.42	43.73	56.52	48.76							
C10:0	P	71.22	54.88	68.29	56.84	12.828	0.042	0.537	0.001	0.470	0.349	0.494
	M	115.65	89.50	117.96	99.29							
C12:0	P	78.28	62.50	74.52	67.07	12.532	0.028	0.381	0.003	0.426	0.283	0.457
	M	124.73	99.39	129.83	110.71							
C13:0	P	3.54	2.83	3.10	2.87	0.680	0.032	0.717	0.004	0.247	0.180	0.401
	M	5.98	4.69	6.19	5.22							

[1]PE = Parity effects; [2]TE = Type effects, [3]LE = Level effects, PE x TE = Parity x Type, PE x LE = Parity x Level, TE x LE = Type x Level
[4]P = Primiparous cows. [5]M = Multiparous cows.

Table 7B. Total fatty acid yieldof milk fat (g/d) from primiparous and multiparous cows supplemented with either safflower oil or a combination of safflower and fish oils

Fatty Acid	Parity	Safflower Oil		Safflower/Fish Oil		SE	PE[1]	TE[2]	LE[3]	PE X TE	PE X LE	TE X LE
		400	800	400	800							
							P					
C14:0	P[4]	341.07	285.22	337.71	314.68	32.881	0.013	0.321	0.019	0.800	0.759	0.379
	M[5]	468.97	404.52	476.39	440.71							
C14:1 *cis*	P	20.99	19.09	20.45	20.99	3.032	0.015	0.280	0.394	0.508	0.667	0.406
	M	32.73	29.27	34.09	33.48							
C15:0	P	32.53	27.99	32.24	29.76	2.643	0.019	0.227	0.010	0.533	0.994	0.369
	M	41.96	37.25	43.04	40.68							
C16:0	P	1050.46	916.12	1080.67	1039.31	58.157	0.014	0.025	0.011	0.964	0.933	0.330
	M	1252.39	1142.79	1315.56	1238.73							
C16:1 *trans*	P	16.64	18.35	19.60	24.98	2.651	0.941	0.049	0.126	0.672	0.780	0.630
	M	17.22	19.71	20.40	22.89							
C16:1 *cis*	P	44.69	39.37	48.80	53.39	5.423	0.779	0.012	0.961	0.619	0.855	0.261
	M	45.58	44.87	50.58	52.55							
C17:0	P	25.17	20.40	27.52	23.79	1.813	0.407	0.010	0.003	0.992	0.386	0.654
	M	26.02	23.12	28.55	26.38							

[1]PE = Parity effects; [2]TE = Type effects, [3]LE = Level effects, PE x TE = Parity x Type, PE x LE = Parity x Level, TE x LE = Type x Level
[4]P = Primiparous cows
[5]M = Multiparous cows

Table 7C. Total fatty acid yieldof milk fat (g/d) from primiparous and multiparous cows supplemented with either safflower oil or a combination of safflower and fish oils

Fatty Acid	Parity	Safflower Oil		Safflower/Fish Oil		SE	PE[1]	TE[2]	LE[3]	PE X TE	PE X LE	TE X LE
		400	800	400	800							
							P					
C18:0	P[4]	651.75	667.98	615.96	516.18	65.200	0.247	0.003	0.763	0.922	0.086	0.062
	M[5]	710.75	821.47	663.74	669.99							
C18:1 t5	P	1.36	1.71	1.22	1.58	0.185	0.108	0.279	0.0001	0.797	0.099	0.915
	M	1.54	2.27	1.48	1.15							
C18:1 t7	P	1.28	1.71	1.10	1.69	0.156	0.213	0.380	0.0004	0.795	0.202	0.229
	M	1.42	2.00	1.66	1.65							
C18:1 *trans* 6&8	P	19.55	26.88	19.06	30.52	1.940	0.233	0.019	<.0001	0.359	0.206	0.132
	M	19.89	30.84	22.33	35.02							
C18:1 t9	P	19.29	26.62	19.01	27.95	1.753	0.217	0.172	<.0001	0.441	0.050	0.930
	M	19.03	31.31	21.51	32.47							
C18:1 t10	P	35.02	72.75	27.44	61.77	8.951	0.959	0.766	0.007	0.542	0.813	0.420
	M	32.57	53.68	33.65	48.64							
C18:1 t11	P	72.64	128.79	88.99	240.31	19.063	0.909	<.0001	<.0001	0.559	0.560	0.0003
	M	64.96	120.37	102.07	232.65							
C18:1 t12	P	39.51	55.88	38.43	52.27	3.834	0.124	0.688	<.0001	0.122	0.011	0.599
	M	38.93	66.00	43.61	69.12							

Table 7C. (Continued)

Fatty Acid	Parity	Safflower Oil		Safflower/Fish Oil		SE	PE[1]	TE[2]	LE[3]	PE X TE	PE X LE	TE X LE
		400	800	400	800							
							------------------P------------------					
C18:1 *trans* 13&14	P	64.96	88.87	59.69	76.67	7.532	0.053	0.409	<.0001	0.182	0.039	0.327
	M	70.80	113.33	77.35	111.07							

[1]PE = Parity effects; [2]TE = Type effects, [3]LE = Level effects, PE x TE = Parity x Type, PE x LE = Parity x Level, TE x LE = Type x Level
[4]P = Primiparous cows
[5]M = Multiparous cows

Table 7D. Total fatty acid yield of milk fat (g/d) from primiparous and multiparous cows supplemented with either safflower oil or a combination of safflower and fish oils

Fatty Acid	Parity	Safflower Oil		Safflower/Fish Oil		SE	PE[1]	TE[2]	LE[3]	PE X TE	PE X LE	TE X LE
		400	800	400	800							
							------------------P------------------					
C18:1 *cis* 9&10	P[4]	964.06	974.29	973.99	872.30	59.480	0.912	0.213	0.594	0.992	0.085	0.218
	M[5]	916.67	1034.96	905.69	955.28							
C18:1 c11	P	19.85	17.67	21.06	21.00	1.209	0.473	0.004	0.507	0.728	0.043	0.101
	M	17.25	17.88	18.56	22.14							
C18:1 c12	P	44.76	80.79	35.11	43.08	5.352	0.149	<.0001	<.0001	0.019	0.023	0.002
	M	44.99	86.60	41.40	71.18							

Fatty Acid	Parity	Safflower Oil		Safflower/Fish Oil		SE	PE[1]	TE[2]	LE[3]	PE X TE	PE X LE	TE X LE
		400	800	400	800		P					
C18:1 c13	P	4.01	4.81	3.88	5.31	0.327	0.484	0.335	<.0001	0.977	0.070	0.672
	M	3.64	5.66	3.99	5.71							
C18:1 t16	P	27.50	33.73	25.39	23.18	2.350	0.067	0.001	0.002	0.579	0.033	0.008
	M	28.79	41.74	28.42	32.60							
C18:2	P	141.41	145.15	132.75	129.87	10.603	0.238	0.584	0.065	0.039	0.078	0.240
	M	139.30	162.85	153.59	163.03							
C18:3	P	21.55	17.98	22.11	17.07	2.163	0.221	0.159	0.002	0.114	0.356	0.597
	M	23.17	20.76	26.18	23.34							
RA[6]	P	25.39	43.12	32.02	85.68	7.619	0.650	<.0001	<.0001	0.752	0.408	0.003
	M	23.29	40.13	33.47	73.69							

[1]PE = Parity effects; [2]TE = Type effects, [3]LE = Level effects, PE x TE = Parity x Type, PE x LE = Parity x Level, TE x LE = Type x Level.
[4]P = Primiparous cows.
[5]M = Multiparous cows.
[6]RA = Rumenic acid, *cis*-9, *trans*-11 C18:2.

Table 7E. Total fatty acid yield of milk fat (g/d) from primiparous and multiparous cows supplemented with either safflower oil or a combination of safflower and fish oils

Fatty Acid	Parity	Safflower Oil		Safflower/Fish Oil		SE	PE[1]	TE[2]	LE[3]	PE X TE	PE X LE	TE X LE
		400	800	400	800							
							--------	--------	P	--------	--------	--------
C20:4	P[4]	5.16	4.37	4.64	3.25	0.368	0.005	0.020	0.002	0.505	0.582	0.240
	M[5]	6.19	5.69	6.02	4.90							
C20:5	P	1.67	1.46	2.40	2.09	0.196	0.183	<.0001	0.247	0.838	0.428	0.429
	M	1.96	1.65	2.44	2.65							
C22:5	P	3.83	3.43	4.43	3.38	0.333	0.741	0.092	0.091	0.758	0.059	0.711
	M	3.51	3.38	3.73	3.96							
C22:6	P	0.78	0.63	1.32	1.20	0.150	0.411	<.0001	0.579	0.187	0.097	0.104
	M	0.71	0.60	1.20	1.81							
SCFA[6]	P	283.40	243.73	285.58	233.37	30.687	0.042	0.820	0.011	0.592	0.539	0.830
	M	374.56	332.96	372.73	354.98							
MCFA[7]	P	1684.57	1446.75	1712.91	1633.68	112.100	0.009	0.063	0.011	0.828	0.776	0.325
	M	2131.23	1895.11	2222.58	2072.00							
LCFA[8]	P	2159.05	2363.09	2121.86	2209.67	131.890	0.312	0.263	0.0002	0.660	0.030	0.318
	M	2163.17	2657.54	2184.73	2551.37							

[1]PE = Parity effects; [2]TE = Type effects, [3]LE = Level effects, PE x TE = Parity x Type, PE x LE = Parity x Level, TE x LE = Type x Level
[4]P = Primiparous cows
[5]M = Multiparous cows
[6]Short chain fatty acids = sum of C4:0 to C8:0
[7]Medium chain fatty acids = sum of C10:0 to C17:0
[8]Long chain fatty acids = sum of C18:0 to C20:4

Table 7F. Total fatty acid yieldof milk fat (g/d) from primiparous and multiparous cows supplemented with either safflower oil or a combination of safflower and fish oils

Fatty Acid	Parity	Safflower Oil		Safflower/Fish Oil		SE	PE[1]	TE[2]	LE[3]	PE X TE	PE X LE	TE X LE
		400	800	400	800							
							P					
Ratio 1[6]	P[4]	2.69	4.38	2.38	3.28	0.245	0.096	0.004	<.0001	0.289	0.118	0.039
	M[5]	2.75	4.88	2.70	4.21							
Ratio 2[7]	P	0.63	0.42	0.81	0.40	0.055	0.629	0.050	<.0001	0.921	0.796	0.333
	M	0.71	0.39	0.77	0.50							
Total *trans* FA[8]	P	1894.87	1873.24	1895.76	1976.88	112.780	0.059	0.227	0.099	0.940	0.293	0.610
	M	2145.11	2276.39	2209.60	2329.84							
Total unsat *cis* FA[9]	P	1636.44	1678.78	1627.74	1579.17	85.182	0.073	0.594	0.130	0.604	0.117	0.478
	M	1721.68	1914.56	1748.92	1885.91							
PUFA[10]	P	204.94	231.26	201.63	247.46	16.664	0.343	0.056	<.0001	0.239	0.286	0.510
	M	203.04	255.45	227.68	281.11							

[1]PE = Parity effects; [2]TE = Type effects, [3]LE = Level effects, PE x TE = Parity x Type, PE x LE = Parity x Level, TE x LE = Type x Level
[4]P = Primiparous cows.
[5]M = Multiparous cows.
[6]Ratio 1 = *cis*-9&10 C18:1 / C18:0.
[7]Ratio 2 = RA / *trans*-11 C18:1
[8]Sum of unsaturated *cis* fatty acids.
[9]Sum of *trans* fatty acids.
[10]Polyunsaturated fatty acids.

Table 8A. Fatty acid composition of milk fat phospholipid (g/100 g of phospholipid) from primiparous and multiparous cows supplemented with either safflower oil or a combination of safflower and fish oils

Fatty Acid	Parity	Safflower Oil		Safflower/Fish Oil		SE	PE[1]	TE[2]	LE[3]	PE X TE	PE X LE	TE X LE
		400	800	400	800							
							P					
C4:0-C8:0	P[4]	-	-	-	-	-	-	-	-	-	-	-
	M[5]	-	-	-	-							
C10:0	P	0.36	0.32	0.40	0.28	0.099	0.202	0.020	0.119	0.021	0.717	0.637
	M	0.75	0.52	0.32	0.30							
C12:0	P	0.51	0.49	0.58	0.47	0.105	0.105	0.027	0.124	0.012	0.465	0.460
	M	1.03	0.72	0.52	0.51							
C13:0	P	-	-	-	-	-	-	-	-	-	-	-
	M	-	-	-	-							
C14:0	P	3.36	3.12	3.78	3.61	0.275	0.189	0.334	0.056	0.004	0.358	0.399
	M	4.65	3.82	3.55	3.28							
C14:1 *cis*	P	0.08	0.11	0.11	0.18	0.021	0.656	0.949	0.175	0.006	0.056	0.620
	M	0.14	0.13	0.09	0.08							
C15:0	P	0.52	0.45	0.53	0.51	0.030	0.535	0.927	0.002	0.074	0.513	0.108
	M	0.59	0.49	0.53	0.49							

[1]PE = Parity effects; [2]TE = Type effects, [3]LE = Level effects, PE x TE = Parity x Type, PE x LE = Parity x Level, TE x LE = Type x Level
[4]P = Primiparous cows.
[5]M = Multiparous cows.

Table 8B. Fatty acid composition of milk fat phospholipid (g/100 g FA) from primiparous and multiparous cows supplemented with either safflower oil or a combination of safflower and fish oils

Fatty Acid	Parity	Safflower Oil		Safflower/Fish Oil		SE	PE[1]	TE[2]	LE[3]	PE X TE	PE X LE	TE X LE
		400	800	400	800							
							P					
C16:0	P[4]	17.15	15.35	18.63	18.51	0.355	0.588	0.010	<.0001	<.0001	0.041	0.001
	M[5]	19.26	16.09	17.27	16.34							
C16:1 *trans*	P	0.27	0.30	0.33	0.47	0.459	0.361	0.001	0.012	0.478	0.444	0.114
	M	0.24	0.27	0.30	0.37							
C16:1 *cis*	P	0.71	0.64	0.75	0.91	0.089	0.789	0.043	0.597	0.074	0.097	0.041
	M	0.78	0.65	0.75	0.71							
C17:0	P	0.48	0.41	0.53	0.49	0.018	0.0002	0.019	<.0001	0.057	0.474	0.293
	M	0.46	0.37	0.45	0.39							
C18:0	P	19.04	18.73	18.37	15.57	0.629	0.067	0.011	0.201	0.078	0.022	0.024
	M	18.86	20.12	19.22	18.94							
C18:1 t5	P	-	-	-	-	-	-	-	-	-	-	-
	M	-	-	-	-							
C18:1 t7	P	-	-	-	-	-	-	-	-	-	-	-
	M	-	-	-	-							

[1]PE = Parity effects; [2]TE = Type effects, [3]LE = Level effects, PE x TE = Parity x Type, PE x LE = Parity x Level, TE x LE = Type x Level
[4]P = Primiparous cows.
[5]M = Multiparous cows.

Table 8C. Fatty acid composition of milk fat phospholipid (g/100 g of phospholipid) from primiparous and multiparous cows supplemented with either safflower oil or a combination of safflower and fish oils

Fatty Acid	Parity	Safflower Oil		Safflower/Fish Oil		SE	PE[1]	TE[2]	LE[3]	PE X TE	PE X LE	TE X LE
		400	800	400	800		P					
C18:1 *trans* 6&8	P[4]	0.21	0.34	0.28	0.44	0.032	0.090	0.026	<.0001	0.086	0.392	0.582
	M[5]	0.20	0.30	0.20	0.32							
C18:1 t9	P	0.55	0.72	0.55	0.81	0.037	0.712	0.023	<.0001	0.585	0.805	0.306
	M	0.51	0.71	0.58	0.78							
C18:1 t10	P	0.64	0.87	0.63	1.58	0.356	0.392	0.574	0.112	0.421	0.513	0.610
	M	0.53	0.88	0.56	0.72							
C18:1 t11	P	1.27	2.25	1.71	4.79	0.361	0.203	<.0001	<.0001	0.198	0.044	0.02
	M	1.08	1.85	1.64	3.21							
C18:1 t12	P	0.58	0.83	0.65	0.90	0.049	0.702	0.107	<.0001	0.593	0.269	0.951
	M	0.54	0.86	0.58	0.90							
C18:1 *trans* 13&14	P	1.14	1.52	1.28	1.43	0.090	0.724	0.830	<.0001	0.838	0.137	0.205
	M	1.10	1.63	1.16	1.57							
C18:1 *cis* 9&10	P	31.26	30.04	29.95	26.17	0.956	0.099	0.014	0.003	0.094	0.398	0.114
	M	28.20	27.31	28.26	26.13							

[1]PE = Parity effects; [2]TE = Type effects, [3]LE = Level effects, PE x TE = Parity x Type, PE x LE = Parity x Level, TE x LE = Type x Level
[4]P = Primiparous cows.
[5]M = Multiparous cows.

Table 8D. Fatty acid composition of milk fat phospholipid (g/100 g of phospholipid) from primiparous and multiparous cows supplemented with either safflower oil or a combination of safflower and fish oils

Fatty Acid	Parity	Safflower Oil		Safflower/Fish Oil		SE	PE[1]	TE[2]	LE[3]	PE X TE	PE X LE	TE X LE
		400	800	400	800							
							--------	--------	P	--------	--------	--------
C18:1 c11	P[4]	0.62	0.49	0.61	0.63	0.053	0.023	0.048	0.104	0.859	0.874	0.083
	M[5]	0.44	0.34	0.48	0.46							
C18:1 c12	P	1.41	2.42	1.24	1.38	0.122	0.816	<.0001	<.0001	0.018	0.265	0.004
	M	1.27	2.10	1.15	1.81							
C18:1 c13	P	0.10	0.11	0.11	0.13	0.008	0.251	0.031	0.001	0.782	0.244	0.857
	M	0.08	0.11	0.10	0.12							
C18:1 t16	P	0.43	0.50	0.46	0.37	0.032	0.752	0.017	0.031	0.529	0.015	0.028
	M	0.41	0.56	0.37	0.45							
C18:2	P	14.28	14.61	13.27	12.98	0.633	0.139	0.895	0.929	0.001	0.972	0.119
	M	13.88	14.70	16.06	15.32							
C18:3	P	0.80	0.55	0.69	0.55	0.045	0.043	0.888	<.0001	0.020	0.444	0.003
	M	0.88	0.64	0.85	0.77							
RA[6]	P	1.03	1.89	1.21	4.11	0.376	0.294	0.001	<.0001	0.493	0.071	0.009
	M	0.90	1.50	1.41	2.75							

[1]PE = Parity effects; [2]TE = Type effects, [3]LE = Level effects, PE x TE = Parity x Type, PE x LE = Parity x Level, TE x LE = Type x Level
[4]P = Primiparous cows.
[5]M = Multiparous cows.
[6]RA = Rumenic acid, *cis*-9, *trans*-11 C18:2

Table 8E. Fatty acid composition of milk fat phospholipid (g/100 g of phospholipid) from primiparous and multiparous cows supplemented with either safflower oil or a combination of safflower and fish oils

Fatty Acid	Parity	Safflower Oil		Safflower/Fish Oil		SE	PE[1]	TE[2]	LE[3]	PE X TE	PE X LE	TE X LE
		400	800	400	800		P					
C20:4	P[4]	1.53	1.43	1.37	1.05	0.124	0.288	0.045	0.008	0.002	0.171	0.007
	M[5]	1.44	1.55	1.70	1.44							
C20:5	P	0.28	0.23	0.37	0.35	0.027	0.964	<.0001	0.111	0.825	0.357	0.169
	M	0.27	0.24	0.35	0.37							
C22:5	P	0.86	0.73	0.89	0.69	0.064	0.512	0.004	0.003	0.003	0.085	0.732
	M	0.67	0.60	0.86	0.83							
C22:6	P	0.37	0.36	0.48	0.47	0.075	0.266	0.011	0.469	0.463	0.352	0.383
	M	0.25	0.25	0.35	0.52							
SCFA[6]	P	-	-	-	-	-	-	-	-	-	-	-
	M	-	-	-	-							
MCFA[7]	P	23.44	21.19	25.62	25.42	0.779	0.604	0.435	0.001	<.0001	0.092	0.019
	M	27.89	23.06	23.79	22.42							
LCFA[8]	P	74.89	77.30	72.38	72.91	0.885	0.566	0.276	0.001	0.0001	0.161	0.026
	M	70.32	75.15	74.30	75.61							

[1]PE = Parity effects; [2]TE = Type effects, [3]LE = Level effects, PE x TE = Parity x Type, PE x LE = Parity x Level, TE x LE = Type x Level
[4]P = Primiparous cows.
[5]M = Multiparous cows.
[6]Short chain fatty acids = sum of C4:0 to C8:0.
[7]Medium chain fatty acids = sum of C10:0 to C17:0.
[8]Long chain fatty acids = sum of C18:0 to C20:4.

Table 8F. Fatty acid composition of milk fat phospholipid (g/100 g of phospholipid) from primiparous and multiparous cows supplemented with either safflower oil or a combination of safflower and fish oils

Fatty Acid	Parity	Safflower Oil		Safflower/Fish Oil		SE	PE[1]	TE[2]	LE[3]	PE X TE	PE X LE	TE X LE
		400	800	400	800							
							P					
Ratio 1[6]	P[4]	1.64	1.60	1.65	1.68	0.077	0.039	0.613	0.217	0.672	0.228	0.469
	M[5]	1.51	1.36	1.48	1.39							
Ratio 2[7]	P	0.81	0.84	0.73	0.85	0.064	0.655	0.847	0.354	0.431	0.403	0.374
	M	0.84	0.82	0.84	0.86							
Total *trans*[8]	P	5.09	7.32	5.88	10.80	0.771	0.291	0.003	<.0001	0.217	0.322	0.095
	M	4.61	7.06	5.39	8.29							
Unsat *cis* FA[9]	P	50.90	49.31	48.59	44.11	1.151	0.541	0.132	0.004	0.002	0.420	0.070
	M	47.02	46.50	49.84	46.65							
PUFA[10]	P	18.13	17.91	17.06	16.09	0.725	0.126	0.485	0.340	0.001	0.622	0.170
	M	17.38	17.96	20.17	19.20							

[1]PE = Parity effects; [2]TE = Type effects, [3]LE = Level effects, PE x TE = Parity x Type, PE x LE = Parity x Level, TE x LE = Type x Level
[4]P = Primiparous cows.
[5]M = Multiparous cows.
[6]Ratio 1 = *cis*-9&10 C18:1 / C18:0.
[7]Ratio 2 = RA / *trans*-11 C18:1
[8]Sum of unsaturated *cis* fatty acids.
[9]Sum of *trans* fatty acids.
[10]Polyunsaturated fatty acids.

TVA and rumenic acid concentrations in the phospholipid fractions increased dramatically as the level of oil increased. Increasing the level of dietary oil regardless of parity, increased both rumenic acid and TVA concentrations two-fold. Primiparous cows receiving the high level of lipid supplementation showed the greatest increase of TVA with a ~2.8 fold increase (P=0.04; PE x LE).

Type Effects

Proportion of MCFA and LCFA were not affected by type of oil fed (Table 8). However, there were significant (P = 0.0001) parity by type effects for PL LCFA. Primiparous cows showed decreasing PL LCFA as fish oil was included in the diet while multiparous cows showed an increase. However, including fish oil in the supplement significantly increased C16:0, *trans*-C16:1, *cis*-C16:1, and C17:0 in milk PL. Total PL *trans*-FA increased as fish oil was included in the diet. The cows on the HFS diet tended (P=0.09; TE x LE) to show the greatest increases in *trans* FA. Phospholipid *trans*-C16:1 increased, on average, 44% when fish oil was included in the diet and by 22% when the level increased from low to high.The concentration of phospholipid *cis*-C16:1 increased (P=0.04) with the addition of fish oil. Individual PL LCFA demonstrated differences for type of dietary lipid (Table 8). PL content of C18:0 and *trans*-16 C18:1 significantly decreased while the content of *trans*-6 and -8 C18:1, *trans*-9 C18:1, and *trans*-11 C18:1 increased with the addition of fish oil. Concentration of *cis*-9 and -10 C18:1 and *cis*-12 C18:1 concentration decreased and *cis*-11 C18:1 and *cis*-13 C18:1 increased with fish oil inclusion. Type of oil did not affect the concentrations of C18:2 n6 and C18:3 n3 in the PL fraction. However, primiparous cows fed the fish oil treatment showed decreased concentrations of C18:3 n3 (P < 0.02; PE x TE). *Trans* vaccenic and rumenic acid concentrations in the PL fraction increased dramatically when fish oil was added to the diet. Level by type interactions for TVA and rumenic acid were significant because primiparous cows received a higher proportion of lipid in their diet. Those cows receiving the FS treatments, regardless of parity, showed the greatest increases (P<0.0001; TE) in TVA concentrations.

Parity Effects

Parity alone had little effect on most FA, however there were a few marked changes. Primiparous cows tended (P<0.09) to have greater concentrations of *trans*-6 and -8 C18:1 in their PL fractions. Primiparous cows showed significantly (P=0.02) higher concentrations of *cis*-11 C18:1 and lower concentrations (P=0.04) of PL C18:3 n3 when compared to multiparous cows. Primiparous cows showed a higher (P<0.04) concentration of ratio 1 (ratio 1 = *cis*-9 and -10 C18:1/C18:0) while ratio 2 (ratio 2 = rumenic acid/TVA) was unaffected by parity.

Plasma

Plasma concentration of NEFA was not different for type or level of oil (Table 9). Parity by type was significant and probably reflected the higher concentration of dietary lipid for primiparous cows. Plasma glucose did not differ for treatments. Plasma urea N was lower for the higher level of oil feeding, and the response was greater (P<0.03; PE x LE) for multiparous than primiparous cows. Plasma TG concentration increased (P < 0.008, PE x TE) when multiparous cows received fish oil.

Table 9. Blood plasma constituent concentrations for primiparous and multiparous cows supplemented with either safflower oil or a combination of safflower and fish oils

		Safflower Oil		Safflower/Fish Oil									
	Parity	400	800	400	800	SE	PE	TE	LE	PE X TE	PE X LE	TE X LE	Time
							P						
NEFA[3], uM	P[4]	276.6	287.8	306.6	457.5	49.84	0.050	0.371	0.403	0.017	0.062	0.097	<.0001
	M[5]	271.4	210.4	195.1	193.3								
Glucose, g/dl	P	73.9	74.6	75.5	71.4	2.50	0.517	0.205	0.468	0.446	0.076	0.619	0.003
	M	72.3	75.3	68.3	73.1								
PUN[6], mg/dl	P	14.7	14.7	14.8	14.4	0.70	0.865	0.074	0.006	0.038	0.030	0.663	<.0001
	M	15.0	12.6	15.9	14.6								
TG[7] mg/dl	P	35.8	37.1	35.4	35.1	1.65	0.961	0.266	0.490	0.008	0.168	0.49	0.013
	M	35.1	33.8	38.3	36.5								

[1]PE = Parity effects; [2]TE = Type effects, [3]LE = Level effects, PE x TE = Parity x Type, PE x LE = Parity x Level, TE x LE = Type x Level
[3]NEFA = Non-esterified fatty acid.
[4]P = Primiparous cows.
[5]M = Multiparous cows.
[6]PUN = Plasma urea nitrogen.
[7]TG = Triglycerides

Table 10. Fatty acids (mg FA/100 ml plasma) in the phospholipid (PL) fraction of plasma from primiparous and multiparous cows supplemented with either safflower oil or a combination of safflower and fish oils

Fatty Acid	Parity	Safflower Oil		Safflower/Fish Oil		SE							
		400	800	400	800		PE[1]	TE[2]	LE[3]	PE X TE	PE X LE	TE X LE	PE X TE X LE
							P						
C12:0	P[4]	No data											
	M[5]												
C14:0	P	0.186	1.050	0.215	0.243	0.306	0.645	0.461	0.247	0.320	0.412	0.421	0.291
	M	0.271	0.291	0.272	0.408								
C15:0	P	0.580	1.944	0.628	0.515	0.507	0.659	0.433	0.332	0.266	0.450	0.314	0.307
	M	0.629	0.704	0.747	0.832								
C16:0	P	23.218	24.242	26.592	23.425	1.891	0.917	0.482	0.337	0.674	0.978	0.100	0.900
	M	23.658	24.335	25.794	22.849								
C16:1 *trans*	P	0.361	0.419	0.440	0.586	0.046	0.991	0.001	0.083	0.312	0.026	0.629	0.194
	M	0.412	0.418	0.507	0.472								
C16:1 *cis*	P	0.559	0.500	0.547	0.698	0.282	0.099	0.561	0.369	0.894	0.496	0.319	0.618
	M	0.846	0.863	0.682	1.323								
C17:0	P	1.183	1.196	1.364	0.916	0.141	0.193	0.642	0.003	0.844	0.593	0.040	0.388
	M	1.503	1.306	1.583	1.186								
C18:0	P	38.989	41.499	41.066	34.122	4.013	0.199	0.077	0.265	0.575	0.963	0.112	0.510
	M	48.217	47.846	45.313	40.871								
C18:1 t6 e8	P	0.094	0.146	0.150	0.244	0.021	0.299	0.013	<.0001	0.025	0.564	0.257	0.774
	M	0.138	0.216	0.130	0.233								
C18:1 t9	P	0.179	0.288	0.230	0.347	0.023	0.031	0.172	<.0001	0.014	0.829	0.568	0.382
	M	0.272	0.399	0.274	0.362								

Fatty Acid	Parity	Safflower Oil		Safflower/Fish Oil		SE	PE[1]	TE[2]	LE[3]	PE X TE	PE X LE	TE X LE	PE X TE X LE
		400	800	400	800								
							---P---						
C18:1 t10	P	0.153	0.244	0.180	0.430	0.080	0.987	0.586	0.042	0.200	0.434	0.616	0.382
	M	0.222	0.324	0.200	0.258								
C18:1 t11	P	1.085	2.124	1.505	5.158	0.333	0.581	<.0001	<.0001	0.486	0.012	0.001	0.027
	M	1.095	1.981	2.179	3.807								
C18:1 t12	P	0.974	1.599	1.141	1.872	0.094	0.532	0.005	<.0001	0.193	0.631	0.667	0.141
	M	0.918	1.644	1.103	1.640								
C18:1 t13 e14	P	1.216	1.934	1.328	1.707	0.165	0.171	0.162	<.0001	0.444	0.460	0.087	0.867
	M	1.651	2.212	1.601	1.881								
C18:1 c9 e10	P	8.697	7.736	9.916	8.193	1.107	0.962	0.790	0.045	0.097	0.834	0.610	0.874
	M	9.594	8.691	8.655	7.350								
C18:1 t15	P	0.267	0.410	0.300	0.361	0.042	0.249	0.105	0.001	0.180	0.787	0.049	0.715
	M	0.356	0.503	0.341	0.372								
C18:1 c11	P	0.530	0.434	0.545	0.523	0.064	0.940	0.174	0.118	0.752	0.749	0.316	0.832
	M	0.518	0.454	0.526	0.511								
C18:1 c12	P	1.806	3.567	1.707	2.696	0.206	0.412	0.001	<.0001	0.736	0.609	0.002	0.989
	M	1.929	3.797	1.902	2.992								
C18:1 c13	P	0.079	0.083	0.090	0.091	0.016	0.058	0.406	0.143	0.931	0.207	0.723	0.630
	M	0.092	0.115	0.093	0.135								
C18:1 t16	P	0.374	0.537	0.380	0.359	0.073	0.397	0.065	0.032	0.864	0.562	0.069	0.731
	M	0.431	0.612	0.422	0.476								

Table 10.(Continued)

Fatty Acid	Parity	Safflower Oil		Safflower/Fish Oil		SE	PE[1]	TE[2]	LE[3]	PE X TE	PE X LE	TE X LE	PE X TE X LE
		400	800	400	800								
							P						
C18:2	P	47.279	51.392	50.152	45.791	4.122	0.105	0.547	0.882	0.983	0.840	0.041	0.696
	M	54.314	61.196	58.915	53.671								
C18:3	P	1.701	1.470	1.857	1.419	0.234	0.123	0.369	0.004	0.696	0.922	0.050	0.285
	M	2.072	2.043	2.531	1.850								
C18:2 c9 t11	P	0.133	0.218	0.177	0.401	0.030	0.466	<.0001	<.0001	0.089	0.020	0.002	0.606
	M	0.169	0.188	0.173	0.298								
C20:4	P	3.098	4.552	3.377	2.715	0.581	0.901	0.141	0.778	0.633	0.463	0.110	0.302
	M	3.539	3.604	3.374	2.954								
C20:5	P	0.716	0.733	1.301	1.400	0.125	0.176	<.0001	0.548	0.013	0.705	0.351	0.797
	M	0.688	0.631	0.917	1.001								
C22:5	P	1.346	1.375	1.644	1.538	0.198	0.423	0.312	0.523	0.312	0.755	0.983	0.576
	M	1.381	1.209	1.325	1.277								
C22:6	P	0.402	0.395	0.686	0.906	0.117	0.415	0.000	0.410	0.202	0.409	0.105	0.935
	M	0.421	0.318	0.549	0.652								
Total	P	135.200	150.080	147.520	136.650	12.126	0.314	0.608	0.856	0.674	0.878	0.068	0.837
	M	155.330	165.900	160.100	149.880								

[1]PE = Parity effects; [2]TE = Type effects, [3]LE = Level effects, PE x TE = Parity x Type, PE x LE = Parity x Level, TE x LE = Type x Level, PE x TE x LE = Parity x Type x Level.
[4]P = Primiparous cows.
[5]M = Multiparous cows.

Table 11. Fatty acids (mg FA/100 ml plasma) in the TG fraction of plasma from primiparous and multiparous cows supplemented with either safflower oil or a combination of safflower and fish oils

Fatty Acid	Parity	Safflower Oil		Safflower/Fish Oil		SE							
		400	800	400	800		PE[1]	TE[2]	LE[3]	PE X TE	PE X LE	TE X LE	PE X TE X LE
							P						
C12:0	P[4]	0.017	0.015	0.016	0.018	0.003	0.363	0.890	0.849	0.742	0.890	0.419	0.945
	M[5]	0.019	0.018	0.017	0.020								
C14:0	P	0.169	0.111	0.151	0.135	0.024	0.609	0.051	0.654	0.072	0.017	0.112	0.687
	M	0.111	0.128	0.143	0.228								
C15:0	P	0.165	0.080	0.136	0.090	0.039	0.953	0.295	0.857	0.172	0.043	0.279	0.703
	M	0.078	0.091	0.106	0.203								
C16:0	P	1.240	1.236	1.502	1.372	0.223	0.649	0.148	0.964	0.790	0.650	0.906	0.617
	M	1.278	1.259	1.464	1.646								
C16:1 *trans*	P	0.024	0.023	0.024	0.032	0.006	0.852	0.254	0.931	0.955	0.478	0.693	0.407
	M	0.025	0.024	0.031	0.027								
C16:1 *cis*	P	0.038	0.023	0.035	0.026	0.020	0.441	0.532	0.614	0.527	0.188	0.511	0.398
	M	0.027	0.077	0.031	0.036								
C17:0	P	0.068	0.070	0.083	0.070	0.007	0.154	0.060	0.358	0.538	0.931	0.639	0.356
	M	0.077	0.070	0.088	0.086								
C18:0	P	2.489	2.750	2.731	2.105	0.470	0.786	0.969	0.961	0.535	0.564	0.461	0.588
	M	2.355	2.640	2.653	2.800								
C18:1 t6 e8	P	0.016	0.023	0.018	0.036	0.008	0.958	0.291	0.075	0.920	0.838	0.818	0.513
	M	0.014	0.026	0.022	0.030								
C18:1 t9	P	0.015	0.020	0.016	0.021	0.005	0.829	0.583	0.173	0.695	0.864	0.835	0.762
	M	0.013	0.021	0.019	0.023								

Table 11.(Continued)

Fatty Acid	Parity	Safflower Oil		Safflower/Fish Oil		SE							
		400	800	400	800		PE[1]	TE[2]	LE[3]	PE X TE	PE X LE	TE X LE	PE X TE X LE
							P						
C18:1 t10	P	0.033	0.049	0.033	0.072	0.018	0.878	0.326	0.137	0.913	0.555	0.646	0.670
	M	0.031	0.043	0.045	0.058								
C18:1 t11	P	0.047	0.123	0.079	0.309	0.068	0.472	0.078	0.049	0.732	0.322	0.308	0.599
	M	0.048	0.077	0.978	0.177								
C18:1 t12	P	0.064	0.082	0.072	0.078	0.015	0.891	0.578	0.247	0.702	0.946	0.946	0.539
	M	0.067	0.073	0.070	0.090								
C18:1 t13 e14	P	0.069	0.112	0.083	0.105	0.024	0.890	0.326	0.100	0.421	0.890	0.764	0.782
	M	0.060	0.878	0.092	0.119								
C18:1 c9 e10	P	0.144	0.154	0.160	0.146	0.032	0.705	0.465	0.707	0.574	0.645	0.982	0.607
	M	0.123	0.130	0.141	0.173								
C18:1 t15	P	0.017	0.031	0.023	0.028	0.007	0.906	0.251	0.075	0.320	0.888	0.753	0.598
	M	0.015	0.024	0.025	0.037								
C18:1 c11	P	0.010	0.011	0.016	0.017	0.003	0.227	0.011	0.186	0.681	0.421	0.794	0.681
	M	0.011	0.015	0.018	0.025								
C18:1 c12	P	0.027	0.068	0.033	0.041	0.010	0.731	0.743	0.003	0.285	0.848	0.186	0.441
	M	0.021	0.053	0.031	0.054								
C18:1 c13	P	0.013	0.009	0.012	0.016	0.003	0.388	0.312	0.961	0.840	0.961	0.761	0.052
	M	0.011	0.017	0.019	0.013								
C18:1 t16	P	0.016	0.023	0.029	0.024	0.006	0.729	0.316	0.652	0.450	0.784	0.284	0.784
	M	0.018	0.024	0.022	0.022								

Table 11.(Continued)

Fatty Acid	Parity	Safflower Oil		Safflower/Fish Oil		SE							
		400	800	400	800		PE[1]	TE[2]	LE[3]	PE X TE	PE X LE	TE X LE	PE X TE X LE
							P						
C18:2	P	0.050	0.082	0.103	0.051	0.027	0.650	0.269	0.605	0.554	0.314	0.646	0.107
	M	0.060	0.067	0.070	0.124								
C18:3	P	0.007	0.006	0.010	0.012	0.003	0.396	0.018	0.296	0.754	0.396	0.979	0.444
	M	0.005	0.010	0.012	0.014								
C18:2 c9 t11	P	No data											
	M												
C20:4	P	0.026	0.033	0.034	0.025	0.006	0.932	0.728	0.325	0.771	0.252	0.207	0.686
	M	0.022	0.035	0.028	0.034								
C20:5	P	0.045	0.029	0.068	0.105	0.049	0.778	0.107	0.262	0.784	0.400	0.210	0.585
	M	0.034	0.040	0.038	0.174								
C22:5	P	No data											
	M												
C22:6	P	No data											
	M												
Total	P	4.807	5.160	5.461	4.932	0.880	0.784	0.367	0.619	0.562	0.527	0.855	0.618
	M	4.521	5.049	5.280	6.218								

[1]PE = Parity effects; [2]TE = Type effects, [3]LE = Level effects, PE x TE = Parity x Type, PE x LE = Parity x Level, TE x LE = Type x Level, PE x TE x LE = Parity x Type x Level.

[4]P = Primiparous cows.

[5]M = Multiparous cows.

There was little change in the FA composition of plasma TG with the exception of *cis*-12 C18:1, which increased with increasing oil intake (Table 10). We were not able to detect *cis*-9 *trans*-11 C18:2 and C22:6 n3 in plasma TG, and C20:5 n3 was not different for type and level of oil. Individual FA changed in plasma PL (Table 11). Feeding fish oil at 800 g/d increased the *cis*-9 *trans*-11 C18:2 in PL. Both C20:5 n3 and C22:6 n3 increased in PL in response to the addition of fish oil in the lipid supplement although the changes were small.

Fatty acid composition of plasma TG changed little in response to type and level of oil (Table 10). Adding fish oil to the lipid supplement significantly increased the concentrations of both C20:5 n3 and C22:6 n3 in the PL fraction of plasma. Level of oil feeding had no effect on these FA. The α-Linolenic acid decreased in concentration with increased level of oil feeding. Concentration of various trans FA including *trans*-11 C18:1 and *trans*-12 C18:1 increased with higher oil feeding.

The FA composition of PL was affected by type and level of oil but not by parity (Table 11). *Trans* vaccenic acid and *trans*-12 C18:1 were higher when the lipid supplement contained fish oil. Adding fish oil to the lipid supplement increased the long chain n3 FA, C20:5 and C22:6, in the PL fraction. Increasing the level of oil in the diet increased the concentration of *trans* monoenes of C18 except *trans*-13 and -14 C18:1, *trans*-15 C18:1, and *trans*-16 C18:1. The concentrations of *cis*-9 and -10 C18:1 and *cis*-12 C18:1 were lower with high oil feeding. Increasing the amount of fish oil consumed did not markedly affect the concentrations of C20:5 and C22:6 in PL. The FA composition of plasma TG was not affected by parity and only a few selected FA were altered by type or level of oil (Table 12). Concentration of C20:5 was not impacted by type or level of oil in the diet, and C22:6 was not measurable. Total TG in plasma was not affected by treatments in agreement with total TG concentration (Table 9).

Discussion

In the present study the higher amount of fish oil feeding was 200 g/cow daily. Reduced feed intake was observed with fish oil (Doreau and Chilliard, 1997). Feeding or infusion of fish lipids at high levels depressed milk fat (Opstvedt, 1984). However, based on the literature available, levels below 100g lipid/cow daily did not typically depress milk fat concentration, but levels above 200 g lipid/cow consistently depressed milk fat concentration (Opstvedt, 1984). Firkins et al. (1994) summarized data from numerous studies and suggested that as the percentage of unsaturated FA increased in the diet, feed intake would decrease. The objective of this study was to perturb rumen biohydrogenation without affecting productive performance including milk fat concentration and yield and therefore, 200 g fish oil daily was the upper limit. In agreement, milk yield and composition (Tables 3A and 3B) were not affected by either type or level of oil feeding. Likewise, dry matter intake was also not influenced by dietary treatments (Table 4).

Proportion of *trans*-10 C18:1 increased in milk fat as the level of fish oil increased (Table 7C) in agreement with previous work (Loor et al., 2004) who fed fish oil at 2.5% DM. Increased *trans*-10 C18:1 was associated with milk fat depression (Chouinard et al., 1999). Other studies indicate that the *trans*-10 *cis*-12 isomer of CLA was responsible for milk fat depression (Baumgard et al., 2000, 2001, 2002). Milk fat depression was observed in previous research that involved feeding fishmeal or fish oil (AbuGhazaleh et al., 2001a; Donovan et al., 2000; AbuGhazaleh, 2002; Whitlock et al., 2002). In the current study *trans*-10 *cis*-12C18:2 was not found in milk fat (Table 6), and milk fat percentages averaged 3.6% or higher (Table 3) with no indication of milk fat depression in diets based on alfalfa hay and adequate dietary fiber.

The proportion of SCFA and MCFA decreased and LCFA increased with increasing amount of oil feeding. This change was anticipated as the availability of LCFA increases, either due to increased dietary intake, as with the current study, or with increased body lipid mobilization, so does the incorporation of these FA into milk fat resulting in more LCFA (Chilliard et al., 2004). Palmquist et al. (1993) reported that increased dietary LCFA increased the concentration of LCFA in milk fat and decreased de novo synthesis of SCFA and MCFA. The SCFA and MCFA are synthesized de novo in the mammary gland, while LCFA are removed from blood and supplied through the diet or mobilization of adipose tissue (Grummer, 1991; Palmquist, 1993).

The decrease in C18:0 in both primiparous and multiparous cows with the inclusion of fish oil might indicate incomplete biohydrogenation (Table 6C). Recent studies also observed decreased concentrations of C18:0 in milk fat and increased concentrations of C18:1 when fish oil was added to the diet (Loor et al., 2004; Shingfield et al., 2004). Jones et al. (1998) observed a linear decrease in C18:0 concentrations in milk fat and increased concentrations of C18:1 and C18:3 n3 when fish oil increasingly replaced tallow in treatment diets. Increasing levels of TVA and rumenic acid in milk fat corresponded with decreases in C18:0 and C18:2 n6 (Kepler et al., 1966). Feeding 800 g compared to 400 g and the inclusion of fish oil both increased TVA (Table 6C) and rumenic acid (Table 6D) concentrations in milk fat. In addition, a large proportion of the individual C18: isomers in milk fat were affected by type of oil and level of oil feeding. The concentrations of *cis*-9 and -10 C18:1 in milk fat decreased by 7.5% for both parities as dietary fish oil was added. Franklin et al. (1999) also observed a decreased concentration of *cis*-9 C18:1 when marine algae products were included in the dairy ration of lactating cows.

Concentration of TVA in milk fat increased 1.9-fold with type and 1.8-fold with level of oil supplemented. TVA can be desaturated to rumenic acid by the tissues of humans and cows by the enzyme*delta*-9 desaturase reaction (Griinari et al., 2000; Turpeinen et al., 2002; Mosley et al., 2006). *Delta*-9 desaturase enzyme was responsible for the production of 64 to 78% of the rumenic acid found in milk fat (Griinari et al., 2000; Corl et al., 1999). Concentrations of TVA increased when cows were fed fish oil alone (Donovan et al., 2000) or with extruded soybeans (Whitlock et al., 2002). Concentrations of C20:5 n3 and C22:6 n3 in milk fat were increased when fish oil was included in the supplement, but the increase was small (Table 6E).The higher concentrations of C20:5 n3 and C22:6 n3 in milk fat likely result from their increased intakes. Similar results were found by AbuGhazaleh et al. (2001b, 2002) who showed elevated levels of C20:5 n3 and C22:6 n3 when a source of fish oil was included in the diet of lactating dairy cows.The low concentrations of C20:5 n3 and C22:6 n3 in milk fat could be due to the preferential deposition in body tissues (Ashes et al., 1992), incorporation into phospholipid fractions (Offer et al., 1999) or by extensive biohydrogenation in the rumen (Wachira et al., 2000). While C20:5 n3 and C22:6 n3 are present in triglycerides to a small extent, their concentration is greatest in cholesterol esters and phospholipid fractions of plasma (Kitessa et al., 2001a). Fish oil contains high levels of C20:5 n3 and C22:6 n3 but low concentrations of C22:5 n6. In the current study, C22:5 n6 only accounted for 2.9 g/100g of fish oil (Table 2). However it has a high transfer rate (0.30-0.39) when compared to C20:5 n3 and C22:6 n3 (Chilliard et al., 2001). C22:5 n6 is hydrogenated to a lesser extent in the rumen and is readily used by the mammary gland, thereby increasing its concentration in milk fat.

A majority of the FA in milk fat are derived primarily from plasma triglycerides and non-esterified FA extracted by the mammary gland (Palmquist, 1976). In this study there was no significant effect on plasma concentration of TG and non-esterified FA for either type or level of oil supplementation (Table 9). These results contrast with those of Loor et al. (2002), who showed the concentrations of both TG and non-esterified FA increased with the dietary inclusion of high levels of unsaturated FA. Loor et al. (2002) suggested that the increase in the concentration of plasma TG and non-esterified FA was in response to the increased absorption of unsaturated FA by the small intestine. When plasma lipid levels were elevated due to high levels of dietary unsaturated fats, the mammary gland could hydrolyze up to 70% of triglycerides making their FA available for milk fat synthesis. One possible explanation for the lack of effect by level and type of oil supplemented could be due to the high removal rate of triglycerides by the mammary gland (Palmquist, 1976). Finally, glucose levels were not affected by either type or level of dietary oil in the current trial, which supports the findings in other feeding trials (Gagliostro, 1991). Phospholipids surround the core triglyceride in the milk fat globule and make up approximately 1% of total lipids in milk fat (Kurtz, 1974).Phospholipids are synthesized de novo in the mammary gland (Easteret al., 1971; Bitman et al., 1990). Fatty acids taken up by the mammary gland are used to synthesize PL. We observed elevated levels of TVA and rumenic acid in the phospholipid fraction of milk fat (Tables 8C and 8D) and plasma (Table 10) relative to total milk FA (Tables 6C and 6D).

Conclusion

Type of oil and level of oil feeding can be used to modify the FA composition of milk fat. Fish oil affects rumen biohydrogenation and can increase the TVA and RA concentration in milk fat particularly in the PL compared to the TG fraction of milk fat. Feeding fish oil increased the concentrations of C20:5 and C22:6 in plasma PL as well as milk fat, but the differences were small. Offer et al. (2001) proposed a poor transfer of C22:6 from lipoproteins into the mammary gland. Since we found no C20:5 and C22:6 in plasma TG and only small amounts of these FA in plasma PL, the small change in their concentration in milk fat might be a result of the poor transfer from lipoproteins as suggested by Offer et al. (2001) although this aspect of substrate availability deserves further study.

Acknowledgements

The authors thank J. Pareas, S. Burgos, and S. Juchem for their technical assistance; W. Paroczai and S. Amato for collecting milk samples; S. Cunningham and D. Gisi for care of the animals; and F. Sauers and V. Vieu for preparation of the diets. Financial support was provided by the USDA, the California Dairy Research Foundation (Davis, CA) through support of the Dairy Milk Components Laboratory, and the California Agricultural

Experiment Station (Davis). This research is a contribution of Multistate Research Project W-1181.

References

Abayasekara, D.R.E., Wathes, D.C. (1999). Effects of altering dietary fatty acid composition on prostaglandin synthesis and fertility. *Prostag., Leukotrien., Essent. Fatty Acids* 61, 275-287.

AbuGhazaleh, A.A., Shingoethe, D.J., Hippen, A.R. (2001). Conjugated linoleic acidand other beneficial fatty acids in milk fat from cows fed soybean and/or fishmeals. *J. Dairy Sci.* 84: 1845-1850.

AbuGhazaleh, A.A., Shingoethe, D.J., Hippen, A.R., Whitlock, L.A. (2002). Feeding fish meal and extruded soybeans enhances the conjugated linoleic acid(CLA) content of milk. *J. Dairy Sci.* 85: 624-631.

AbuGhazaleh, A.A., Shingoethe, D.J., Hippen, A.R. (2003). Milk conjugated linoleic acid response to fish oil supplementation of diets differing in fatty acid profiles. *J. Dairy Sci.* 86: 944-953.

AbuGhazaleh, A.A, Jenkins, T.C. (2004). Disappearance of docosahexaenoic and eicosapentaenoic acids from cultures of mixed ruminal microorganisms. *J. Dairy. Sci.* 87: 645-651.

A.O.A.C. (1995). Official Methods of Analysis, 16th Ed. Association of Official Analytical Chemists, Arlington, Virginia, USA

A.O.A.C. (2000). Official Methods of Analysis of AOAC International, 17th Ed. Official method #990.20.Gaithersburg, MD, USA

Ashes, J.R., Siebert, B.D., Gulati, S.K., Cuthbertson, A.Z., Scott, T.W. (1992). Incorporation of n-3 fatty acids of fish oil into tissue and serum lipids of ruminants. *Lipids* 27, 629-631.

Barber, M.C., Clegg, R.A., Travers, M.T., Vernon, R.G. (1997). Review. Lipid metabolism in the lactating mammary gland. *Biochim. Biophys. Acta* 1347, 101-126.

Bauman, D.E., Baumgard, L.H., Corl, B.A., Griinari, J.M. (1999). Biosynthesis of conjugated linoleic acidin ruminants. Proc. Am. Soc. Anim. Sci. Available at http://www.asas.org/jas/symposia/proceedings/0937.pdf (accessed April 04, 2005).

Bauman D.E., Griinari, J.M. (2003). Nutritional regulation of milk fat synthesis. *Amer. Rev. Nutr.* 23, 203-227.

Baumgard, L.H., Corl, B.A., Dwyer, D. A., Sæbø, A., Bauman, D.E. (2000). Identification of the conjugated linoleic acidisomer that inhibits milk fat synthesis. Amer. J. Physiol. 278, R179-184.

Baumgard, L.H., Sangster, J.K., Bauman, D.E. (2001). Milk fat synthesis in dairy cows is progressively reduced by increasing supplemental amounts of *trans*-10, *cis*-12 conjugated linoleic acid(CLA). *J. Nutr.* 131, 1764-1769.

Baumgard, L.H., Matitashvili, E., Corl, B.A., Dwyer, D.A., Bauman, D.E. (2002). *Trans*-10, *cis*-12 conjugated linoleic aciddecreases lipogenic rates and expression of genes involved in milk lipid synthesis in dairy cows. *J. Dairy Sci.* 85, 2155-2163.

Belury, M.A. (1995). Conjugated Dienoic Linoleate: A polyunsaturated fatty acid with unique chemical properties. *Nutr. Rev.* 53, 83-89.

Bickerstaffe, R., Noakes, D.E., Annison, E.F. (1972). Quantitative aspects of fatty acid biohydrogenation, absorption and transfer into milk fat of the lactating goat, with special reference to the *cis*- and *trans*-isomers of octadecenoate and linoleate. Biochem. J. 130, 607-617.

Bitman, J., Wood, D.L., Mehta, N.R., Hamosh, P., Hamosh, M. (1984). Comparison of the phospholipid composition of breast milk from mothers of term and preterm infants during lactation. *Amer. J. Clin. Nutr*. 40, 1103-1119.

Cant, J.P., Fredeen, J.H., MacIntyre, T., Gunn, J., Crowe, N. (1997). Effect of fish oil and monensin on milk composition in dairy cows. *Can. J. Anim. Sci*. 77, 125-131.

Chilliard, Y., Chardigny, J.M., Chabrot, J., Ollier, A., Sébédio, J.L., Doreau, M. (1999). Effects of ruminal and postruminal fish oil supply on conjugated linoleic acid(CLA) content of cow milk fat. *Proc. Nutr. Soc*. 58: 70A (Abstr.).

Chilliard, Y., Ferlay, A., Doreau, M. (2001).Effect of different types of forages, animal fat or marine oils in cow's diet on milk fat secretion and composition, especially conjugated linoleic acid(CLA) and polyunsaturated fatty acids. *Livestock Prod. Sci*. 70, 31-48.

Chilliard, Y.A., Ferlay, A., Mansbridge, R.M., Doreau, M. (2000). Ruminant milk fat plasticity: nutritional control of saturated, polyunsaturated, trans and conjugated fatty acids. *Ann. Zootech*. 49, 181-205.

Chilliard, Y., Ferlay, A. (2004). Dietary lipids and forages interactions on cow and goat milk fatty acid composition and sensory properties. Reprod. Nut. Dev. 44, 467-492.

Chin, S.F., Liu, W., Storkson, J.M., Ha, Y.L, Pariza, M.W. (1992). Dietary sources of conjugated dienoic isomers of linoleic acid, a newly recognized class of anticarcinogens. *J. Food Compos. Anal*. 5, 185-197.

Chin, S.F., Storkson, J.M., Albright, K.J., Cook, M.E., Pariza, M.W. (1994). Conjugated linoleic acidis a growth factor for rats as shown by enhanced weight gain and improved feed efficiency. *J. Nutr*. 124, 2344-2349.

Chouinard, P.Y., Corneau, L., Bauman, D.E., Butler, W.R., Chilliard, Y., Drackley, J.K. (1998). Conjugated linoleic acidcontent of milk from cows fed different sources of dietary fat. *J. Dairy Sci*., 81 (Suppl. 1): 223 (Abstr.)

Chouinard, P.Y., Corneau, L., Barbano, D., Metzger, L.E., Bauman, D.E. (1991). Conjugated linoleic acids alter milk fatty acid composition and inhibit milk fat secretion in dairy cows. *Amer. Soc. Nutr. Sci*, 1579-1584.

Collumb, M, Sieber, R., Bütikofer, U. (2004). CLA isomers in milk fat from cows fed diets with high levels of unsaturated fatty acids. *Lipids* 39, 355-363.

Cook, M.E., Miller, C.C., Park, Y., Pariza, M. (1993). Immune modulation by altered nutrient metabolism: Nutritional control of immune-induced growth depression. *Poult. Sci*. 72, 1301-1305.

Cook, M.E., Jerome, D.L., Crenshaw, T.D., Buege, D.R., Pariza, M.W., Schmidt, S.P., Scimeca, J.A., Lofgren, P.A., Hentges, E.J. (1998). Feeding conjugated linoleic acidimproves feed efficiency and reduces fat in pigs. *FASEB J*. 12, 4843 (Abstr.).

Consumer and Food Economics Institute, Agricultural Research Service, U.S. Department of Agriculture. (1976). Composition of foods, dairy and egg products, raw, processed, prepared. Agriculture Handbook 8-1. Washington D.C.: U.S. Government Printing Office.

Cordain, L., Watkins, B.A., Florant, G.L., Kelher, M., Roger, L., Li, Y. (2002). Fatty acid analysis of wild ruminant tissues: Evolutionary implications for reducing diet related chronic disease. *Eur. J. Clin. Nutr.* 56,181-191.

Corl, B.A., Chouinard, Y.P., Bauman, D.E., Dwyer, D.A., Griinari, J.M., Nurmela, K.V. (1998). Conjugated linoleic acid in milk fat of dairy cows originates in part by endogenous synthesis from trans-11 octadecadienoic acid. *J. Dairy Sci.* 81 (Suppl. 1), 233 (Abstr.).

Corl, B.A., Lacy, S.H., Baumgard, L.H., Dwyer, D.A., Griinari, J.M., Phillips, B.S., Bauman, D.E. (1999). Examination of the importance of Δ^9-desaturase and endogenous synthesis of CLA in lactating cows. *J. Anim. Sci.* 77 (Suppl. 1), 118 (Abstr.).

Corl, B.A., Baumgard, L.H., Dwyer, D.A., Griinari, J.M., Phillips, B.S., Bauman, D.E. (2001). The role of Δ^9-desaturase in the production of cis-9, trans-11 CLA. *J. Nutr. Biochem,* 12,622-630.

Crocker, L.M., DePeters, E.J., Fadel, J.G., Perez-Monti, H., Taylor, S.J., Wyckoff, J.A., Zinn, R.A. (1998). Influence of processed corn grain in diets of dairy cows on digestion of nutrients and milk composition. *J. Dairy Sci.* 81, 2394-2407.

Cunningham, D.C., Harrison, L.Y., Shultz, T.D. (1997). Proliferative responses of normal human and MCF-7 breast cancercells to linoleic acid, conjugated linoleic acid and eicosanoids synthesis inhibitors in culture. *Anticancer Res.* 17 (IA),197-203.

Dawson, R.M.C., Kemp, P. (1970). Biohydrogenation of dietary fats in ruminants. In: A.T. Phillipson (Ed.), Physiology of digestion and metabolism in the ruminant. Pp. 504-518. Oriel Press, Newcastle-upon-Thyne, United Kingdom.

DePeters, E.J., German, J.B., Taylor, S.J., Essex, S.T., Perez-Monti, H. (2001). Fatty acid and triglyceride composition of milk fat from lactating Holstein cows in response to supplemental canola oil. *J. Dairy Sci.* 84, 929-936.

Dhiman, T.R., Satter, L.D., Pariza, M.W., Galli, M.P., Albright, K., Tolosa, M.X. (2000). Conjugated linoleic acid (CLA) content of milk from cows offered diets rich in linoleic and linolenic acid. *J. Dairy Sci.* 83, 1016-1027.

Doll, R. (1992). The lessons of life: keynote address to the nutrition and cancer conference. *Cancer Res.* 52, 2024-2029.

Donovan, C.D., Shingoethe, D.J., Baer, R.J., Ryali, J., Hippen, A.R., Franklin, S.T. (2000). Influence of dietary fish oil on conjugated linoleic acidand other fatty acids in milk fat from lactating dairy cows. *J. Dairy Sci.* 83, 2620-2628.

Doreau, M., Chilliard, Y. (1997). Effects of ruminal or postruminal fish oil supplementation on intake and digestion in dairy cows. *Reprod. Nutr. Dev.* 37, 113-124.

Firkins, J.L., Eastridge, M.L. (1994). Assessment of the effects of iodine value on fatty acid digestibility, feed intake, and milk production. *J. Dairy Sci.* 77, 2357-2366.

Franklin, S.T., Martin, K.R., Baer, R.J., Shingoethe, D.J., Hippen, A.R. (1999). Dietary marine algae (*Schizochytrium* sp.) increase concentration of conjugated linoleic acid, docosahexanoic acid, and trans vaccenic acid of milk in dairy cows. *J. Nutr.* 129, 2048-2052.

Gil, J.L., Hafs, H.D. (1971). Analysis of repeated measurements of animals. *J. Anim. Sci.* 33, 331-336.

Griinari, J.M., Dwyer, D.A., McGuire, M.A., Bauman, D.E. (1996). Partially hydrogenated fatty acids and milk fat depression. *J. Dairy Sci.* 79 (Suppl.1), 177 (Abstr.).

Griinari, J.M., Nurmela, K.V.V., Corl, B.A., Chouinard, P.Y., Bauman, D.E. (1998). The endogenous synthesis of milk fat conjugated linoleic acid (CLA) from absorbed vaccenic acid in dairy cows. 89th AOCS Annual Meeting Abstr. May 10-13.

Griinari J.M., Bauman, D.E. (1999). Biosynthesis of conjugated linoleic acid and its incorporation into meat and milk in ruminants. In: Advances in conjugated linoleic acid. P. Yurawez, M.M. Mossaba, J.K.G. Kramer, G. Nelson, and M.W. Pariza (eds.), Vol. 1., Pp. 180-200 American Oil Chemists Society (AOCS) Press, Champaign, Ill.

Griinari, J.M., Corl, B.A., Lacy, S.H., Chouinard, P.Y., Nurmela, K.V.V., Bauman, D.E. (2000). Conjugated linoleic acidis synthesized endogenously in lactating dairy cows by Δ^9-desaturase. *J. Nutr.* 130, 2285-2291.

Grummer, R.R. (1991). Effect of feed on the composition of milk fat. *J. Dairy Sci.* 74, 3244-3257.

Gulati, S.K., Ashes, J.R., Scott, T.W. (1997). In vitro assessment of fat supplements for ruminants. *Anim. Feed Sci. Tech.* 64, 127-132.

Gulati, S.K., Ashes, J.R., Scott, T.W. (1999). Hydrogenation of eicosapentaenoic and docosahexaenoic acids and their incorporation into milk fat. *Animal Feed Sci. Technol.* 79, 57-64.

Gulati, S.K., May, C., Wynn, P.C., Scott, T.W. (2002). Milk fat enriched in n-3 fatty acids. *Anim. Feed Sci. Technol.* 98, 143-152.

Gulati, S.K., McGrath, S., Wynn, P.C., Scott, T.W. (2003). Preliminary results on the relative incorporation of docosahexaenoic and eicosapentaenoic acids into cows milk from two types of rumen protected fish oil. *Int. Dairy J.* 13, 339-343.

Harfoot, C.G. (1978). Lipid metabolism in the rumen. *Prog Lipid Res* 17,21-54.

Harfoot, C.G., Hazelwood, G.P. (1988). Lipid metabolism in the rumen. In: The rumen microbial ecosystem (Hobson, P.N., Ed.), pp. 285-322. Elsevier Science Publishers B.V., Amsterdam, The Netherlands.

Harfoot, C.G., Noble, R.C., Moore, J.H. (1973). Food particles as a site of biohydrogenation of unsaturated fatty acids in the rumen. Biochem. J. 132, 829-832.

Hoffman, W. S. (1937). A rapid photoelectric method for the determination of glucose in blood and urine. *J. Biol. Chem.* 120,51-55.

Howe, G.R. (1994). Dietary fat and breast cancerrisks. Cancer 74 (Suppl. 3),1078-1084.

Hu, F.B., Manson, J.E., Willett, W.C. (2001). Types of dietary fat and risk of coronary heart disease: A critical review. *J. Amer. College Nutr.* 20, 5-19.

Ip, C., Singh, M., Johnson, H.J., Scimeca, J.A. (1994). Conjugated linoleic acidsuppresses mammary carcinogenesis and proliferative activity of the mammary gland in the rat. *Canc. Res.* 54, 1212-1215.

Ip C., Banni, S., Angioni, E., Carta, G., McGinley, J., Thompson, H.J., Barbano, D., Bauman, D. (1999). Conjugated linoleic acid-enriched butter fat alters mammary gland morphogenesis and reduces cancer risk in rats. *J. Nutr.* 129, 2135-2142.

Jayan, G., Herbein, J. (2000). "Healthier" dairy fat using *trans*-vaccenic acid. *Nutr. Food Sci.* 30, 304-309.

Johnson, A.H. (1974). The composition of milk. In: Fundamentals of dairy chemistry. B.H. Webb, A.J. Johnson, and J.A. Alford (eds.), pp. 1-57. The Avi Publishing Company, Inc., Westport, CT, USA.

Jones, D.F., Weiss, W.P. (1998). Effects of feeding dairy cows differing concentrations of tallow and fish oil in milk yield and composition. In: The Ohio State University Extension Research Bulletin. Special Circular: pp. 163-169.

Kelly, M.L., Berry, J.R., Dwyer, D.A., Griinari, J.M., Chouinard, P.Y., Vanamburgh, M.E., Bauman, D.E. (1998a). Dietary fatty acid sources affect CLA concentrations in milk from lactating dairy cows. *J. Nutr*. 128, 881-885.

Kelly, M.L., Kolver, E.S., Bauman, D.E., Vanamburgh, M.E., Muller, L.D. (1998b). Effect of intake of pasture on concentrations of CLA in milk of lactating cows. *J. Dairy Sci.* 81, 1630-1636.

Kemp, P., White, R., Lander, D. (1975). The hydrogenation of unsaturated fatty acids by five bacterial isolates from the sheep rumen, including a new species. *J. Gen. Microbial.* 90, 100-114.

Kepler, C.R., Hirons, K.P., McNeill, J.J., Tove, S.B. (1966). Intermediates and products of the biohydrogenation of linoleic acidby *Butyrivibrio fibrisolvens*. J. Biol. Chem. 241, 1350-1354.

Kepler, C.R., Tove, S.B. (1967). Biohydrogenation of unsaturated fatty acids: III. Purification and properties of a linoleate Δ^{12}-cis, Δ^{11}-*trans*-isomerase from *Butyrivibrio fibrisolvens. J. Biol. Chem.* 242, 5686-5692.

Khanal, R.C. Dhiman, T.R. (2004). Biosynthesis of conjugated linoleic acid(CLA): A Review. *Pakistan J. Nutr*. 3, 72-81.

Kinsella, J.E. (1972). Steroyl CoA as a precursor for oleic acidand glycerolipids in mammary microsomes from lactating bovine: possible regulatory step in milk triglyceride synthesis. *Lipids* 7, 349-355.

Kitessa, S.M., Gulati, S.K., Ashes, J.R., Fleck, E., Scott, T.W., Nichols, P.D. (2001a). Utilisation of fish oil in ruminants I. Fish oil metabolism in sheep. *Anim. Feed Sci. Technol*. 89, 189-199.

Kitessa, S.M., Gulati, S.K., Ashes, J.R., Fleck, E., Scott, T.W., Nichols, P.D. (2001b). Utilisation of fish oil in ruminants II. Transfer of fish oil fatty acids into goats' milk. *Anim. Feed Sci. Technol*. 89, 201-208.

Knapp, H.R. (1999). Conjugated linoleic acid as a neutraceutical: Observations in the context of 15 years of n-3 polyunsaturated fatty acid research. In: P. Yurawez, M.M. Mossaba, J.K.G. Kramer, G. Nelson, and M.W. Pariza (Eds.), Advances in conjugated linoleic acid, Vol. 1. American Oil Chemists Society (AOCS) Press, Champaign, Ill.

Kraml, M. (1966). Semi-automated determination of phospholipids. *Clin. Chimica Acta.* 13, 442-448.

Kramer, J.K.G., Parodi, P.W., Jensen, R.G., Mossoba, M.M., Yurawecz, M.P., Adlof, R.O. (1998). Rumenic Acid: A proposed Common Name for the Major Conjugated Linoleic Acid Isomer Found in Natural Products. Lipids 33, 835.

Kurtz, E.F. (1974). The lipids of milk: Composition and properties. In: Fundamentals of Dairy Chemistry, Webb, B.H. and A.H. Johnson (Eds.). Pp. 91-169. Westport, CT. AVI Publishing.

Lacasse, P., Kennelly, J.J., Delbecchi, L., Ahnadi, C.E. (2002). Addition of protected and unprotected fish oil to diets for dairy cows. I. Effects on the yield, composition and taste of milk. *J. Dairy Res*. 69, 511-520.

Lands, W.E. (1986). Renewed questions about polyunsaturated fatty acids. *Nutr. Rev.* 44, 189-195.

Lee, K.N., Kritchevsky, D., Pariza, M.W. (1994). Conjugated linoleic acidand atherosclerosis in rabbits. *Atherosclerosis* 108, 19-25.

Lessard, M., Gagnon, N., Petit, H.V. (2003). Immune response of postpartum dairy cows fed flaxseed. *J. Dairy Sci.* 86, 2647-2657.

Li, Y., Watkins, B.A. (1998). Conjugated linoleic acidalters bone fatty acids composition and reduces *ex vivo* prostaglandin E_2 biosynthesis in rats fed n-6 or n-3 fatty acids. *Lipids* 33, 417-425.

Loor, J.J., Doreau, M., Chardigny, J.M., Ollier, A., Sebedio, J.L., Chilliard, Y. (2005). Effects of ruminal or duodenal supply of fish oil on milk fat secretion and profiles of *trans*-fatty acids and conjugated linoleic acidisomers in dairy cows fed maize silage. *Anim. Feed Sci. Technol.* 119, 227-246.

Loor, J.J., Quinlan, L.E., Bandara, A.B.P.A., Herbein, J.H. (2002). Distribution of *trans*-vaccenic acid and *cis*9, *trans*11-conjugated linoleic acid (rumenic acid) in blood plasma lipid fractions and secretion in milk fat of Jersey cows fed canola or soybean oil. *Anim. Res.* 51, 119-134.

Loor, J.J., Ueda, K., Ferlay, A., Chilliard, Y., Doreau, M. (2004). Short communication: Diurnal profiles of conjugated linoleic acids and trans fatty acids in ruminal fluid from cows fed a high concentrate diet supplemented with fish oil, linseed oil, or sunflower oil. *J. Dairy Sci.* 87, 2468-2471.

Mahfouz, M.M., Valicenti, A.J., Holman, R.T. (1980). Desaturation of isomeric trans-octadecadienoic acids by rat liver microsomes. Biochem. Biophys. Acta. 618, 1-12.

Marsh, W.H., Fingerhut, B., Kirsch,E. (1957). Urea nitrogen autoanalyzer methodology. *Amer. J. Clin. Path.* 28,681-688.

Martin, G.S., Lunt, D.K., Britain, K.G., Smith, S.B. (1999). Postnatal development of Stearoyl coenzyme A desaturase gene expression and adiposity in bovine subcutaneous adipose tissue. *J. Anim. Sci.* 77, 630-636.

McCaughey, K.M. (2003). Effect of iron sulfate supplementation of the diet on plasma gossypol concentration and productivity of lactating Holstein cows fed cracked Pima cottonseed. M.S. Thesis, University of California, Davis, CA.

McGuire, M.A., McGuire, M.K., Guy, M.A., Sanchez, W.K., Shultz, T.D., Harrison, L.Y., Bauman, D.E., Griinari, J.M. (1996). Effect of dietary lipid concentration on content of conjugated linoleic acid(CLA) in milk from dairy cattle. *J. Anim. Sci.* 74 (Suppl. 1), 266 (Abstr.).

McGuire, M.A., McGuire, M.K., Ritzenthaler, K., Shultz, T.D. (1999). Dietary sources and intakes in humans. In: P. Yurawez, M.M. Mossaba, J.K.G. Kramer, G. Nelson, and M.W. Pariza (Eds.), Advances in conjugated linoleic acid, Vol. 1. American Oil Chemists Society (AOCS) Press, Champaign, Ill.

Mosley, E.E., McGuire, M.K., Williams, J.E., McGuire, M.A. (2006). *Cis*-9, *trans*-11 conjugated linoleic acid is synthesized from vaccenic acid in lactating women. *J. Nutr.* 136, 2297-2301.

Murphy, J.J. (1999). Synthesis of milk fat and opportunities for nutritional manipulation. Teagasc, Dairy Production Research Department, Moorepark, Fermoy, Co Cork. www.bsas.org.uk/publs/milkcomp/07.pdf.

National Dairy Council. (1988). An interpretive review of recent nutrition research. *Dairy Council Digest* 59,1-6.

National Research Council. (1996). Carcinogens and Anticarcinogens in the Human Diet. National Academy of Science, Washington D.C.

Ney, D. (1991). Symposium: The role of nutritional and healthy benefits in the marketing of dairy products. Potential for enhancing the nutritional properties of milk. *J. Dairy Sci.* 74, 4002-4012.

Offer, N.W., Marsden, M., Dixon, J., Speake, B.K., Thacker, F.E. (1999). Effect of dietary fat supplements on levels of n-3 polyunsaturated fatty acids, trans acids, and conjugated linoleic acidin bovine milk. *Anim. Sci.* 69, 613-625.

Offer, N.W., Speake, B.K., Dixon, J., Marsden, M. (2001). Effect of fish-oil supplementation on levels of (n-3) poly-unsaturated fatty aicds in the lipoprotein fractions of bovine plasma. Anim. Sci. 73, 523-531.

Opstvedt, J. (1984). Fish fats. In Fats in Animal Nutrition. J. Wiseman (Ed.). Pp. 53-81. Butterworths.

Palmquist, D. (2001). Milk fat – It's good for you! *In*: Research and Review: Dairy. Ohio State University, *Department of Animal Science*, Special Circular: 182-201.

Palmquist, D.L., Griinari, J.M. (2006). Milk fatty acid composition in response to reciprocal combinations of sunflower and fish oils in the diet. *Anim. Feed Sci. Technol.* 131, 358-369.

Parodi, P.W. (1994). Conjugated linoleic acid: an anticarcinogenic fatty acid present in milk fat. *Aust. J. Dairy Technol.* 49, 93-97.

Parodi, P.W. (1997a). Cow's milk fat components as potential anticarcinogenic agents. *J. Nutr.* 127, 1055-1066.

Parodi, P.W.(1997b). Milk fat conjugated linoleic acid: can it help prevent breast cancer? *Proc. Nutr. Soc.* N.Z. 22, 137-149.

Parodi, P.W. (1999). Conjugated linoleic acidand other anticarcinogenic agents of bovine milk fat. *In*: Symposium: A Bold New Look At Milk Fat. *J. Dairy Sci.* 82, 1339-1349.

Pennington, J.A., Davis, C.L. (1975). Effects of intraruminal and intra-abomasal additions of cod-liver oil on milk fat production in the cow. *J. Dairy Sci.* 58, 49-55.

Petit, H.V., Dewhurst, R.J., Proulx, J.G., Khalid, M., Haresign, W., Twagiramungu, H. (2001). Milk production, milk composition, and reproductive function of dairy cows fed different fats. *Can. J. Anim. Sci.* 81, 263-271.

Petit, H.V., Dewhurst, R.J., Scollan, N.D., Proulx, J.G., Khalid, M., Haresign, W., Twagiramungu, H., Mann, G.E. (2002). Milk production and composition, ovarian function, and prostaglandin secretion of dairy cows fed omega-3 fats. *J. Dairy Sci.* 85, 889-899.

Piperova, L.S.; Sampugna, J., Teter, B.B., Kalsheur, K.F., Yurawicz, M.P., Ku, Y., Morehouse, K.M., Erdman, R.A. (2002). Duodenal and milk *trans* octadecadienoic acid and conjugated linoleic acid (CLA) isomers indicate that post absorptive synthesis is the predominant source of *cis*-9-containing CLA in lactating dairy cows. *J. Nutr.* 132, 1235-1241.

Pollard, M.R., Gunstone, F.D., James, A.T., Morris, L.J. (1980). Desaturation of positional and geometric isomers of monoenoic fatty acids by microsomal preparations from rat liver. *Lipids* 15, 306-314.

Polon, C.E, McNeill, J.J., Tove, S.B. (1964). Biohydrogenation of unsaturated fatty acids by rumen bacteria. *J. Bacteriol.* 88, 1056-1064.

Rego, O.A., Rosa, H.J.D., Portugal, P., Cordeiro, R., Borba, A.E.S., Vouzela, C.M., Bessa, R.J.B. (2005). Influence of dietary fish oil on conjugated linoleic acid, omega-3 and other fatty acids in milk fat from grazing dairy cows. *Live. Prod. Sci.* 95, 27-33.

Riel, R.R. (1963). Physico-chemical characteristics of Canadian milk fat: Unsaturated fatty acids. *J. Dairy Sci.* 46, 102-106.

Reiser, R. (1951). Hydrogenation of polyunsaturated fatty acids by the ruminant. Fed. Proc. 10, 236 (Abstr.).

Robertson, J.B., Van Soest, P.J. (1981). The detergent system of analysis and its application to human foods. In: W.P.T. James O. and Theander (Eds.), The analysis of dietary fiber in foods. Pp. 123-130. Marcel Dekker, New York, NY.

Robinson, R.S., Pushpakumara, P.G.A., Cheng, Z., Peters, A.R., Abayasekara, D.R.E., Wathes, D.C. (2002). Effects of dietary polyunsaturated fatty acids on ovarian and uterine function in lactating dairy cows. *Reprod.* 124, 119-131.

Ryder, J.W., Portocarrero, C.P., Song, X.M., Cui, L., Yu, M., Combatsiaris, T., Galuska, D., Bauman, D.E., Barbano, D.M., Charron, M.J., Zierath, J.R., Houseknecht, K.L. (2001). Improved glucose tolerance, skeletal muscle insulin action and UCP-2 gene expression. *Diabetes* 50, 1149-1157.

SAS. (1999). SAS User's Guide: Statistics (Version 8 Ed.). SAS Inst. Inc., Cary, NC.

Salminen, I., Mutanen, M., Jauhiainen, M., Aro, A. (1998). Dietary trans fatty acids increase conjugated linoleic acidin human serum. *J. Nutr. Biochem.* 9, 93-98.

Shantha, N.C., Ram, L.N., O'Leary, J., Hicks, C.L., Decker, E.A. (1995). CLA concentrations in dairy products as affected by processing and storage. *J. Food Sci.* 60, 695-697.

Shennan, D.B., Peaker, M. (2000). Transport of milk constituents by the mammary gland. *Physiol. Rev.* 80, 925-951.

Shingfield, K.J., Ahvenjärvi, S., Toivonen, V., Ärölä, A., Nurmela, K.V.V., Huhtanen, P., Griinari, J.M. (2004). Effect of dietary fish oil on biohydrogenation of fatty acids and milk fatty acid content in cows. *Anim. Sci.* 77, 165-179.

Simopoulas, A.P. (1999a). New products from the agri-food industry: The return of n-3 fatty acids into the food supply. *Lipids* 34, S297-S301.

Simopoulas, A.P. (1999b). Evolutionary Aspects of omega-3 fatty acids in the food supply. *Prostaglandins Leuko.Essent. Fatty Acids* 60, 421-429.

Spain, J. (2002). Essentiality of specific fatty acids in reproductive performance of high producing dairy cows. *Adv. Dairy Tech.* 14,185-192.

Stanton, C., Lawless, F., Murphy, J., Connolly, B. (1997). CLA- A health promoting component of dairy fats, I. Biological Properties of CLA. Farm Food 7, 19-20.

St. John, L.C., Lunt, D.K., Smith, S.B. (1991). Fatty acid elongation and Desaturation activities of bovine liver and subcutaneous adipose tissue microsomes. *J. Anim. Sci.* 69, 1064-1073.

Sumathipala, R.N., Travers, M.T., Thomas, P., Munday, M.R., Clegg, R.A. (1998). Expression of FAT, the putative fatty acid translocator protein, is developmentally regulated in rat mammary tissue. *Biochem. Soc. Trans.* 26, S232.

Thompson, G.E., Christie, W.W. (1991). Extraction of plasma triacylglycerols by the mammary gland of the lactating cow. *J. Dairy Res.* 58, 251-255.

Turpeinen, A.M., Mutanen, M., Aro, A., Salminen, I., Basu, S., Palmquist, D.L., Griinari, J.M. (2002). Bioconversion of vaccenic acid to conjugated linoleic acid in humans. *Am. J. Clin. Nutr.* 76, 504-510.

Van Soest, P.J., Robertson, J.B. (1991). Methods of dietary fiber, neutral detergent fiber, and nonstarch polysaccharides in relation to animal nutrition. *J. Dairy Sci.* 74, 3583-3597.

Van de Vossenberg, J.L.C.M., Joblin, K.N. (2003). Biohydrogenation of C18 unsaturated fatty acids by a strain of *Butyrivibrio hungatei* from the bovine rumen. *Letters Appl. Microbiol.* 37, 424-428.

Viviana, R. (1970). Metabolism of long-chain fatty acids in the rumen. *Adv. Lipid Res.* 8, 367-346.

Wachira, A.M., Sinclair, L.A., Wilkinson, R.G., Hallet, K., Enser, M. Wood, J.D. (2000). Rumen biohydrogenation of n-3 polyunsaturated fatty acids and their effects on microbial efficiency and nutrient digestibility in sheep. *J. Agric. Sci.* (Camb.) 135, 419-428.

Ward, R.J., Travers, M.T., Richards, S.E., Vernon, R.G., Salter, A.M., Buttery, P.J., Barber, M.C. (1998). Stearoyl CoA desaturase mRNA is transcribed from a single gene in the ovine genome. *Biochem. Biophys. Acta.* 1391, 145-156.

Watkins, B.A., Shen, C.L., McMurtry, J.P., Xu, H., Bain, S.D., Allen, K.G.D., Seifert, M.F. (1997). Dietary lipids modulate prostaglandin E_2 production, Insulin-like Growth Factor-I concentration and formation rate in chicks. *J. Nutr.* 127, 1084-1091.

Whigam, L.D., Cook, M.E., Atkinson, R.L. (2000). Conjugated linoleic acid: Implications for human health. *Pharmacol. Res.* 42, 503-510.

Whitlock, L.D., Shingoethe, D.J., Hippen, A.R., Kalscheur, K.F., Baer, R.J., Ramaswamy, N., Kasperson, K.M. (2002). Fish oil and extruded soybeans fed in combination increase CLA in milk of dairy cows more than when fed separately. *J. Dairy Sci.* 85, 234-243.

Wildman, E.E., Jones, G.M., Wagner, P.E., Bowman, R.L., Trout, H.F., Lesch, T.N. (1982). A dairy cow body condition scoring system and its relationship to selected production characteristics. *J. Dairy Sci.* 65, 495-501.

Wonsil, B.J., Herbein, J.H., Watkins, B. (1994). Dietary and ruminally derived *trans*-18:1 fatty acids alter bovine-milk lipids. J. Nutr. 124, 556-565.

World Health Organization and FAO Joint Consultation. (1995). Fats and oils in human nutrition. *Nutr. Rev.* 53, 202-205.

Yurawecz, M.P., Roach, J.A.G., Sehat, N., Mossoba, M.M., Kramer, J.K.G., Fritsche, J.,Steinhart, H., Ku, Y. (1998). A new conjugated linoleic acid isomer, 7 trans, 9 cis-octadecadienoic acid in cow milk, cheese, beef and human milk and adipose tissue. *Lipids* 33, 803-809.

Part IV: Milk Products and Valorization Techniques

In: Milk Production
Editor: Boulbaba Rekik, pp. 245-261

ISBN 978-1-62100-061-7

Chapter XI

Genetic Factors and Dairy-Technological Valorisation of Milk

Andrea Summer[1, 2*], Massimo Malacarne[1], Paolo Formaggioni[1], Piero Franceschi[1] and Primo Mariani[1, 2]

[1]Department of Animal Production, Veterinary Biotechnologies, Food Quality and Safety. University of Parma, Via del Taglio 10, 43126 Parma, Italy

[2]MILC Centre. University of Parma, Via del Taglio 10, 43126 Parma, Italy

Abstract

Among the factors of variation, the genetic ones play a fundamental role in the determination of milk characteristics for cheese making. The genetic factors are more important the more the coagulation is of rennet type. In these conditions, the micelle structure of milk can totally express its intrinsic properties of genetic origin (e.g. polymorphism of caseins). The micelle maintains its structure almost unchanged in all the phases of the whole cheesemaking process, influencing markedly the rheological properties of the curd.

Role of the phenotypic combinations of caseins α_{s1}, β, k – In this part of the chapter the phenotypes (electrophoresis in starch-urea gel pH 8.6) of α_{s1}, β and k caseins of 222 individuals' milk samples from Italian Brown cows were determined. The aim was to estimate the effect of the composed phenotype on the physico-chemical and dairy-technological characteristics of milk. The statistical analysis put in evidence significant differences among the 5 considered phenotypes, regarding both the contents of some constituents and the rennet-coagulation characteristics

Comparison among genetic types – The aim of this part of the chapter was to study the physico-chemical, chemical and technological characteristics of milk of some genetic types present in the production area of the Parmigiano-Reggiano cheese, in comparison with those of the Italian Friesian cow milk. For this aim, the following four comparisons were carried out: Italian Brown *vs* Italian Friesian, Jersey*vs* Italian Friesian; Ayrshire *vs* Italian Friesian. The comparison between Italian Brown and Italian Friesian was made at

*Email: andrea.summer@unipr.it

the level of vat milk; while the other comparisons were carried out at the level of herd milk.

Dairy technological parameters of the Italian Brown cow milk in relation to types A and B of the β-lactoglobulin – The genetic type of the β-lactoglobulin in individual milk samples from Italian Brown cows was determined. The aim of this part of the chapter was to estimate the relationships between types of β-lactoglobulin and characteristics of milk. The β-lactoglobulin BB milk showed particularly interesting technological parameters, i.e. higher casein content and higher casein number.

Keywords: genetic factors, cattle breed, protein polymorphism, milk quality, dairy-technological valorisation.

Introduction

The processing quality of milk is intended as its capacity to give rise to a good cheese (from chemical and sensorial point of views) with an acceptable cheese yield. In the evaluation of the processing properties of milk destined to hard long ripened cheese (e.g. Parmigiano-Reggiano cheese or Grana Padano cheese) production, the rennet-coagulation aptitude represents the most important requisite. Good reactivity of milk with the rennet is associated to discrete rheological properties of the curd and, consequently, to suitable structural characteristics of the cheese paste. Among various factors, those of genetic nature play a fundamental role in determining the characteristics of milk for cheesemaking [1-3]. The importance of these factors is particularly relevant if the milk is coagulated by rennet [4-8]. In this condition, the casein micelle structure can express its intrinsic characteristics of genetic origin (e.g. casein polymorphism) [9-13]. In fact, the casein micelle structure remains almost unaltered throughout the whole cheesemaking process, affecting the rheological properties of the curd. Relevant are differences at the level of species, breeds, and population in fat and protein content of milk as well as in its organic and inorganic components. Several studies focused on the effect of cattle breed on the chemical, physico-chemical, and processing qualities of milk [14-19]. The milk protein polymorphism plays a particular aspect in the definition of the chemical, physico-chemical characteristics, and processing properties of milk. Numerous scientific acquisitions evidenced a discrete applicative interest of milk protein polymorphism, at zootechnical and dairy levels [4, 7, 20-22].In this chapter are summarised the results of different studies focusing on the role of genetic factors (breed and genetic polymorphism) on quantitative and qualitative characteristics of milk used for cheese production.

Comparison among Genetic Types

The objective of this part of the chapter was to compare physico-chemical, chemical and processing characteristics of milk of some cattle breeds reared in the Parmigiano-Reggiano production area. Milk characteristics of each cattle breed was compared with those of Italian Friesian milk, the most diffused cattle breed in the Parmigiano-Reggiano district. To this aim, 3 distinctive comparisons were carried out: Jersey*vs* Italian Friesian, Ayrshire *vs* Italian

Friesian, Italian Brown *vs* Italian Friesian. In the case of Italian Brown *vs* Italian Friesian, the comparison was carried out at vat milk level, while other comparisons were conducted at herd milk level.Productive characteristics of milk of some cattle breeds reared in Italy are reported in table 1 [23]. Differences in milk yield are quite apparent among breeds. Values ranged from 8961 kg/lactation in Italian Friesian to 5930 inJersey. Other breeds showed intermediate values. On the other hand, Jersey milk was characterised by the highest values both for the percentages of protein and fat (3.99% and 5.09%, respectively) while Italian Friesian milk had the lowest protein and fat contents (3.30% and 3.65%protein and fat, respectively).In the following paragraphs, comparisons of each breed with the Italian Friesian are reported.

Jersey vs Italian Friesian

The research was carried out in a single herd located in the province of Reggio Emilia (Northern Italy) in which about 280 Jersey cows and about 280 Italian Friesian cows were present. Eight herd milk samples for each breed were collected. Milk yield was on average 6993 kg/lactation (Jersey) and 9757 kg/lactation (Italian Friesian). Compared to Italian Friesian milk, Jersey milk was characterised by higher values of protein (3.80 *vs* 3.39 g/100g), fat (4.18 *vs* 3.24 g/100g) and casein (2.98 *vs* 2.62 g/100g) (Table 2). Furthermore, Jersey milk showed a higher casein number (78.42 *vs* 77.17 %) and paracasein number (72.96 *vs* 71.49 %) than Italian Friesian milk. The paracasein is the part of the casein that constitutes the coagulum, i.e. the casein without the hydrophilic moiety named glycomacropeptide that is hydrolysed by the chymosin during the rennet-coagulation process. The paracasein number is the percent of the paracasein on total protein.

Milk minerals (in particular calcium, phosphorus and magnesium) are linked to the casein (colloidal phase) or in the solution (soluble phase). The salt equilibria, i.e. the distribution of these minerals between the two phases, are very important for the rennet-coagulation characteristics of the milk; for this reason, it is better to study this distribution rather than only the total content of the minerals. As far as minerals in this comparison, Jersey milk was characterised by higher contents of total calcium (135 *vs* 112 mg/100g), colloidal calcium (97 *vs* 78 mg/100g) and soluble calcium (38 *vs* 34 mg/100g) than Italian Friesian milk (Table 3). Even total phosphorus (110 *vs* 98 mg/100g), colloidal inorganic phosphorus (31 *vs* 24 mg/100g) and soluble phosphorus (50 *vs* 49 mg/100g) were higher in Jersey milk than in Italian Friesian milk (Table 3).

Table 1. Comparison among milks of different breeds: milk yield, protein and fat

		Italian Friesian	Angler	Ayrshire	Italian Brown	Jersey
Milk yield	kg/lactation	8961	7071	7101	6816	5930
Protein	g/100g	3.30	3.63	3.39	3.52	3.99
Fat	g/100g	3.65	4.33	3.81	3.96	5.09

AIA, 2011 [23].

The contents both in colloidal calcium (3.24 *vs* 2.98 g/100g) and inorganic phosphorus (1.04 *vs* 0.91 g/100g) per 100 g of casein were higher in Jersey than in Italian Friesian milk, demonstrated the higher mineralisation degree of the casein micelle of Jersey cows' milk.The somatic cell score was higher in Jersey milk than Italian Friesian (4.39 *vs* 3.29). Jersey milk evidenced a lower chloride value than Italian Friesian milk.

A method for evaluating the milk rennet-coagulation properties is the lactodynamography. In this analytical method, a rennet solutionwas added to milk samples. The coagulation characteristics, milk clotting time (r), curd firming time (k_{20}) and curd firmness (a_{30}), were measured at 35 °C. Milk clotting time is the time from the addition of rennet to the onset of gelation. Curd firming time is the time from the onset of gelation till the signal attains a width of 20 mm. Curd firmness is the width of the signal 30 min (a_{30}) after the addition of rennet. In this way are highlighted the salient points of the rennet cheesemaking process. Other important aspects are the rheological properties, i.e. the resistance to compression and resistance to cut off the coagulum that can be measured 30 min after the beginning of coagulation, using the Gel Tester apparatus. These are important, in particular, to assess the process of curd cutting and curd syneresis.

Table 2. Comparison between Jersey and Italian Friesian herd milks: gross composition and nitrogen fractions. No. obs. 8 for each breed

		Jersey		Italian Friesian		P
Fat	g/100g	4.18	± 0.27	3.24	± 0.15	**
Protein	g/100g	3.80	± 0.11	3.39	± 0.07	**
Casein	g/100g	2.98	± 0.07	2.62	± 0.04	**
Casein number	%	78.42	± 0.68	77.17	± 0.69	*
Paracasein number	%	72.96	± 0.87	71.49	± 0.86	*

Mean±SE.
* P≤0.05; ** P≤0.01.

Table 3. Comparison between Jersey and Italian Friesian herd milks: calcium and phosphorus distribution, chloride, and somatic cells. No. obs. 8 for each breed

		Jersey		Italian Friesian		P
Total calcium	mg/100g	134.63	±2.80	111.92	±2.82	**
Colloidal calcium	mg/100g	96.59	±2.69	78.00	±2.21	**
Soluble calcium	mg/100g	38.04	±1.37	33.92	±2.01	**
Total phosphorus	mg/100g	110.18	±3.86	98.13	±1.01	**
Casein phosphorus	mg/100g	26.81	±5.23	24.28	±1.24	NS
Colloidal inorganic phosphorus	mg/100g	31.01	±3.97	23.70	±1.25	**
Soluble phosphorus	mg/100g	50.45	±2.23	48.65	±0.71	*
Colloidal calcium/casein	g/100g	3.24	±0.05	2.98	±0.11	**
Inorg. coll. phosphorus/casein	g/100g	1.04	±0.13	0.91	±0.04	*
Chloride, Cl^-	mg/100g	80.03	±2.20	91.23	±3.73	**
Somatic cell score	—	4.39	±0.99	3.29	±0.36	*

Mean±SE.
NS, P>0.05; * P≤0.05; ** P≤0.01.

Concerning rennet-coagulation properties, Jersey milk showed lower clotting time (r: 14.70 *vs* 20.14 min), lower curd firming time (k_{20}: 2.07 *vs* 6.69 min), higher curd firmness measured 30 minutes after rennet addition (a_{30}: 46.74 *vs*25.63 mm) than Italian Friesian milk (Table 4). Besides, better rennet-coagulation properties, Jersey milk evidenced better (higher) rheological properties of the curd than Italian Friesian milk: resistance to compression (35.67 *vs*23.50 g) and resistance to cut (80.83 *vs*47.00 g) (Table 4).In general, milk from Jersey cows showed aptitude to rennet-coagulation and rheological properties clearly better than those of milk from Italian Friesian cows. According to these results, the basis of these differences is associated to chemical, physico-chemical, and structural peculiarities that characterise Jersey milk. Among these, the principal role seems to be exerted by the elevated casein content.

Ayrshire vs Italian Friesian

This research was conducted in two herds of Ayrshire cows and two herds of Italian Friesian cows. On a whole, 11 herd milk samples were collected for each breed. Milk samples were analysed for rennet-coagulation characteristics and rheological properties. No significant differences between Ayrshire and Italian Friesian milk were observed for clotting time (r, 21.7 *vs* 21.6 min), curd firming time (k_{20}, 4.3 *vs* 4.8 min), and curd firmness measured 30 minutes after rennet addition (a_{30}, 23.63 *vs*20.00 mm). On the other hand, rheological property values were clearly higher – and thus better – for Ayrshire milk than Italian Friesian milk: resistance to compression (23.63 *vs*20.00 g) and resistance to cut (49.63 *vs*39.38 g) (Table 5).

Table 4. Comparison between Jersey and Italian Friesian herd milks: rennet-coagulation parameters and rheological properties. No. obs. 8 for each breed

		Jersey		Italian Friesian		P
Clotting time, r	min	14.70	±1.07	20.14	±2.17	**
Curd firming time, k_{20}	min	2.07	±0.22	6.69	±2.75	**
Curd firmness, a_{30}	mm	46.74	±1.99	25.63	±7.67	**
Resistance to compression	g	35.67	±7.31	23.50	±4.46	**
Resistance to cut	g	80.83	±3.25	47.00	±8.07	**

Mean±SE.
** P≤0.01.

Table 5. Comparison between Ayrshire and Italian Friesian herd milks: curd rheological properties. No. obs. 11 for each breed

		Ayrshire		Italian Friesian		P
Resistance to compression	g	23.63	±1.85	20.00	±1.85	***
Resistance to cut	g	49.63	±9.94	39.38	±10.60	*

Mean±SE.
* P≤0.05; *** P≤0.001.

Italian Brown vs Italian Friesian

Thirteen cheesemaking trials were performed at 10 different cheese factories producing Parmigiano-Reggiano cheese. Each trial involved the processing, in parallel, of a vat (1100 kg of milk) containing only Italian Brown milk and a vat of only Italian Friesian milk.In the Parmigiano-Reggiano cheese production, the raw material was obtained by commingling (approximately 1:1) the partially skimmed evening milk and the full-cream morning milk, and this is named vat milk. Each vat milk was constituted by milk from at least 3 herds.On each vat milk sample, nitrogen fractions, physico-chemical, and processing properties, cheese yield and cheesemaking losses in the cooked whey, were determined. The content of nitrogen fractions was significantly different between Italian Brown and Italian Friesian vat milks. Italian Brown vat milk was characterised by higher content of crude protein (3.49 *vs* 3.07 g/100g), casein (2.71 *vs* 2.37 g/100g), wheyprotein (0.61 *vs* 0.54 g/100g) and non protein nitrogen (NPN x 6.38: 0.18 *vs* 0.16 g/100g) (Table 6). Titrable acidity was higher in Italian Brown than in Italian Friesian vat milk (3.42 *vs* 3.24 °SH/50mL), whereas no significant differences were observed for pH values (Table 7). As far for rennet-coagulation properties, Italian Brown vat milk evidenced a lower curd firming time (k_{20}, 6.6 *vs* 10.1 min) and higher curd firmness measured 30 minutes after rennet addition (a_{30}, 31.3 *vs*24.2 mm), whereas no significant differences were observed for clotting time (Table 7).

Table 6. Comparison between Italian Brown and Italian Friesian vat milks: nitrogen fractions. No. obs. 13 for each breed

		Italian Brown		Italian Friesian		P
Protein	g/100g	3.49	± 0.17	3.07	± 0.17	**
Casein	g/100g	2.71	± 0.12	2.37	± 0.14	**
Wheyprotein	g/100g	0.61	± 0.04	0.54	± 0.04	**
NPN x 6.38	g/100g	0.18	± 0.02	0.16	± 0.03	*

Mean±SE.
* P≤0.05; ** P≤0.01.

Table 7. Comparison between Italian Brown and Italian Friesian vat milks: pH, titrable acidity, rennet-coagulation parameters and rheological properties. No. obs. 13 for each breed

		Italian Brown		Italian Friesian		P
pH	units	6.72	±0.05	6.71	±0.09	NS
Titratable acidity	°SH/50mL	3.42	±0.20	3.24	±0.19	**
Clotting time, r	min	18.62	±2.92	18.44	±2.62	NS
Curd firming time, k_{20}	min	6.57	±2.31	10.09	±3.61	**
Curd firmness, a_{30}	mm	31.32	±8.36	24.23	±5.95	**
Resistance to compression	g	32.64	±3.85	23.93	±4.32	**
Resistance to cut	g	73.45	±11.28	49.60	±3.14	**

Mean±SE.
NS, P>0.05; ** P≤0.01.

Table 8. Comparison between Italian Brown and Italian Friesian vat milks: 24h Parmigiano-Reggiano cheese yield and cheesemaking losses (%) in the cooked whey. No. obs. 13 for each breed

		Italian Brown		Italian Friesian		P
Cheese yield	kg/100 kg	8.92	± 0.67	7.92	± 0.46	**
Fat % losses	%	15.54	± 2.11	18.23	± 4.00	**
Calcium % losses	%	32.62	± 1.85	35.92	± 1.89	**

Mean±SE.
** P≤0,01.

Forthe rheological properties of the curd, Italian Brown vat milk was characterised by higher value of both resistance to compression (32.6 *vs*23.9 g) and resistance to cut (73.5 *vs*49.6 g) of the curd than Italian Friesian vat milk (Table 7). The cheese yield (measured 24 h after cheesemaking) of Italian Brown vat milk was significantly higher than Italian Friesian vat milks (8.92 *vs* 7.92 kg/100 kg of vat milk). Losses of both fat (15.5% *vs* 18.2%) and calcium (32.6% *vs* 35.9%) resulted in lower cooked whey for Italian Brown than for Italian Friesian vat milks (Table 8). In general, Italian Brown vat milk was characterised by higher casein content, better rennet-coagulation characteristics, and better rheological properties than Italian Friesian vat milk.

Genetic Polymorphism of Milk Proteins

The current knowledge on relationships among genetic polymorphism of milk proteinsand milk processing characteristics shows that genetic variants of β-lactoglobulin and k-casein play a major role on these characteristics [24]. The alleles of β-lactoglobulin exert a complex action towards the protein composition of milk: milk characterised by the presence of the B variant of β-lactoglobulin resulted in high contents of milk in casein and thus milk is particularly suitable for cheesemaking giving higher cheese yield, when compared to milk β-lactoglobulin A [25-27]. The genetic variants of k-casein play a marked role in defining the reactivity of milk toward rennet, and the whole physico-mechanical behaviour of the curd during cheesemaking. In this concern, B variant of k-casein was clearly more favourable than the A variant [26, 28-32]. The effects of the variants A and B of k-casein were widely investigated [28, 33-39], while the effects of the A and B variant of β-lactoglobulin was less characterised.

Processing Parameters of Italian Brown Milk According to the Variants A and B of β-lactoglobulin

The aim of this research was to evaluate the relationships between β-lactoglobulin genotypes and milk characteristics. The β-lactoglobulin genotypewas assessed in individual milk samples of Italian Brown cows. Then, two groups of cows were constructed according to their β-lactoglobulin genetic type: a group of cows heterozygous AB and a group of cows homozygous BB. The casein nitrogen content was higher in BB milk than in AB milk (446 *vs* 430 mg/100g) (Table 9). On the contrary, soluble nitrogen was lower in BB milk than in AB milk (119 *vs* 129 mg/100g) (Table 9). As a consequence, total nitrogen was not different between the two groups. The casein number resulted higher for the BB than for AB phenotype(79.04 *vs* 76.93 %). On minerals analysed (Ca, P, Mg, Na, K and chloride, Table 10), only Mg content evidenced significant differences between the two groups of milk: BB milk was characterised by lower values than AB milk (10 *vs* 11 mg/100g). Concerning rennet-coagulation properties, significant differences were recorded only for the curd firmness adjusted for the clotting time ($a_{1/2r}$): this parameter was higher in β-lactoglobulin BB milk (Table 11). This effect on rennet-coagulation is related to the higher content of casein in β-lactoglobulin BB milk than in β-lactoglobulin AB milk.

Table 9. Comparison between individual milks of AB and BB β–lactoglobulin phenotypes in Italian Brown breed: nitrogen fractions. No. obs.: 139 for β–Lg AB class and 79 for β–Lg BB class

		AB β–lactoglobulin		BB β–lactoglobulin		P
Total N	mg/100g	557.3	±3.60	564.2	±4.76	NS
Soluble N at pH 4.6	mg/100g	129.4	±1.12	118.6	±1.51	***
Casein N	mg/100g	430.4	±2.92	446.0	±3.87	**
Casein number	%	76.93	±0.13	79.04	±0.17	***

Mean±SE.
NS, P>0.05; ** P≤0.01; *** P≤0.001.

Table 10. Comparison between individual milks of AB and BB β–lactoglobulin phenotypes in Italian Brown breed: mineral components. For Ca, P and Mg: No. obs. 98 for β–Lg AB class and 62 for β–Lg BB class; for Na, K and Cl: No. obs. 120 for β–Lg AB class and 78 for β–Lg BB class

		AB β–lactoglobulin		BB β–lactoglobulin		P
Calcium	mg/100g	121.3	±0.80	120.2	±1.00	NS
Phosphorus	mg/100g	105.3	±0.91	105.2	±1.15	NS
Magnesium	mg/100g	10.8	±0.13	10.4	±0.16	*
Sodium	mg/100g	41.5	±0.66	42.6	±0.82	NS
Potassium	mg/100g	171.5	±1.45	171.3	±1.72	NS
Chloride, Cl^-	mg/100g	92.0	±0.94	89.3	±1.18	NS

Mean±SE.
NS, P>0.05; * P≤0.05.

Effect of the Phenotypic Combinations of Caseins in Italian Brown Cows Milk

The phenotypes of the caseins α_{s1}-, β- and k-casein were determined (by means urea-starch gel electrophoresis at pH 8.6) in 222 individual milk samples collected from Italian

Brown cows. The objective of this research was to evaluate the effect of phenotypic combinations on physico-chemical and processing properties of milk.

The phenotypic combinations of milk caseins considered (in the order α_{s1}-β-k) were the following: 1, BB-AA-AA (n. = 43); 2, BB-AA-AB (n. = 64); 3, BB-BB-AA (n. = 44); 4, BB-BB-AB (n. = 45); 5, BC-AA-AB (n. = 26). In the following paragraphs, combinations will be indicated with the number, from 1 to 5, as reported above. Significant differences among the 5 phenotypic combinations considered for the contents of some constituents and rennet-coagulation properties were evidenced by statistical analysis. The content of total nitrogen was lower in the combinations 3 and 4 than in other combinations (Table 12). As reported in table 12, soluble nitrogen evidenced the lowest values for combinations 3 and 4 and the highest ones for combinations 1 and 5. The same trend was observed for the content of casein nitrogen. As far as minerals are concerned, total calcium was lower in the combinations 4 and 5 than in other combinations (Table 13). Also the ratio Ca/P was lowest in combinations 4 and 5 and highest in combinations 1, 2 and 3. Potassium (K) showed the highest values in combination 5. Concerning the rennet-coagulation properties, the lowest clotting times (r, minutes) were observed in the combinations 3 and 4 and the lowest curd firming time (k_{20}, minutes) in the combinations 3, 4 and 5 (Table 14).

Table 11. Comparison between individual milks of AB and BB β–lactoglobulin phenotypes in Italian Brown breed: rennet-coagulation properties. No. obs.: 139 for β–Lg AB class and 79 for β–Lg BB class

		AB β–lactoglobulin		BB β–lactoglobulin		P
Clotting time, r	min	18.9	±0.32	19.0	±0.43	NS
Curd firming time, k_{20}	min	8.9	±0.21	8.5	±0.28	NS
Curd firmness, a_{30}	mm	24.2	±0.78	25.3	±1.05	NS
Curd firmness, $a_{1/2}r$	mm	22.1	±0.22	23.3	±0.29	**

Mean±SE.
NS, P>0.05; ** P≤0.01.

Table 12. Comparison between individual milks of the main casein phenotype combinations present in the Italian Brown breed: nitrogen fractions. Mean±SE

Phenotypic combinations												
	α_{s1}	BB		BB		BB		BB		BC		P
	β	AA		AA		BB		BB		AA		
	k	AA		AB		AA		AB		AB		
No. observations		43		64		44		45		26		
Total N	mg/100g	576.84	b	563.22	b	541.92	a	536.04	a	581.81	b	***
		6.71		5.52		6.72		6.71		8.67		
Soluble N at pH 4.6	mg/100g	132.62	c	125.81	b	122.00	ab	118.16	a	126.91	bc	**
		2.31		1.91		2.31		2.31		3.04		
Casein N	mg/100g	444.18	bc	437.59	b	420.03	a	420.29	a	456.45	c	***
		5.45		4.48		5.40		5.44		7.03		

** P≤0.01; *** P≤0.001; different letters in the same row differ significantly for P<0.05.

Table 13. Comparison between individual milks of the main casein phenotype combinations present in the Italian Brown breed: mineral fractions

Phenotypic combinations												
	α_{s1}	BB		BB		BB		BB		BC		P
	β	AA		AA		BB		BB		AA		
	k	AA		AB		AA		AB		AB		
No. observ.		43		64		44		45		26		
Calcium	mg/100g	122.55	b	122.62	b	122.09	b	116.35	a	116.23	a	**
		1.29		1.18		1.303		1.61		1.75		
Phosphorus	mg/100g	106.27		105.27		102.67		105.54		103.36		NS
		1.66		1.52		1.67		2.07		2.25		
Magnesium	mg/100g	10.95		10.66		10.61		10.50		9.98		NS
		0.22		0.20		0.22		0.27		0.29		
Sodium	mg/100g	40.75		42.86		42.20		41.29		38.46		NS
		1.16		0.97		1.15		1.12		1.66		
Potassium	mg/100g	168.37	a	170.15	a	168.28	a	166.44	a	190.70	b	***
		2.12		1.69		2.11		1.94		2.92		
Chloride, Cl^-	mg/100g	88.96		91.60		93.48		90.11		89.82		NS
		1.61		1.39		1.63		1.60		2.36		
Ca/P	—	1.16	bc	1.17	bc	1.20	c	1.11	a	1.14	ab	**
		0.02		0.02		0.02		0.02		0.02		

Mean±SE.
NS, P>0.05; ** P≤0.01; *** P≤0.001.
Different letters in the same row indicate significant differences at P<0.05.

Table 14. Comparison between individual milks of the main casein phenotype combinations present in the Italian Brown breed: rennet-coagulation properties

Phenotypic combinations												
	α_{s1}	BB		BB		BB		BB		BC		P
	β	AA		AA		BB		BB		AA		
	k	AA		AB		AA		AB		AB		
No. observations		43		64		44		45		26		
Clotting time	min	20.3	bc	21.1	c	15.4	a	16.2	a	19.3	b	***
		0.5		0.4		0.5		0.5		0.6		
Curd firming time	min	10.0	b	9.8	b	7.9	a	7.5	b	8.4	b	***
		0.4		0.3		0.4		0.4		0.5		
Curd firmness, a_{30}	mm	21.1	ab	20.1	a	30.6	c	30.3	c	24.8	b	***
		1.2		1.0		1.2		1.2		1.6		
Curd firmness, $a_{1/2}r$	mm	21.9	a	22.4	a	21.6	b	22.1	b	23.7	b	***
		0.4		0.3		0.4		0.4		0.5		

Mean±SE.
*** P≤0.001. Different letters in the same row indicate significant differences at P<0.05.

The combinations 3 and 4 registered the highest values of curd firmness (a_{30}, mm), while combination 5 was characterised by the highest values of curd firmness adjusted for the clotting time ($a_{1/2r}$, mm). In general, combinations 3, 4, and 5 were characterised by better rennet-coagulation properties.

Frequency of k-Casein Genetic Variants in Different Cattle Breeds and their Effects on Cheese Yield

Influence of k-Casein Genetic Variants on Dairy Quality of Milk

Several studies evidenced the influence of k-casein genetic variants on the dairy quality of milk [26, 28-30]. Milk with k-casein B is characterised by a higher presence of k-casein compared to milk with k-casein A. This leads to the presence of a higher number of casein micelles, with lower size and uniform distribution [31, 40]. The differences at the level of casein micelles size are the basis for differin behaviour of k-casein B milk towards rennet. Compared to k-casein, the A and k-casein B milks showed lower clotting times and gave rise to firming faster curds resulting in higher firmness of the final product [35, 37, 41]. This leads to better processing of milk, with consistent curds, that showed a uniform and more adequate drainage leading to higher cheese yields from lowered losses of fat and protein in the cooked whey [29, 33, 36, 42, 43].In Parmigiano-Reggiano cheesemaking, Mariani *et al.* [28] observed an increase of 6 kg of cheese per vat (1000 kg of milk) processing k-casein B milk in place of k-casein A milk. According to FitzGerald *et al.* [44], if a cheese factory produced 20,000 tons of cheese yearly processing only k-casein A milk, the cheese yield would improve to 21,780 tons, in the case of Mozzarella cheese, and to 21,180, in the case of Cheddar cheese, processing only k-casein B milk.

Frequency of k-Casein B in Different Cattle Breeds

The frequency of k-casein A and B in different cattle breeds are reported in table 15 [45-48]. The Italian Friesian was characterised by the highest frequency of k-casein A (82% A *vs* 18% B). On the contrary, the Italian Brown was characterised by the highest frequency of the B variant (37% A *vs* 63% B), the most favourable for cheesemaking as, presumably, a result of selection within the Italian Brown breed. The k-casein B frequency in other cattle breeds was intermediate, within the positioned within the (37% A *vs* 63% B) interval.

Table 15. Percent frequency of k-casein A and B variants in different cattle breeds

	Italian Friesian[1]	Rendena[2]	Italian Red Pied[3]	Reggiana[4]	Alpine Grey[5]	Modenese[6]	Italian Brown[7]
k-casein A	82	65	63	53	52	50	37
k-casein B	18	35	37	47	48	50	63

[1]Associazione Nazionale Allevatori Frisona Italiana, ANAFI, 2004. Personal communication.

[2]Mariani and Russo, 1975 [45]

[3]Messina et al., 2002 [46]

[4]Mariani et al., 2002 [47]

[5]Di Stasio and Merlin, 1979 [48]

[6]Mariani et al., 2002 [47]

[7]Italian Brown Cattle Breeders' Association, ANARB, 2003 Personal communication.

Frequencies of k-Casein A and B and Cheese Yield

The cheese yields reported in several studies compared to that from the Italian Friesian milk are illustrated in figure 1.In each study, the cheese yield of the milk of particular cattle breed was compared with the one obtained with Friesian milk. As evidenced in figure 1 [14, 16, 18, 49, 50], the cattle breed play a major role in defining cheese yield. Differences among studies can be ascribed to the different kind of cheeses analysed, as fresh or ripened cheeses. As reported in figure 2, a k-casein BB milk gave rise to a higher cheese yield, than k-casein AA milk, in the case of Parmigiano-Reggiano cheese (7.07% BB *vs* 6.47% AA, +6 kg of cheese every 1000 kg of milk processed) [28], in the case of Cheddar cheese (9.91% BB *vs* 9.36% AA, +5.5 kg of cheese every 1000 kg of milk processed) [31], and in the case of Mozzarella cheese (10.05% BB *vs* 9.23% AA, +8.2 kg of cheese every 1000 kg of milk processed) [43].

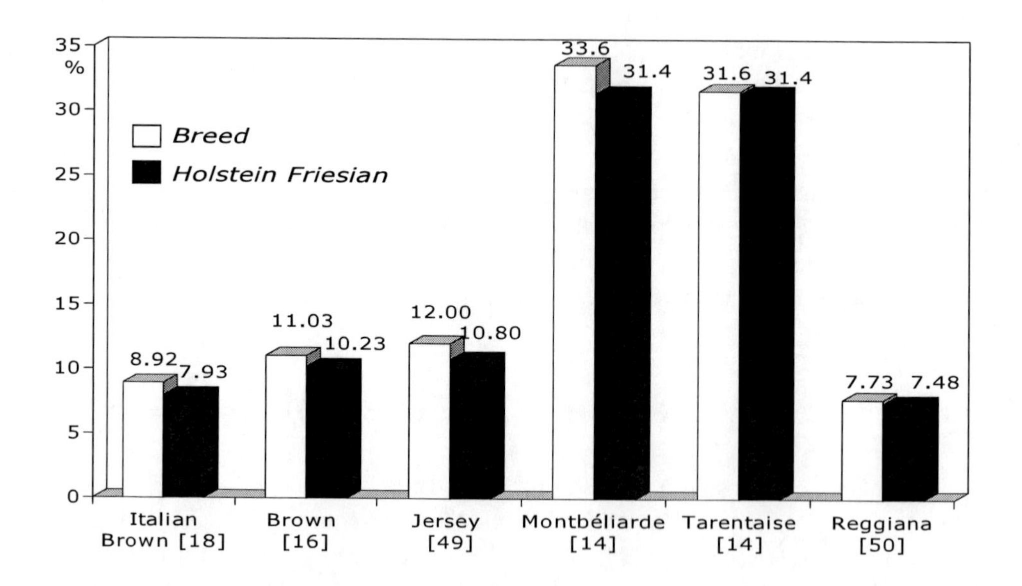

Figure 1. Cheese yield of milk from different cattle breeds, compared to that from the Italian Friesian milk, in the production of different cheeses.

Determination of k-Casein B in Bulk Milk Samples: Set Up of Test k

An immuno-enzymatic test for the determination of the content of k-casein B in bulk milk samples was ideated and constructed. The "*test kappa*" is the result of collaboration among the University of Parma (Italy), Schweizer Braunviehzuchtverband SBVZ (Switzerland) and Italian Brown Cattle Breeders' Association (ANARB, Italy).

The conception of the "*test kappa*" had to satisfy an easy routine use that does not necessitate skilled personnel. It was based on an immuno-enzymatic test, based on specific recognition of bovine k-casein B by a monoclonal antibody. A monoclonal antibody (antik-B) against an oligopeptide of 23 AA corresponding to the region 131–153 of bovine k-casein B was generated using the Human Combinatorial Antibody Library (HuCAL) technology. Both AA substitutions distinguishing k-casein A and B are located in that region (positions 136 and

148). In this study, the reactivity of antik-B to milk samples collected from cows previously genotyped as *CSN3*AA*, *CSN3*AB,* and *CSN3*BB* was tested. According to Western blot results, antik-B recognized k-casein B and it showed no cross-reactivity toward k-casein A and other milk proteins. Furthermore, a modified Western blot method, urea-PAGE Western blot, was set up to assess the reactivity of antik-B toward all isoforms of k-casein B. In conclusion, antik-B was specific to k-casein B in milk and it seemed to be reactive toward all its isoforms [51].Once realized, antik-B was employed to set up a competitive indirect ELISA quick test, called "*test kappa*".

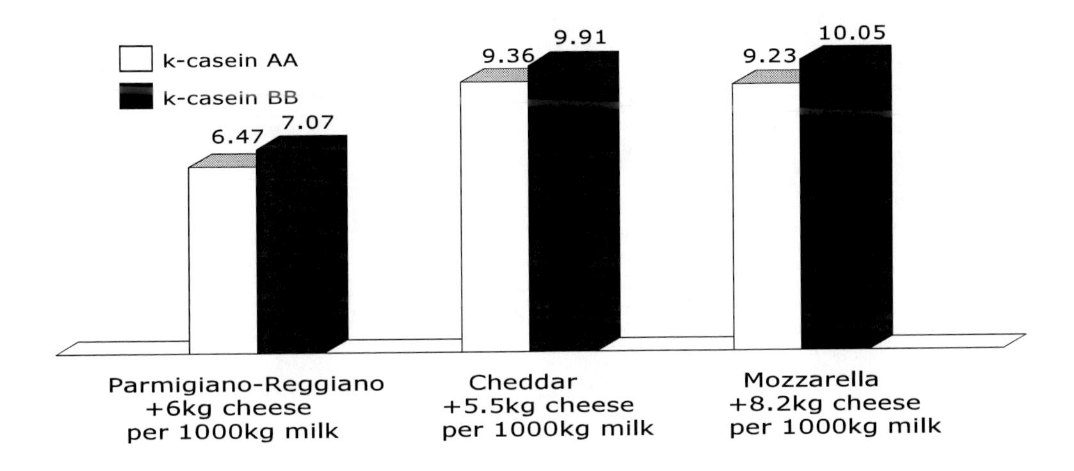

Figure 2. Cheese yields (%) of herd milks of k-casein AA and BB phenotypes in the production of different cheeses.

Conclusion

The genetic factors influence the characteristics of milk for cheesemaking. In fact, they are more important, particularly if the coagulation is of rennet type. Milk from Jersey cows showed better aptitude to rennet-coagulation and rheological properties than milk from Italian Friesian cows. According to results reported in this chapter, differences are related to chemical, physico-chemical and structural peculiarities that characterise Jersey milk. Among these, the principal role seems to be exerted by the elevated casein content. Milk from Ayrshire cows showed better rheological properties in comparison to that from Italian Friesian cows.On the contrary, the rennet-coagulation characteristics did not show significant differences between the two breeds.The Italian Brown vat milk was characterised by higher casein content, better rennet-coagulation characteristics and better rheological properties than Italian Friesian vat milk. The cheese yield of milk from Italian Brown cows is higher in comparison to that from Italian Friesian cows. In the Italian Brown, milk from cows with the β-lactoglobulin BB genotypehad higher casein content, casein number and better curd firmness aptitude than milk from cows with β-lactoglobulin AB genotype.

The α_{s1}-casein, β-casein and k-casein combinations affect cheesemaking properties. In particular the combinations (in the order α_{s1}-β-k) BB-BB-AA, BB-BB-AB and BC-AA-AB are characterised by better rennet-coagulation properties.

The k-casein polymorphism affected markedly the cheesemaking properties. In terms of cheese yield, differences between milks with k-casein B and A are evident for different cheese varieties. In particular, in Parmigiano-Reggiano cheesemaking, cheese yield was increasedby6 kg per 1000 kg of processed k-casein B milk compared to k-casein A milk. Similar results were found for Cheddar and Mozzarella cheeses (5.5 kg and 8.2 kg more, respectively).

The frequency of k-casein A and B in different cattle breeds are markedly different. In particular, the frequency of k-casein A was the highest in the Italian Friesian milk; on the contrary, the Italian Brown milk was characterised the highest frequency of the most favourable k-casein alleles for cheesemaking, i.e the k-casein B. This was due to the genetic selection index adopted by the Italian Brown Breed Association, but now most of the best Italian Friesian bulls are k-casein BB or AB.

It is possible to determine the content of k-casein B in bulk milk by means of a fast and economic method: the "*test kappa*". In this way is possible to pay the milk on the base of the content of k-casein B.

References

[1] Mariani, P., & Pecorari, M. (1987). Fattori genetici, attitudine alla caseificazione e resa del latte in formaggio. *Sci. Tecn. Latt*, 38, 286-326.

[2] Dalgleish, D. G. (1998). Casein micelles as colloids: surface structures and stabilities. *J. Dairy Sci.*, 81, 3013-3018.

[3] De Kruif, C. G. (1998). Supra-aggregates of casein micelles as a prelude to coagulation. *J. Dairy Sci.*, 81, 3019-3028.

[4] Losi, G., & Mariani, P. (1984). Significato tecnologico del polimorfismo delle proteine del latte nella caseificazione a formaggio grana. *Industria Latte*, 20(1), 23-53.

[5] Mariani, P. (1987). Fattori genetici e variazioni quantitative e qualitative della caseina. *Atti Società Italiana Buiatria*, 19, 59-63.

[6] Puhan, Z., & Jakob, E. (1994). Genetic variants of milk proteins and cheese yield. *IDF, International Dairy Federation, Brussels, Belgium*, 9402, 111-122.

[7] Mariani, P., & Summer, A. (1999). Polimorfismo delle proteine ed attitudine tecnologico-casearia del latte. *Sci. Tecn. Latt.-cas.*, 50, 197-230.

[8] Mariani, P., Summer, A., Formaggioni, P., Malacarne, M., & Battistotti, B. (2001). Rilievi sui principali requisiti tecnologico-caseari del latte per la produzione di formaggio grana. *Sci. Tecn. Latt.-cas.*, 52, 49-91.

[9] Corradini, C, Pettinau, M (1972). Possibili riflessi tecnologici del polimorfismo delle caseine. Sci. Tecn. Latt.-cas., 23, 243-253.

[10] Hossain, M. A. (1974). Genetic variants of milk protein and their importance in dairyng. Kieler Milchwirtschaft. Forschung., 26, 17-65.

[11] Losi, G., Castagnetti, G. B., & Morini, D. (1979). Le varianti genetiche della caseina k e attitudine del latte alla coagulazione presamica. *Il Latte*, 4, 1062-1068.

[12] Schaar, J. (1981). Casein stability and cheesemaking properties of milk: effects of handling, mastitis and genetic variation. Swed. *Univ. Agric. Sci., Uppsala, Rep.*, 52, pp, 38.

[13] Mariani, P. (1983). Genetica e qualità del latte per la caseificazione. *Ob. Doc. Vet.*, 4 (9), 25-30.

[14] Macheboeuf, D., Coulon, J. B., & D'Hour, P. (1993). Effect of breed, protein genetic variants and feeding on cows' milk coagulation properties. *J. Dairy Res.*, 60, 43-54.

[15] Marletta, D., Summer, A., Bordonaro, S., Mariani, P., & D'Urso, G. (1998). Composizione chimica, ripartizione percentuale delle caseine e caratteristiche di coagulazione del latte di massa delle razze bovine Modicana e Frisona italiana allevate in provincia di Ragusa. *Zoot. Nutr. Anim.*, 24, 185-192.

[16] Mistry, VV, Brouk, MJ, Kasperson, KM, Martin, E (2002). Cheddar cheese from milk of Holstein and Brown Swiss cows. Milchwissenschaft, 57,19-23.

[17] Auldist, M., Mullins, C., O'Brien, B., O'Kennedy, B. T., & Guinee, T. (2002). Effect of cow breed on milk coagulation properties. *Milchwissenschaft*, 57, 140-143.

[18] Malacarne, M., Summer, A., Fossa, E., Formaggioni, P., Franceschi, P., Pecorari, M., & Mariani, P. (2006). Composition, coagulation properties and Parmigiano-Reggiano cheese yield of Italian Brown and Italian Friesian herd milks. *J. Dairy Res.*, 73, 171-177.

[19] De Marchi, M., Dal Zotto, R., Cassandro, M., & Bittante, G. (2007). Milk coagulation ability of five dairy cattle breeds. *J. Dairy Sci.*, 90, 3986-3992.

[20] Russo, V. (1975). Polimorfismo delle proteine del latte e possibilità di utilizzazione zootecnica. *Il Mondo del Latte*, 29, 296-304.

[21] Russo, V., & Mariani, P. (1978). Polimorfismo delle proteine del latte e relazioni tra varianti genetiche e caratteristiche di interesse zootecnico, tecnologico e caseario. *Riv. Zoot. Vet.*, 6 (5, 6), 289-384 e 365-379.

[22] Mariani, P. (1988). Fattori genetici, proprietà tecnologico-casearie e resa del latte in formaggio grana. Proc. Seminar “La trasformazione del latte in formaggio grana: tecnologie e qualità”, pp. 9-30. Gonzaga (MN), Italy, Quaderni Agropolis, Arti Grafiche Chiribella, MN, Italy.

[23] Associazione Italiana Allevatori. Milk recording activity. Official Statistics 2010. 2011, February 14. Available from: www.aia.it/bollettino/bollettino.htm

[24] Grosclaude, F. (1988). Le polymorphisme génétique des principales lactoprotéines bovines. Relations avec la quantité, la composition et les aptitudes fromagères du lait. *INRA Prod. Anim.*, 1, 5-17.

[25] Mariani, P., Morini, D., Castagnetti, G. B., & Losi, G. (1982). La consistenza del coagulo in latti caratterizzati dalle varianti A e B della beta lattoglobulina. *Industria Latte*, 18 (2), 9-17.

[26] van den Berg, G., Escher, J. T. M., de Koning, P. J., & Bovenhuis, H. (1992). Genetic polymorphism of k-casein and β-lactoglobulin in relation to milk composition and processing properties. *Neth. Milk Dairy J.*, 46, 145-168.

[27] Lundén, A., Nilsson, M., & Janson, L. (1997). Marked effect of β-lactoglobulin polymorphism on the ratio of casein to total protein in milk. *J. Dairy Sci.*, 80, 2996-3005.

[28] Mariani, P., Losi, G., Russo, V., Castagnetti, G. B., Grazia, L., Morini, D., & Fossa, E. (1976). Prove di caseificazione con latte caratterizzato dalle varianti A e B della k-caseina nella produzione del formaggio Parmigiano-Reggiano. *Sci. Tecn. Latt.-cas.*, 27, 208-227.

[29] Schaar, J., Hansson, B., & Pettersson, H. E. (1985). Effects of genetic variants of k-casein and β-lactoglobulin on cheesemaking. *J. Dairy Res.*, 52, 429-437.

[30] Ng-Kwai-Hang, K. F., Hayes, J. F., Moxley, J. E., & Monardes, H. G. (1986). Relationships between milk protein polymorphism and major milk constituents in Holstein-Friesian cows. *J. Dairy Sci.*, 69, 22-26.

[31] Walsh, C. D., Guinee, T. P., Reville, W. D., Harrington, D., Murphy, J. J., O'Kennedy, B. T., & FitzGerald, R. J. (1998). Influence of k-casein genetic variant on rennet gel microstructure, Cheddar cheesemaking properties and casein micelle size. *Int. Dairy J.*, 8, 707–714.

[32] Bobe, G., Beitz, D. C., Freeman, A. E., & Lindberg, G. L. (1999). Effect of milk protein genotypes on milk protein composition and its genetic parameter estimates. *J. Dairy Sci.*, 82, 2797–2804.

[33] Marziali, A. S., & Ng-Kwai-Hang, K. F. (1986). Relationships between milk protein polymorphism and cheese yielding capacity. *J. Dairy Sci.*, 69, 1193-1201.

[34] Davoli, R., Dall'Olio, S., & Russo, V. (1990). Effect of k-casein genotype on the coagulation properties of milk. *J. Anim. Breed. Gen.*, 107, 458-464.

[35] Mariani, P., & Pecorari, M. (1991). Il ruolo delle varianti genetiche della k-caseina nella produzione del formaggio. Sci. Tecn. Latt.-cas., 42, 255-285.

[36] Rahali, V., & Ménard, J. L. (1991). Influence des variants génétiques de la β-lactoglobuline et de la k-caséine sur la composition du lait et son aptitude fromagère. *Lait*, 71, 275-297.

[37] Lodes, A., Krause, I., Buchberger, J., Aumann, J., & Klostermeyer, H. (1996). The influence of genetic variants of milk proteins on the compositional and technological properties of milk. II. Rennet coagulation time and firmness of the rennet curd. Milchwissenschaft, 51, 543-548.

[38] Amigo, L., Martin Alvarez, P. J., Garcia Muro, E., & Zarazaga, I. (2001). Effect of milk protein haplotypes on the composition and technological properties of Fleckvieh bovine milk. Milchwissenschaft, 56, 488-491.

[39] Comin, A., Cassandro, M., Chessa, S., Ojala, M., Dal Zotto, R., De Marchi, M., Carnier, P., Gallo, L., Pagnacco, G, & Bittante, G. (2008). Effects of composite β- and κ-casein genotypes on milk coagulation, quality, and yield traits in Italian Holstein cows. *J. Dairy Sci.,* 91, 4022-4027.

[40] Morini, D., Losi, G., Castagnetti, G. B., Benevelli, M., Resmini, P., & Volonterio, G. (1975). L'influenza delle varianti genetiche della k-caseina sulla dimensione delle micelle caseiniche. *Sci. Tecn. Latt.-cas.*, 26, 437-444.

[41] Losi, G., Capella, P., Castagnetti, G. B., Grazia, L., Zambonelli, C., Mariani, P., & Russo, V. (1973). Influenza delle varianti genetiche della caseina k sulla formazione e sulle caratteristiche della cagliata. *Sci. Tecnol. Alim.*, 3, 373-374.

[42] Hartung, H., & Gernand, E. (1997). Investigation about cheese yielding capacity in relation to casein-polymorphism. *Arch. für Tierzucht*, 40, 305-308.

[43] Walsh, C. D., Guinee, T. P., Harrington, D., Mehra, R., Murphy, J., & FitzGerald, R. J. (1998). Cheesemaking, compositional and functional characteristics of low-moisture part-skim Mozzarella cheese from bovine milks containing k-casein AA, AB, or BB genetic variants. *J. Dairy Res.*, 65, 307-315.

[44] FitzGerald, R. J., Walsh, D., Guinee, T. P., Murphy, J. J., Mehra, R., Harrington, D., & Connolly, J. F. (1998). Genetic variants of milk proteins and their association with milk

production and processing properties. In: RJ FitzGerald (ed.), Genetic variants of milk proteins: Relevance to milk composition and cheese production. Dairy Products Research Centre, Report n. 19, 2-11.

[45] Mariani, P., & Russo, V. (1975). Varianti genetiche delle proteine del latte nella razza Rendena. *Riv. Zoot. Vet.*, 3 (4), 345-348.

[46] Messina, M., Maier, S., Corradini, C., Renaville, R., & Prandi, A. (2002). Indagine preliminare su polimorfismi genetici in bovine e tori di Pezzata Rossa Italiana per la produzione di latte da trasformazione. *Sci. Tecn. Latt.-Cas.*, 53 (5), 371-381.

[47] Mariani, P., Summer, A., Formaggioni, P., & Malacarne, M. (2002). La qualità tecnologico-casearia del latte di differenti razze bovine. La Razza Bruna, 1, 7-13.

[48] Di Stasio, L., & Merlin, P. (1979). Polimorfismi biochimici del latte nella razza bovina Grigio Alpina. *Riv. Zoot. Vet.*, 2, 64-67.

[49] Auldist, M. J., Johnston, K. A., White, N. J., Fitzsimons, W. P., & Boland, M. J. (2004). A comparison of the composition coagulation characteristics and cheesemaking capacity of milk from Friesian and Jersey dairy cows.*J. Dairy Res.*, 71, 51-57.

[50] Castagnetti, G. B., Grazia, L., Losi, G., & Mariani, P. (1986). Il latte della Reggiana nella produzione del Parmigiano-Reggiano. II. Prove di caseificazione. Industria Latte, 22 (2), 3-22.

[51] Summer, A., Santus, E., Casanova, L., Joerg, H., Rossoni, A., Nicoletti, C., Donofrio, G., Mariani P., & Malacarne, M. (2010). Characterisation of a monoclonal antibody for k-casein B of cow's milk. *J. Dairy Sci.*, 93, 796-800.

Part V: Animal Welfare

In: Milk Production
Editor: Boulbaba Rekik, pp. 263-269
ISBN 978-1-62100-061-7

Chapter XII

Animal Welfare in Dairy Operations: A Postmodern View

Zenobia C. Y. Chan[1] and Wing-Fu Lai[2]
[1]School of Nursing, The Hong Kong Polytechnic University,
Hong Kong Special Administrative Region, People's Republic of China
[2]Department of Chemistry, The University of Hong Kong,Hong Kong Special
Administrative Region, People's Republic of China

Overview

Consumption of milk products increased sharply in Asia, which is widely believed to potentially be a significant export market in the future. For this reason, Asia becomes a target of export expansion among countries having powerful dairy production (Dong, 2005). Under the prevailing moral discourse, animals are simply a cash cow and appear to be exploitable for human wants. Given this situation, so far extensive efforts have been devoted to optimize the dairy operations to increase the quality and quantity per yield, at the expense of animal welfares. By focusing on the context of milk production, this article aims to challenge, with a perspective of postmodernism, the existing monistic anthropocentric conception on the use of animals in agricultural practices. It is hoped that this article can raise the awareness of ethical animal use, and offer an insight into the development of a more moral environment in dairy production in which both animal welfare and industrial development could be secured.

Milk Production in China

From 1985 to 2000,there was a triple increase in milk consumption per capita (including all products on a whole milk basis) in China. Several factors have contributed to

[1] E-mail: hszchan@inet.polyu.edu.hk (Z. C. Y. Chan)
[2] E-mail: rori0610@graduate.hku.hk

thistremendous demand growth for dairy products. Examples of these include rapid income growth, change of urban lifestyles, governmental promotion of the dairy industry, and improved marketing channels(Fuller et al., 2005). With the growing demand in milk consumption, an increase in milk production follows, assuredly. The Food and Agricultural Policy Research Institute (FAPRI) projects that China's total production of milk will be escalated to 29.1 million tons in 2014, and the total consumption of milk (whole milk equivalent basis) will become 31.9 million tons. It is estimated that a net import of 2.8 million tonswill likely be required.

According to Simpson (2005), China has four main types of cow milk production systems nowadays. The most elementary one is part of the grassland animal production system. It provides milk for suckling calves and herder families. The second one is the low-input, low-cost operation based on crossbred cow, which is commonly found in urban areas. This kind of system is usually run by small-size producers. Most of the milk produced by this system is sold to nearby urban residents as fresh dairy products. The third type is the traditional medium to large-scale operations. They were derived initially from state farms. Regarding to the ownership, management and modernization level, this type of system is in a period of flux. The fourth and the most important type of cow milk production system in modern dairy industries is a kind of practice made up of operations owned by individuals, partnerships and private or semi-public corporations. Even there are four types of milk production systems, facing the rapid economic growth, the rising demand on milk products, and the large consumer population, there is no doubt that China is encountering a challenge to meet the projected demand and also other livestock commodities. Under such a context, some of the milk producers chose to sacrifice the quality of their products in order to boost their yield and profits. A good example is the previous tainted milk event which is one of the most influential foodsafety crises being reported in recentyears.

The event surfaced on 11th September, but its preliminarysymptoms could be traced back three months earlier when some parents of the affected children complained about the Sanlu milk powder to General Administration of Quality Supervision, Inspectionand Quarantine (AQSIQ). Since July 2008, there was an abnormally high prevalenceof renal stone cases amongmainland infants. However, no action was taken at thattime (Ma, 2008). Only when the event came public, the mainland authorities attempted to mitigate possible public health hazards and penalize the relevant milk powder manufacturers. Barring this, medical care was provided free-of-charge by the Minister of Health to affected infants, and a large-scale food investigation was launched to comfortpanicked parents on the safety of Chinese food. Unfortunately,the panic did not end. By late September, morethan 52,857 children had been remedied for renal complicationsand at least 4 had died.

Apart from the life of humans, the welfare of animals can also be harmed by the unethical profit-driven practice in milk production. As a matter of fact, in order to meet the increasing milk demand, some manufacturers have turned cows into tools, rather than respecting them as lives, during the process of milk production. Dairy cows are often crammed into a huge area covered in mud and their own feces. They are also fed with different kinds of drugs to keep them producing enough "artificial" milk. This nonstop milk production often causes those milk cows to result in swollen udders and udder infection. Such a terrifying situation does not only limit to China, but also occur overseas. It has been narrated that one-third of California's cows suffer from painful udder infections, and more than half are enduring other painful infections and illnesses (PETA, n.d.). According to *People for the Ethical Treatment of*

Animals (n.d.), every year dairy cows are forced to be pregnant, and male baby cows are often killed by farm workers, who chain their necks in veal crates before slaughtering when those baby cows reach just 16 weeks old. Furthermore, more than a fourth of California's dairy cows were slaughtered in a brutal, terrifying and barbarian way simply because they become crippled from foot infections or calcium depletion (PETA, n.d.).

A Postmodern Perspective

There is no denying that, under the prevailing moral discourse, human beings predominate over other species and the nature. sing animals for the good of the human society is regarded as being "rational" and "moral". However, it is high time for us to revisit this taken-for-granted notion, and the first thing we have to ask is, how should "rationality" and "morality" be determined? Till now, "rationality" is regarded as being objective, universal and singular. Given this, any paradigms other than the dominant one were seen as being "irrational" or absent. Under the prevailing anthropocentric paradigm, striving for animal welfare at the expense of human benefits is highly condemned and deemed "irrational". This notion appears to be a universal truth across all political systems. However, it is this so-called "universality of rationality" that is challenged by postmodernism.

Postmodernism is a controversial theme in philosophy and social sciences (Abma 2002; Anderson 1999; Chambon *et al* 1999; Foucault 1990). It is an inspiring thinking approach (Chan & Ma 2005), and its capacity of tolerating difference (Moules 2000) and creating multiple meanings via deconstruction (Dumm 1996; Laird 2000) render it especially suitable to examine the kernel of "knowledge" by challenging the monovision of meaning and interpretation (Chan & Ma 2005). Under a postmodern perspective, the acceptability of truth and rationality are flexible, relative, contextual and also dependent on a pre-set paradigm (Putnam 1988).

In another word, the rationality of compromising the welfare of animals simply for the social good is more subjective than objective, and after putting down our egoism, we should start reconsidering the righteousness of our exploitation of animals, such as milk cows, in our daily practice.

In fact, if the multiple paradigms in determining "rational" practices can be recognized some day in the future, it is hoped that the conflicts between animal rights and human needs can also advance to higher enlightenment.

As far as animal use is concerned, the concept of "morality" should also be discussed. "Morality" is defined as a set of social norms and value judgments differentiating the good from the bad. According to Beck (1995), "morality" is a cultural and social product. Its meaning changes over time, and differs from one culture to another. In fact, moral standards are usually created by different interest groups to suit their own needs and cultures. The existing utilitarian ideology and the ethical acceptability of animal use for the sake of human betterment are merely a cultural construction rather than a universal truth. They are thought to be "moral", "rational" or "infallible" just because they are judged under an artificial belief system in which values are constructed based solely on the interests of man (Beck 1995).

Conclusion

Milk production is an important industrial component in dairy operations and agricultural practices.While so far much of our efforts have anthropocentrically focused on optimization of the operational practice to increase the yield and its quality, little efforts have ever been paid to the welfare of animals. In this chapter, we have revisited the welfare of animals with a postmodern perspective. However, this chapter is not a research paper. The extent of its analysismay be limited by the availability of information sources. For this, the content of this chapter can only be a lead-in to the topic.

More endeavorsfrom readers, especially scholars and dairy workers, are required so that our philosophical understanding of the topic could be put into practice. It is hoped that with the awareness of ethical animal use, a moral environment in dairy production could be developed, and both the animal welfare and industrial development could be secured simultaneously in the future.

Acknowledgements

The authors would like to thank Shelley Man-Hoi Tsang and Dorothy Kwok for their help during the writing of this chapter.

References

Abma, T. A. (2002). Emerging narrative forms of knowledge representation in the health sciences: Two texts in a postmodern context. *Qualitative Health Research*,12, 5-27.

Anderson, H. (1999). Reimagining family therapy: Reflections on Minuchin's visible family. *Journal of Marital and Family Therapy*, 25, 1-8.

Beck, C. (1995). Postmodernism, ethics, and moral education. In W. Kohli (Ed.), *Critical conversations in philosophy of education* (pp. 127-136). New York: Routledge.

Chambon, A. S., Irving, A., and Epstein, L. (1999). *Reading Foucault for social work.* Columbia University Press: New York.

Chan, Z. C. Y., and Ma, J. L. C. (2005). Through the lens of postmodernism: Uniqueness of the anorectic families. *The Qualitative Report*, 10, 246-256.

Dong, F. (2005). The outlook for Asian dairy markets: The role of demographics, income, and prices. *CARD working paper 05-WP*, USA: Iowa State University.

Dumm, T. L. (1996). *Michel Foucault and the politics of freedom.* London: Sage Publications.

Foucault, M. (1990). *The history of sexuality* (Vol 1). New York: Vintage Books.

Fuller, F.H.,Huang, J., Ma, H., Rozelle, S. (2005). The rapid rise of China's dairy sector: Factors behind the growth in demand and supply.*Working paper 05-WP 394*, USA: Center for Agricultural and Rural Development (CARD), Iowa State University.

Laird, J. (2000). Gender in lesbian relationships: Cultural, feminist, and constructionist reflections. *Journal of Marital and Family Therapy*, 26, 455-467.

Ma, J. (2008, September 13). Complaints were received in June but nothing was done, *South China Morning Post*, p. 4.

Moules, N. J. (2000). Postmodernism and the sacred: Reclaiming connection in our greater-than-human worlds. *Journal of Marital and Family Therapy*, 26, 229-240.

People for the Ethical Treatment of Animals (PETA) (n.d.). Happy cows? You decide. Retrieved November 18, 2010, from http://www.peta.org/features/unhappy- cows.aspx

Putnam, H. (1988). *Representation and reality*. Cambridge: MTT Press.

Simpson, J. R. (2005). Future of the dairy industries in China, Japan and the United States: Conflict resolution in the Doha Round of WTO agricultural trade negotiations. *Working paper series no. 1*, Japan: Afrasian Centre for Peace and Development Studies.

Index

L

M

N

O

P

R

S

T

U

V

W

B